MW00851939

Combinatorics of Genome Rearrangements

Computational Molecular Biology

Sorin Istrail, Pavel Pevzner, and Michael Waterman, editors

Computational molecular biology is a new discipline, bringing together computational, statistical, experimental, and technological methods, which is energizing and dramatically accelerating the discovery of new technologies and tools for molecular biology. The MIT Press series on Computational Molecular Biology is intended to provide a unique and effective venue for the rapid publication of monographs, textbooks, edited collections, reference works, and lecture notes of the highest quality.

Computational Molecular Biology: An Algorithmic Approach
Pavel A. Pevzner, 2000

Computational Methods for Modeling Biochemical Networks
James M. Bower and Hamid Bolouri, editors, 2001

Current Topics in Computational Molecular Biology
Tao Jiang, Ying Xu, and Michael Q. Zhang, editors, 2002

Gene Regulation and Metabolism: Postgenomic Computation Approaches
Julio Collado-Vides, editor, 2002

Microarrays for an Integrative Genomics
Isaac S. Kohane, Alvin Kho, and Atul J. Butte, 2002

Kernel Methods in Computational Biology
Bernhard Schölkopf, Koji Tsuda, and Jean-Philippe Vert, editors, 2004

An Introduction to Bioinformatics Algorithms
Neil C. Jones and Pavel A. Pevzner, 2004

Immunological Bioinformatics
Ole Lund, Morten Nielsen, Claus Lundegaard, Can Keşmir, and Søren Brunak, 2005

Ontologies for Bioinformatics
Kenneth Baclawski and Tianhua Niu, 2005

Biological Modeling and Simulation: A Survey of Practical Models, Algorithms, and Numerical Methods
Russell Schwartz, 2008

Combinatorics of Genome Rearrangements
Guillaume Fertin, Anthony Labarre, Irena Rusu, Éric Tannier, and Stéphane Vialette, 2009

Combinatorics of Genome Rearrangements

Guillaume Fertin, Anthony Labarre, Irena Rusu, Éric Tannier and Stéphane Vialette

The MIT Press
Cambridge, Massachusetts
London, England

For information about special quantity discounts, please email special_sales@mitpress.mit.edu

This book was set in Times New Roman and Syntax on 3B2 by Asco Typesetters, Hong Kong.
Printed and bound in the United States of America.

Library of Congress Cataloging-in-Publication Data

Combinatorics of genome rearrangements / Guillaume Fertin ... [et al.].
p. cm. — (Computational molecular biology)
Includes bibliographical references and index.
ISBN 978-0-262-06282-4 (hardcover : alk. paper) 1. Translocation (Genetics)—Mathematical models. 2. Translocation (Genetics)—Data processing. 3. Combinatorial analysis. 4. Genomics—Mathematics. I. Fertin, Guillaume, 1972– II. Series.
[DNLM: 1. Gene Rearrangement. 2. Genome. 3. Models, Genetic. QU 470 C731 2009]
QH462.T7C66 2009
572.8′77—dc22 2008042152

10 9 8 7 6 5 4 3 2 1

Contents

Preface xiii
Acknowledgments xv

1 **Introduction 1**
1.1 A Minimalist Introduction to Molecular Evolution 1
1.2 Birth of the Combinatorics of Genome Rearrangements 4
1.3 Statement of the Problem 6
1.4 Scope of This Survey 7
1.5 Overview of the Models 7
1.6 Organization of the Book 8

I **DUPLICATION-FREE MODELS: PERMUTATIONS 11**

2 **Genomes as Permutations 13**
2.1 The Symmetric Group 13
2.2 The Cycles of a Permutation 14
2.3 Signed Permutations 15
2.4 Distances on Permutation Groups 15
2.4.1 Rearrangements as Generators 16
2.4.2 Invariant Distances 17
2.5 Circular Permutations 18
2.5.1 Classical Circular Permutations 19
2.5.2 Genomic Circular Permutations 19
2.6 First Measures of Similarity between Permutations 20
2.6.1 Breakpoints 20
2.6.2 Common Intervals and Semipartitive Families 21

3 **Distances between Unsigned Permutations 25**
3.1 Transposition Distance 25
3.1.1 Lower Bounds on the Transposition Distance 26
3.1.2 Upper Bounds 29
3.1.3 Improving Bounds Using Toric Permutations 32

3.1.4 Easy Cases 33
3.1.5 Approximation Algorithms 34
3.1.6 Conjectures and Open Problems 35
3.2 Prefix Transposition Distance 36
3.2.1 Lower Bounds 37
3.2.2 Upper Bounds 38
3.2.3 Diameter 38
3.2.4 Easy Cases 39
3.2.5 Approximation Algorithms 39
3.2.6 Variant: Insertion of the Leading Element 40
3.3 Reversal Distance 40
3.3.1 Lower Bounds 40
3.3.2 Upper Bounds 43
3.3.3 Easy Cases 43
3.3.4 Computational Complexity 44
3.3.5 Approximation Algorithms 45
3.3.6 Exact Algorithms 46
3.4 Prefix Reversal Distance (Pancake-Flipping) 47
3.4.1 Lower Bounds 47
3.4.2 History 48
3.4.3 Variants 48
3.5 Variants 49
3.5.1 Block Interchange Distance 49
3.5.2 Element Interchange Distances 50
3.5.3 Weighted Reversals 52
3.5.4 Fixed-Length Reversals 54
3.5.5 Bounded Variants 54
3.5.6 Cut-and-Paste 55
3.5.7 Strip Moves 55
3.5.8 Stack-Sorting 56
3.5.9 Tandem Duplications and Random Losses 58
3.5.10 Combined Operations: Reversals and Transpositions 59
3.6 Relations between Distances on Unsigned Permutations 61

4 Distances between Signed Permutations 63
4.1 Conserved Interval Distance 63
4.2 Signed Reversal Distance 64
4.2.1 Reversals 64
4.2.2 The Distance Formula 65
4.2.3 The Scenario of Reversals 67
4.2.4 The Space of All Optimal Solutions 68
4.2.5 Experimental Results 69
4.3 Variants of Sorting by Reversals 69
4.3.1 Perfect Signed Reversal Distance 69
4.3.2 Prefix Reversals (Burned Pancakes) 70
4.3.3 Reversals That Are Symmetric around a Point 70

4.3.4 Weighted Reversals 71
4.3.5 Fixed-Length Reversals 71
4.4 Combined Operations 72
4.4.1 Reversals and Transpositions 72
4.4.2 Reversals, Transpositions, Transreversals, Revrevs 72
4.5 Double Cut-and-Joins 73

5 Rearrangements of Partial Orders 75
5.1 Genomes as Partially Ordered Sets 75
5.2 Partially Ordered Sets 75
5.2.1 Basic Definitions 75
5.2.2 Representing Posets 77
5.2.3 Topological Sorting 77
5.3 Constructing a Poset 78
5.4 Reversal Distance 79
5.5 Breakpoint Distance 80
5.5.1 Exact Algorithms 80
5.5.2 Heuristics for Computing the Breakpoint Distance 81

6 Graph-Theoretic and Linear Algebra Formulations 83
6.1 Simple Permutations and the Interleaving Graph 83
6.2 The Overlap Graph 84
6.3 The Local Complementation of a Graph 85
6.4 The Matrix Tightness Problem 85
6.5 Extension to Sorting by Transpositions 86
6.6 The Intermediate Case of Directed Local Complementation 87

II MODELS HANDLING DUPLICATIONS: STRINGS 89

7 Generalities 91
7.1 Biological Motivations 91
7.2 Strings and Rearrangements on Strings 92
7.3 Balanced Strings 94
7.4 How to Deal with Multiple Copies? 95

8 Distances between Arbitrary Strings 97
8.1 The Match-and-Prune Model 98
8.1.1 Breakpoint Distance 100
8.1.2 Signed Reversal Distance 106
8.1.3 Adjacency Similarity 108
8.1.4 Common Intervals Similarity 111
8.1.5 Conserved Intervals Similarity 113
8.1.6 Conserved Intervals Distance 114
8.1.7 MAD and SAD Numbers 118
8.1.8 Heuristics 119

8.2 The Block Edit Model 123
8.2.1 Block Covering Distance 123
8.2.2 Symmetric Block Edit Distance 126
8.2.3 Large Block Edit Distance 129
8.2.4 String Edit Distance with Transpositions 130
8.2.5 Signed Strings 131

9 Distances between Balanced Strings 133
9.1 Minimum Common String Partition Problems 133
9.1.1 Unsigned MCSP 134
9.1.2 Signed MCSP 135
9.1.3 Reversed MCSP 137
9.1.4 Full Breakpoint Distance 138
9.2 Reversal Distance 138
9.2.1 Unsigned Reversals 138
9.2.2 Signed Reversals 141
9.2.3 Sorting by Reversals with Length-Weighted Costs 142
9.2.4 Prefix Reversals on Unsigned Strings (Pancake-Flipping) 144
9.2.5 Reversals of Length at Most 2 147
9.3 Unsigned Transpositions 147
9.3.1 Unit Cost Transpositions 147
9.3.2 Length-Weighted Transpositions 150
9.3.3 Restricted Length-Weighted Transpositions 150
9.3.4 Prefix Transpositions 152
9.3.5 Adjacent Swaps 153
9.4 Unsigned Block Interchanges 153
9.4.1 Unit-Cost Block Interchanges 153
9.4.2 Character Swaps 155
9.5 Relations between Distances 157

III MULTICHROMOSOMAL MODELS 159

10 Paths and Cycles 161
10.1 Genomes 161
10.2 Breakpoints 162
10.3 Intervals 163
10.4 Translocation Distance 164
10.4.1 Feasibility 166
10.4.2 Unsigned Genomes 166
10.4.3 Signed Genomes 167
10.4.4 Translocations Preserving Centromeres 168
10.4.5 Variants and Special Cases 169
10.5 Double Cut-and-Joins (2-Break Rearrangement) 170
10.6 *k*-Break Rearrangement 171
10.7 Fusions, Fissions, Translocations, and Reversals 172
10.8 Rearrangements with Partially Ordered Chromosomes 174

11 Cycles of a Permutation 175
11.1 A Model for Multichromosomal Circular Genomes 175
11.2 A Generalization to Signed Genomes 178
11.2.1 A Different Kind of Signed Permutation 178
11.2.2 The Operations 179
11.2.3 Some Results 179

12 Set Systems and the Syntenic Distance 181
12.1 Introduction 181
12.2 Structural Properties 182
12.2.1 Compact Representation 182
12.3 Lower Bounds 184
12.4 Diameter 185
12.5 Algorithmic Results 185
12.5.1 Syntenic Distance 185
12.5.2 Easy Cases 186
12.6 Conjectures and Open Problems 189

IV MULTIGENOMIC MODELS 191

13 Median and Halving Problems 193
13.1 Breakpoint Median 194
13.1.1 Complexity 194
13.1.2 Algorithms 195
13.2 Reversal and DCJ Median 197
13.2.1 Complexity 197
13.2.2 Algorithms 197
13.2.3 Variants 198
13.3 Duplicated Genomes 199
13.3.1 The Double Distance 199
13.3.2 Genome Halving 201
13.3.3 Solving Tetraploidy 202
13.3.4 Guided Halving 202
13.3.5 Genome Halving with Unordered Chromosomes 203
13.4 Other Variants, Generalizations, and Discussion 205
13.4.1 Other Operations 205
13.4.2 More Permutations in the Input 205
13.4.3 Medians and Centers 205
13.4.4 Discussion 206

14 Rearrangement Phylogenies 207
14.1 The Large Parsimony Problem 207
14.2 The Large Parsimony Problem with Gene Orders 209
14.2.1 Breakpoint and Reversal Phylogenies on Permutations 209
14.2.2 Variants 211

14.3 Heuristics for the Breakpoint/Reversal Phylogeny Problem 211
14.3.1 Tree Steinerization 212
14.3.2 Sequential Addition 216
14.3.3 Character Encodings 217
14.4 Variants 220

V MISCELLANEOUS 221

15 Software 223
15.1 Pairwise Rearrangements 223
15.1.1 Unichromosomal Models 223
15.1.2 Multichromosomal Models 225
15.2 Phylogeny Reconstruction and Medians 226
15.2.1 BPAnalysis 226
15.2.2 MGR 226
15.2.3 GRIL 226
15.2.4 GRAPPA 227
15.2.5 MedRbyLS 227
15.2.6 rEvoluzer and amGRP 227
15.2.7 GENESIS 228

16 Open Problems 229
16.1 Complexity Issues 229
16.1.1 Hardness 229
16.1.2 Approximability 230
16.1.3 Polynomial Complexity 231
16.2 Diameter 231
16.3 Tightness of Bounds 232

APPENDICES 233

A Graph Theory 235
A.1 Undirected Graphs 235
A.1.1 Basic Definitions 235
A.1.2 Paths and Cycles 236
A.1.3 Connectivity 237
A.1.4 Bipartite Graphs 238
A.1.5 Trees and Forests 238
A.1.6 Matching 238
A.1.7 Adjacency Matrix 239
A.2 Directed Graphs 240
A.2.1 Basic Definitions 240
A.2.2 Paths and Cycles 241
A.2.3 Connectivity 241
A.2.4 Directed Acyclic Graphs 241

B Complexity Theory 243

B.1 The Class **NP** 243

B.1.1 **NP**-Optimization Problems: From **PTAS** to **APX** 246

B.1.2 **NP**-Optimization Problems: Beyond **APX** 250

B.1.3 Parameterized Complexity 250

B.2 Some **NP**-Complete Problems 252

Glossary 257

Bibliography 263

Index 283

Preface

In 1984, at a congress in Paris, François Jacob, one of the most famous evolutionary scientists, stated that "La molécule de l'hérédité est raboutée, modifiée, coupée, rallongée, raccourcie, retournée" (the molecule of heredity is sewed together, modified, cut, lengthened, shortened, reversed) during evolution. From one cell to another, one individual to another, one species to another, the content of the DNA molecules is often similar, but their organization often differs dramatically. The mutations that affect this organization are called *genome rearrangements*, and the structural differences between molecules in two genomes motivate the study of their combinatorics. Indeed, the inference of the evolutionary events that can explain the multiple combinations of observed genomes can often be formalized as combinatorial optimization problems.

The variety of problems that have been raised in this domain is so interesting from a combinatorial point of view that this field has grown and become partly independent of the application, so that it now belongs as much to mathematics as to biology. The mathematics and algorithmics related to genome rearrangements have witnessed a huge expansion over these last years, and this dynamics seems to be continuing at the present time. Due to this success, the field has swallowed other studies that were developed earlier and without biological motivations. For example, many problems about sorting permutations with constraints are now presented as rearrangement problems, without considering the biological relevance of the constraint.

Although molecular biology gave birth to it, combinatorics of genome rearrangements is now a mathematical and algorithmic field that has found its own coherence. It has its own important results, many peripheral developments, and its famous open problems. A great interest of this domain is the simplicity of the formulation of the problems, compared to the sometimes great complexity or even nonexistence of solutions. Moreover, the fact that the subject has now been studied for nearly two decades and has been discussed only in specialized research literature motivates both a thorough survey of the topic and an introduction to a broader audience. This book intends to fulfill both goals.

Acknowledgments

Sèverine Bérard, Gaëlle Giberti, and Julien Moncel participated in the birth of the project that led to the present book. The decision to undertake this work was made during a meeting in Lyon funded by the ACI "Nouvelles Interfaces des Mathématiques," π-vert, led by Marie-France Sagot.

Anthony Labarre was funded by the Fonds pour la Formation à la Recherche dans l'Industrie et dans l'Agriculture (FRIA), by a grant from the Fonds National de la Recherche Scientifique (FNRS), and by Communauté Française de Belgique—Actions de Recherche Concertées. He wishes to thank his family and friends, as well as his adviser Jean-Paul Doignon. Eric Tannier was funded by the French National Research Institute for Computer Science (INRIA) and by the Agence Nationale de la Recherche (ANR), projects JC05_49162 (GENOMICRO) and NT05-3_45205 (REGLIS).

1 Introduction

Although this book is combinatorially inclined and does not devote much discussion to the biological issues, we will start with a short introduction to molecular evolution, for conceptual and historical purposes; indeed, this is where the combinatorial study of genome rearrangements originates, and the invention of most variants of genome rearrangement problems are still driven by biological constraints. This introduction is not necessary to understand the combinatorial problems and their solutions, but it allows us to place them in their context and explain why they are important, independently of their mathematical value.

1.1 A Minimalist Introduction to Molecular Evolution

The "molecules of heredity," the support of genetic information, are present in every cell of all living organisms (bacteria, plants, animals, etc.). Each molecule is called a *chromosome*, and the set of all chromosomes is what we will call the *genome*. Chromosomes are made of *DNA* (**d**eoxyribo**n**ucleic **a**cid), a double-stranded molecule in which each *strand* is a long succession of *nucleotides* (a *sequence*). Nucleotides can be of four types—*A*, *C*, *G*, and *T*—and the two DNA strands are coupled in such a way that an *A* on one strand is always coupled with a *T* on the other strand, and a *C* on one strand is always coupled with a *G* on the other strand. Those strands are said to be *complementary*: the sequence on one strand determines the sequence on the other one. Figure 1.1 shows a representation of the above concepts.

Because of complementarity, a DNA molecule is usually represented as a single sequence (one arbitrary strand), but the organization in two strands will often be crucial for our purpose. Chromosomes can be either circular (the sequence forms a circle and has no ends), which is often the case in bacteria, or linear (the sequence has two ends, called *telomeres*), which is often the case in animals and plants. A *segment* of DNA is a part of this molecule made of consecutive nucleotides. A *gene* is a segment of DNA that contains the information needed to construct the other molecules in the cell.

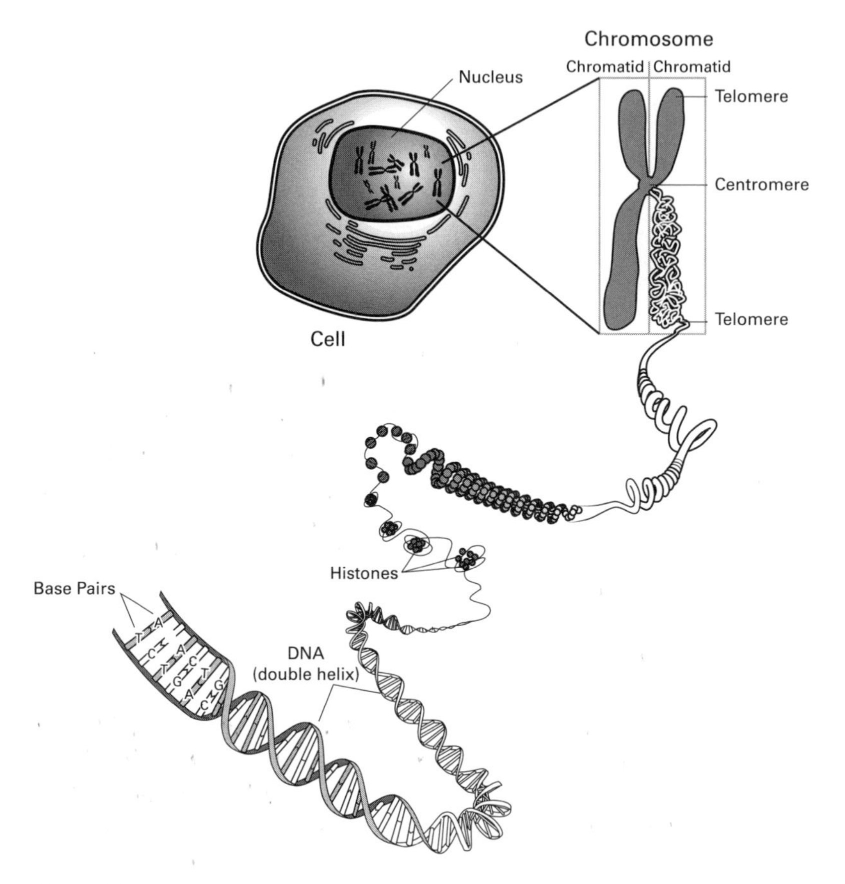

Figure 1.1
A chromosome and a fragment of a DNA molecule
Source: National Institutes of Health, National Human Genome Research Institute, Division of Intramural Research

What accounts for the diversity of living organisms is the possibility for DNA to *replicate* itself with some inaccuracy: one genome is used to produce another, almost identical genome. This inaccuracy is the principle of molecular *evolution*.

A DNA molecule may evolve by *point mutations* (i.e., mutations at the level of nucleotides). There are three different kinds of point mutations: substitutions (one nucleotide is replaced with another), insertions (a nucleotide is added to the sequence), and deletions (a nucleotide is removed from the sequence). Detecting these events is the goal of *sequence alignment* (for a presentation of this topic, see, for example, Setubal and Meidanis [333] or Jones and Pevzner [224]).

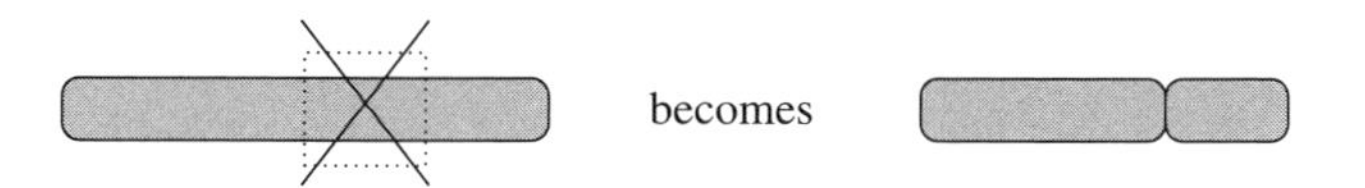

Figure 1.2
Deletion of the dotted region in a chromosome

Figure 1.3
Transposition of the dotted region in a chromosome

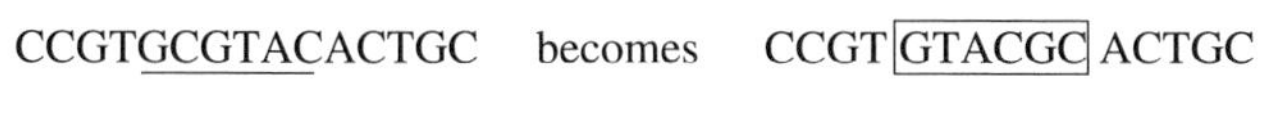

Figure 1.4
Reversal of the underlined segment, resulting in the boxed segment

Figure 1.5
Tandem duplication of the dotted region in a chromosome

However, a sequence may also evolve by modifying its organization at a larger scale. These large-scale mutations are called *rearrangements*, or *structural variations*, and detecting them is the goal of *genome rearrangement* problems. The main rearrangements include the following:

• *Deletions*. A segment of the genome is lost (see figure 1.2).

• *Transpositions*. A segment of the genome moves to another location (see figure 1.3). Transpositions are sometimes referred to as translocations or insertions, but transposition is well adopted in the field of combinatorics of genome rearrangements.

• *Inversions* or *reversals*. A segment of the genome is reversed and the strands are exchanged (see figure 1.4).

• *Duplications*. A segment of DNA is copied and inserted in the genome. There are three main standard types of duplications: *tandem duplications*, illustrated by figure 1.5, which insert the copy next to the original; *retrotranspositions*, which insert a copy of a gene at an arbitrary location in the genome; and *whole genome duplications*, which copy either the whole genome or some of its chromosomes.

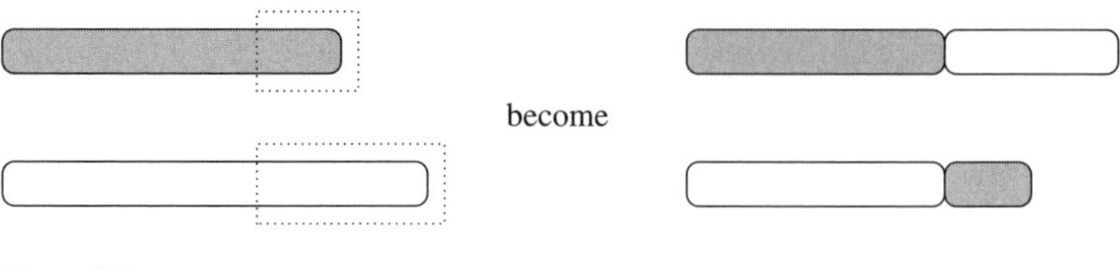

Figure 1.6
Reciprocal translocation of the dotted regions in two chromosomes

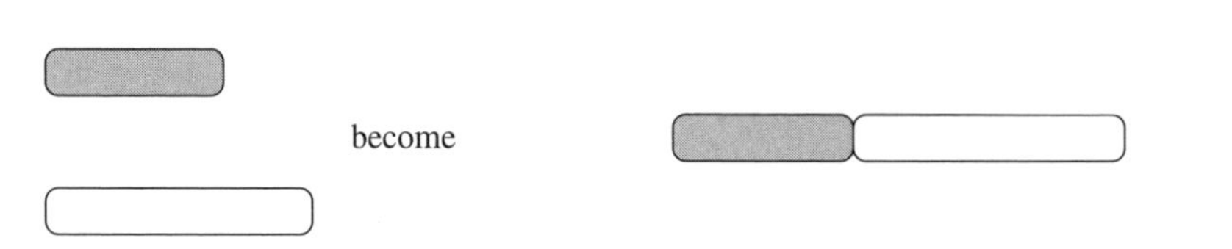

Figure 1.7
Fusion of two chromosomes

- *Reciprocal translocation.* A segment of a chromosome that contains a telomere is exchanged with a segment of another chromosome that also contains a telomere (see figure 1.6).
- *Fusion.* Two chromosomes are joined into one (see figure 1.7).
- *Fission.* One chromosome splits into two (this is the inverse of a fusion).
- *Horizontal*, or *lateral, transfer.* A segment of the genome is copied from one genome to another. This is common mainly in unicellular organisms.

All these operations act on a genome at the level of DNA segments rather than on nucleotides. This is why a genome is often represented by a sequence of segments in that setting: they are the segments that are found in an almost identical state in several species, not cut by rearrangements. Two segments are said to be *homologous* if they derive from a common ancestor and are distinguished by a replication event (they end up in two different genomes) or by a duplication event (they both belong to the same genome).

Genes are often taken as those homologous segments because, due to their functional utility, they are less subject to small mutations and are rarely cut by rearrangements, which is not the case for other parts of the genome.

1.2 Birth of the Combinatorics of Genome Rearrangements

In 1936 two renowned biologists, Dobzhansky (the inventor of the synthetic theory of evolution) and Sturtevant (the discoverer of rearrangement processes in genomes at the beginning of the twentieth century) proposed for the first time to use the degree of disorder between the organization of genes in two different genomes as an indicator of an evolutionary distance between organisms (see Dobzhansky and Sturtevant [145, 146]). They proposed a scenario of inversions to explain chromosomal differences between 17 groups of flies, as well as a reconstruction of putative ancestral gene arrangements and species histories from the observation of the gene order along the chromosomes.

Since rearrangements are relatively rare events, scenarios minimizing their number are more likely to be close to reality. In 1941 Sturtevant and Novitski [343] formulated the problem of minimizing the number of inversions that may explain the differences in arrangements between two species: "...for each such sequence there was determined the minimum number of successive inversions required to reduce it to the ordinal sequence chosen as 'standard.' For numbers of loci above nine the determination of this minimum number proved too laborious, and too uncertain, to be carried out...."

The reconstruction of genome rearrangements from the examination of chromosomes, using techniques such as "chromosome banding" or "in-situ hybridisation" [301] were numerous, all focusing on relatively close species, so that the number of rearrangements was small. All these studies were based on the parsimony criterion, which makes molecular biologists often prefer explanations of differences between genomes that involve as few mutations as possible. This principle makes the connection with combinatorial optimization possible, because the optimization principle meets the parsimony criterion.

As we entered the genome sequencing era, the importance of rearrangements in evolution or illnesses was pointed out by several biologists, such as Palmer and Herbon [289], who examined the differences in the gene order of the mitochondrial genomes of cabbage and turnip, which are very similar in sequence but dramatically different in structure. It was not until 1982 that some researchers working in combinatorial optimization started to formalize and become involved in this problem, in order to overcome the limit of nine genes stated by Sturtevant and Novitski [343]. Watterson et al. [369] proposed to represent the relative positions of genes in different genomes as *permutations*. In order to propose an evolutionary scenario between two species, one had to solve the problem of transforming one circular permutation into another with a minimum number of inversions. The problem was far from being solved after this first article, but it was well stated.

Transforming one permutation into another by means of a minimum number of allowed operations is often equivalent to *sorting* a permutation by means of the same operations (see page 17). Though it took a decisive start in a biological context, the problem of sorting permutations with constraints was not new: a few mathematicians and computer scientists had already tackled that kind of problem in the past. Those problems were not, however, motivated by biology: constraints were related to data structures as stacks, or were simply introduced as games that later turned out to be particular cases of genome rearrangement problems, or found uses in other fields.

New models were later proposed to handle more operations, duplicated segments, and several chromosomes. Shortly after Watterson et al. [369], the field started its dramatic expansion.

1.3 Statement of the Problem

The *genome rearrangement problem* is formulated in its most general form as follows: given a set of genomes and a set of possible evolutionary events, find a shortest set of events transforming those genomes into one another.

What "genome" means here, and what events are, makes the diversity of the problem. Miscellaneous models have been proposed, depending on various parameters, and we briefly review them in section 1.5. "Shortest" usually refers to the number of events, but it may also mean "of least weight" if events are weighted (e.g., according to their probability of occurrence).

The length (or weight) of an optimal sequence of events transforming one genome into another is called the *distance* between the two genomes. We will often require that this distance be a metric on the set of genomes, in the mathematical sense, and we recall its definition here.

Definition 1.1 A **metric** d on a set S is an application

$$d : S \times S \to \mathbb{R} : (s, t) \mapsto r$$

satisfying the following three axioms:

1. For all $s, t \in S$, $d(s, t) \geq 0$ and $d(s, t) = 0$ if and only if $s = t$ (positivity).
2. For all $s, t \in S$, $d(s, t) = d(t, s)$ (symmetry).
3. For all $s, t, u \in S$, $d(s, u) \leq d(s, t) + d(t, u)$ (triangular inequality).

A set S equipped with a metric d is called a **metric space** and is denoted (S, d). Finding an optimal sequence of events between two genomes of course yields the distance between the two genomes, but the converse is not always true. Therefore, most of the time we will examine both aspects of the problem. A related problem we will be interested in is that of determining the *diameter* of a metric space.

Definition 1.2 The **diameter** of a metric space (S, d) is the maximal value the distance can reach, that is,

$$\max_{s, t \in S} d(s, t).$$

1.4 Scope of This Survey

This survey is restricted to the algorithmic and combinatorial aspects of genome rearrangements, but it also encompasses a few problems that are similar in spirit, even if they were not motivated by biology in the first place. The motivation for this is twofold: first, those problems deserve at least to be mentioned here, since they are closely related to genome rearrangement problems; and second, the study of related problems or variants of our problems may provide insight on the original problems we are interested in.

There has been a lot of work on probabilistic models and statistical studies of genome rearrangement problems, which we will not consider in this work. We refer the reader to the surveys of Eriksson [165] and Durrett [150]. As the reader may have guessed by reading section 1.1, we also will not delve much into the biological aspects, and will focus on the mathematical aspects of genome rearrangements, though we will mention applications and biological contexts where appropriate.

Some partial surveys have been published in earlier articles or book chapters. In 1995, Hannenhalli and Pevzner [197] wrote the first survey on the combinatorics of genome rearrangements, mainly based on their success at sorting signed permutations by reversals (see sections 3.3 and 4.2). The chapters by Pevzner [296] and Setubal and Meidanis [333] dedicated to genome rearrangements mainly focus on sorting permutations by reversals. The books edited by Gascuel [182], Sankoff and Nadeau [322], Böckenhauer and Bongartz [74], Jiang et al. [222], and Tseng and Zelkowitz [360] contain chapters that survey part of the field or try to give a quick overview of it. A survey article by Li et al. [247] reveals the importance and popularity of the field, and this book intends to be a more developed version of it.

1.5 Overview of the Models

Depending on the assumptions that are made on the data, or the events we want to study, different models can be used. The basic objects will be *homologous markers* (i.e., segments of genomes that can be found in several species, leading to the belief that they belonged to the common ancestor of these species). Genes are a good example of such markers, though they are not the only ones; but since genes were historically first used as markers for genome rearrangement studies, we often say "genes" for "homologous markers," as a simplification.

We will start with the simplest possible model, and progressively extend it by dropping restrictions. In the case where

1. the order of genes in each genome is known,
2. all genomes share the same set of genes,

3. all genomes contain a single copy of each gene, and
4. all genomes consist of a single chromosome,

genomes are modeled by *permutations* (see page 13): each gene can be assigned a unique number and is found exactly once in each genome.

As explained in section 1.1, a DNA molecule has two strands, and some rearrangements may change the strand that a segment belongs to. Therefore, each segment may be assigned a $+$ or a $-$ sign ($+$ is omitted most of the time) to indicate the strand it resides on, leading to the model of *signed permutations* (see page 15). We have also seen in section 1.1 that chromosomes can be linear or circular, and the latter case can be modeled using circular permutations rather than linear (classical) ones.

In spite of all the technical progress that has been made over the last decades and the large number of genomes that have been completely sequenced, many genomes have been only partially sequenced, which means that we cannot model them using permutations because genes are not totally ordered. In that case, genomes can nevertheless be modeled by *partially ordered sets*, and some studies can still be conducted using that model, as we will see in chapter 5.

In general, however, genes do not appear exactly once in each genome: due to duplications and deletions, there can be several copies of a gene in a given genome, or no copy at all. In that case, genomes are modeled by *strings* (on the alphabet of genes, see page 91) rather than permutations. Of course, it is possible to sign the elements of the string or to deal with circular strings.

A great part of living organisms have a genome that consists of several chromosomes (in a variable number, which can lie between 1 and 100), as is the case for all animals, and permutations as we have presented them are no longer a realistic model in that case. One can use the *disjoint cycle decomposition* of permutations (see page 14) to represent each chromosome using a cycle, in the case where chromosomes are circular, but this concept does not extend to linear chromosomes or strings since it cannot model duplicated genes. We may therefore want to extend our model to disjoint sets of paths and cycles (page 159), where each path or cycle models a chromosome.

Finally, one may not care about or simply not know the order of genes in each chromosome, and care only about whether two genes are in *synteny* (i.e., whether they belong to the same chromosome). In that case, genomes are modeled by *collections* of *sets* of genes (see chapter 12).

1.6 Organization of the Book

The first three parts of this book are organized according to the models presented in the previous section, each part being devoted to a mathematical object that has been used to construct genome rearrangement problems. Part IV is devoted to an important generalization of the basic genome rearrangement problem, known as the *median problem*, which aims at considering more than two genomes at the same time and inferring their common ancestors. It surveys the attempts to reconstruct the kin relationships between genomes by drawing *phylogenetic trees* in which nodes are ancestral configurations and branches (edges) account for evolutionary events.

Part V is a collection of summaries that provide useful additional information on the field, such as a list of available software based on the algorithms that we describe in the book and a list of open problems. This book also includes two appendices: appendix A is devoted to basic concepts of graph theory, and appendix B recalls the basics of the algorithmic theory of complexity, as well as a few **NP**-complete problems.

I DUPLICATION-FREE MODELS: PERMUTATIONS

2 Genomes as Permutations

Permutations were the first mathematical objects to serve as formal models for studying arrangements of DNA fragments among several species. Studies that use permutations in this context make use of a common knowledge (either classical or invented ad hoc) that is worth discussing here before stating our first genome rearrangement problems. More information about permutation group theory can be found in the books by Bóna [76] and Wielandt [370].

2.1 The Symmetric Group

Permutations are functions, which we will also view as orderings of a given finite set.

Definition 2.1 A **permutation** π is a bijection over the set $\{1, 2, \ldots, n\}$. The image of $i \in \{1, 2, \ldots, n\}$ by π is denoted by π_i. The elements π_i of the permutation are called **genes**.

A permutation induces a total order $\prec$ on $\{1, 2, \ldots, n\}$: we write $\pi_i \prec \pi_j$ if $i < j$. A classical notation used in combinatorics to denote a permutation π is the *two-row notation*

$$\begin{pmatrix} 1 & 2 & \cdots & n \\ \pi_1 & \pi_2 & \cdots & \pi_n \end{pmatrix};$$

we will adopt the traditional notation used in the genome rearrangement literature by keeping only the second row, that is,

$$\pi = (\pi_1 \; \pi_2 \; \cdots \; \pi_n).$$

We refer to permutations as defined in definition 2.1 as **linear permutations**, as opposed to other kinds of permutations that will be introduced later.

The classical **multiplication** or **composition** of permutations is applied, as the composition of functions, from right to left: when writing $\pi \circ \sigma$, we first apply σ, then π, which results in the permutation $(\pi_{\sigma_1}\ \pi_{\sigma_2}\ \cdots\ \pi_{\sigma_n})$. For example, if $\pi = (3\ 1\ 4\ 2)$ and $\sigma = (4\ 1\ 3\ 2)$, then $\pi \circ \sigma = (2\ 3\ 4\ 1)$. This operation induces a group structure on the set of all permutations. Indeed, composition is associative; the **identity permutation** $\iota = (1\ 2\ \cdots\ n)$ is the corresponding neutral element; and the **inverse permutation** of π is the permutation π^{-1} obtained by exchanging positions and elements in π, that is, $\pi^{-1}_{\pi_i} = i$ for $1 \leq i \leq n$.

Definition 2.2 The **symmetric group**, denoted by S_n, is the set of all permutations of $\{1, 2, \ldots, n\}$ with the operation $\circ$.

2.2 The Cycles of a Permutation

Orderings are by no means the only possible representation of permutations: another representation is the well-known *disjoint cycle decomposition.*

Definition 2.3 A **cycle** of a permutation π, denoted by $C = (i_1, i_2, i_3, \ldots, i_{k-1}, i_k)$ is a set of elements such that for $1 \leq j \leq k-1$, $\pi_{i_j} = i_{j+1}$ and $\pi_{i_k} = i_1$.

Every permutation decomposes into the product of disjoint cycles, and this decomposition is unique (up to ordering of the cycles and of the elements within each cycle). For example, the permutation (4 8 9 7 6 5 1 3 2 10) decomposes into the cycles (1, 4, 7), (2, 8, 3, 9), (5, 6), and (10). Note the distinction between the two notations: for instance, $(1\ 2\ \cdots\ n)$ is the identity permutation, whereas $(1, 2, \ldots, n)$ is the permutation $(2\ 3\ 4\ \cdots\ n-1\ n\ 1)$. A natural graphical representation of permutations follows from this decomposition.

Definition 2.4 The **graph of a permutation** π in S_n is the directed graph with vertex set $\{1, 2, \ldots, n\}$ and with edge set $\{(i, \pi_i) \mid 1 \leq i \leq n\}$.

The cycles of this graph are the cycles of the permutation π. Figure 2.1 shows a representation of the graph of (4 8 9 7 6 5 1 3 2 10).

Definition 2.5 All permutations that have the same disjoint cycle decomposition form a *conjugacy class.*

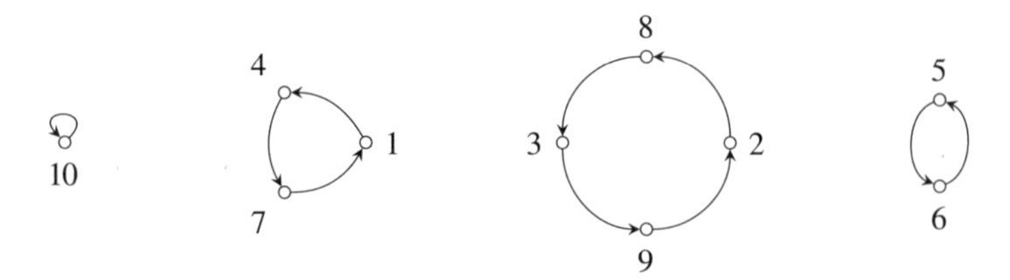

Figure 2.1
Graph of the permutation (4 8 9 7 6 5 1 3 2 10)

For example, $\pi = (1,2,3)(4,5,6)$ and $\sigma = (1,3,5)(2,4,6)$ belong to the same conjugacy class. This is the particular case for the symmetric group of the general notion of a conjugacy class in any group.

Definition 2.6 A permutation is *even* if the number of even cycles in its disjoint cycle decomposition is even or, equivalently, if it can be expressed as a product of an even number of 2-cycles. Otherwise, it is *odd*.

Definition 2.7 The *alternating group* A_n is the subgroup of S_n formed by all even permutations.

2.3 Signed Permutations

Signed permutations model the organization of genomes better than unsigned permutations, because they take into account the double helix structure of DNA. Indeed, given one arbitrary starting point for a chromosome, each DNA strand has an orientation, and by complementarity, the orientation of one strand is the reverse of the orientation of the other. This orientation corresponds to the direction in which genes are transcribed on a given strand.

Definition 2.8 A **signed permutation** on $\{1,2,\ldots,n\}$ is a permutation of the set $\{-n,\ldots,-2,-1,1,2,\ldots,n\}$ that satisfies $\pi_{-i} = -\pi_i$.

The one-row notation that was mentioned in section 2.1 in the context of unsigned permutations is also used for signed permutations. For example, the permutation

$$\begin{pmatrix} -4 & -3 & -2 & -1 & 1 & 2 & 3 & 4 \\ 3 & -1 & 4 & 2 & -2 & -4 & 1 & -3 \end{pmatrix}$$

is simply denoted by

$$(-2\ -4\ 1\ -3),$$

in which we drop the mapping of the negative elements since keeping it would be redundant, according to definition 2.8. Composition and inversion of signed permutations are well defined. The neutral element for the operation $\circ$ remains the identity permutation $\iota = (1\ 2\ \cdots\ n)$.

Definition 2.9 The **hyperoctahedral group**, denoted by $S_n^{\pm}$, is the set of all signed permutations of $\{-n,\ldots,-2,-1,1,2,\ldots,n\}$ with the operation $\circ$.

2.4 Distances on Permutation Groups

In this section, we use $\mathcal{G}$ to denote a permutation group, either signed or unsigned.

2.4.1 Rearrangements as Generators

Permutations may represent not only genomes, but also mutations (rearrangements), in such a way that a mutation ρ will transform a genome π into the genome $\pi \circ \rho$. For example, for a permutation π, the reversal of the segment $\pi_i, \dots, \pi_j$ $(i < j)$ in the permutation π is modeled by the permutation

$$\rho = (1 \cdots i-1 \; j \; j-1 \cdots i+1 \; i \; j+1 \cdots n)$$

or, in the signed case,

$$\rho = (1 \cdots i-1 \; -j \; -(j-1) \cdots -(i+1) \; -i \; j+1 \cdots n),$$

and

$$\pi \circ \rho = (\pi_1 \cdots \pi_{i-1} \; \pi_j \; \pi_{j-1} \cdots \pi_{i+1} \; \pi_i \; \pi_{j+1} \cdots \pi_n)$$

or, in the signed case,

$$\pi \circ \rho = (\pi_1 \cdots \pi_{i-1} \; -\pi_j \; -\pi_{j-1} \cdots -\pi_{i+1} \; -\pi_i \; \pi_{j+1} \cdots \pi_n).$$

Therefore, computing a sequence of rearrangements that transforms a permutation π into a permutation σ comes down to finding a sequence $\rho_1, \dots, \rho_k$ such that $\pi \circ \rho_1 \circ \cdots \circ \rho_k = \sigma$ or, equivalently, $\sigma^{-1} \circ \pi = \rho_k^{-1} \circ \cdots \circ \rho_1^{-1}$.

If the set of allowed rearrangements is such that it is always possible to obtain any permutation in $\mathcal{G}$ by composing those rearrangements, they are said to be **generators** of $\mathcal{G}$. In this case, rearrangement problems on permutations can be formulated as follows:

Given any two permutations π and σ in $\mathcal{G}$ and a set S of generators of $\mathcal{G}$, find a minimum length factorization of $\sigma^{-1} \circ \pi$ that consists only of elements of S.

Problems related to the factorization of permutations had been studied long before mathematicians became interested in genome rearrangement problems. There are some general complexity results that prevent us from hoping for a general solution to the problem: Even and Goldreich [168] have shown that finding such a factorization is **NP**-hard, and Jerrum [221] has shown that the problem is **PSPACE**-complete, even if $|S| = 2$. Some of these problems are easy to solve, however, if the set of generators is fixed, and not part of the instance: well-known examples include the case where S is the set of all exchanges of elements, whether they are adjacent (in which case an optimal sorting algorithm is the well-known "bubble sort"; see Knuth [237]) or not (in which case it corresponds to factorization of permutations into 2-cycles, a problem first solved by Cayley [100]). Jerrum [221] gives a few other examples of such tractable problems.

Generators of a permutation group yield the following natural graphical representation, which is fundamental in group theory.

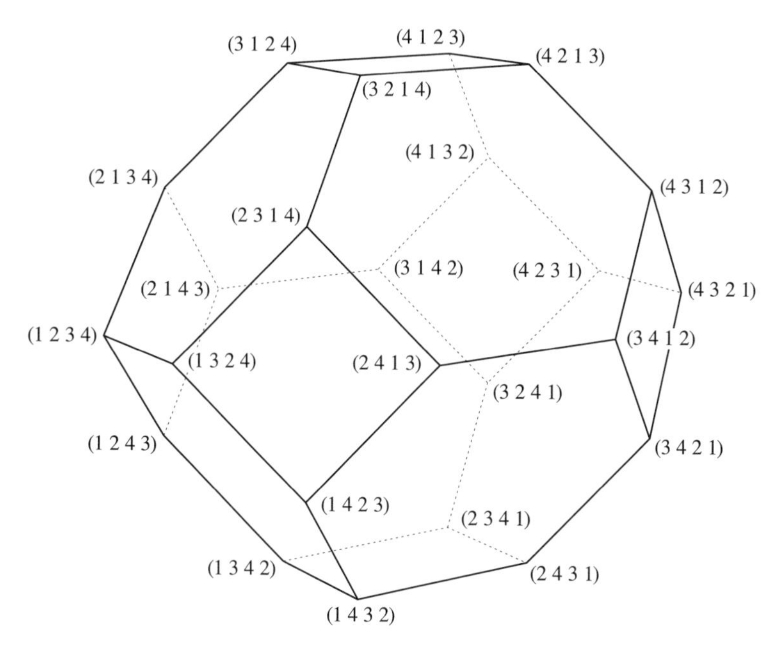

Figure 2.2
The Cayley graph associated with (S, S_4), where S is the set of all exchanges of adjacent elements, also known as the *permutohedron* of order 4
Based on a picture by David Eppstein

Definition 2.10 Given a set S of generators of a permutation group $\mathcal{G}$, the **Cayley graph** associated with $(S, \mathcal{G})$ is the graph whose vertices are the elements of $\mathcal{G}$ and whose edges connect two vertices such that the corresponding elements can be transformed into one another using an element of S.

Figure 2.2 shows the Cayley graph associated with (S, S_4), where S is the set of all exchanges of adjacent elements, and S_4 is the group of all linear unsigned permutations of 4 elements. The notion of diameter introduced in definition 1.2 has a natural interpretation in that setting: it is the length of the "longest shortest path" between any two vertices of the Cayley graph corresponding to the given group and set of generators.

2.4.2 Invariant Distances

Given a set of generators of a permutation group, a distance between two permutations π and σ can be defined as the minimum number of generators $\rho_1, \ldots, \rho_k$ such that $\pi \circ \rho_1 \circ \cdots \circ \rho_k = \sigma$. This is indeed a distance, according to definition 1.1. The distances we consider satisfy an additional property that allows us to reduce genome rearrangement problems to a simpler canonical problem.

Definition 2.11 A distance d on $\mathcal{G}$ is **left-invariant** if for all π, σ, τ in $\mathcal{G}$, we have

$d(\pi, \sigma) = d(\tau \circ \pi, \tau \circ \sigma)$.

Left-invariance can be intuitively explained as follows. Given two genomes that can be represented as permutations of the same set of genes, the number of operations that will be required to transform one genome into the other does not depend on how genes are numbered: one can arbitrarily assign a unique number to each gene and "rename" the genes in the other genome accordingly, without modifying the corresponding distance between the two genomes. Left-invariance is an important underlying concept in genome rearrangements because it is the reason why many problems considered in that field reduce to a sorting problem: indeed, computing the distance of interest between any two permutations π and σ is by left-invariance equivalent to computing the distance between permutations $\sigma^{-1} \circ \pi$ and ι, and we can therefore restrict our attention to computing the distance between a permutation and the identity. An immediate corollary of this property is that for all those distances and any permutation π in $\mathcal{G}$, we have $d(\pi, \iota) = d(\pi^{-1}, \iota)$. Since most of the time we will be considering the distance of a permutation from the identity, we will abbreviate $d(\pi, \iota)$ into $d(\pi)$.

Definition 2.12 Given a permutation π, a sequence of k rearrangements $\rho_1, \ldots, \rho_k$ is called a **sorting sequence** for π if $\pi \circ \rho_1 \circ \cdots \circ \rho_k = \iota$.

As the reader might wonder, there is also a concept known as *right*-invariance, but most distances used in genome rearrangements do not satisfy this property. We note, however, that right-invariant distances between permutations have applications in statistics (see Diaconis [135]) and in the theory of adaptive sorting algorithms (see Estivill-Castro and Wood [166]).

2.5 Circular Permutations

Circular permutations model circular chromosomes. They are especially relevant because most unichromosomal genomes actually consist of a circular chromosome (bacteria are a notable example). They also model the evolution of mitochondrial and chloroplastic genomes, which are small, independent circular chromosomes found in the cells of all animals and plants, as well as the "nuclear genome," which most often contains several linear chromosomes.

We define two kinds of circular permutations: the "classical" ones, found in the mathematical literature, and the "genomic" circular permutations with equivalence classes that are a bit broader than in the classical case: permutations are supposed to model chromosomes, and since linear chromosomes have no prescribed starting point, circular chromosomes have no preferred reading direction (clockwise or counterclockwise). Usually this information is not taken into account in studies of linear chromosomes as permutations. This can be justified by the study of chromosome segments. But it cannot be justified for circular permutations, for which it is necessary to take into account the fact that they may be read in both directions.

2.5.1 Classical Circular Permutations

Let first $\equiv^\circ$ be the equivalence relation between signed or unsigned linear permutations defined as follows: we write $\pi \equiv^\circ \sigma$ if there exists an integer k such that for all i in $\{1, 2, \ldots, n\}$, $\pi_i = \sigma_{(i+k \bmod n)+1}$. In other words, π and σ are equivalent if π can be obtained by rotating the elements of σ (or conversely). An equivalence class under this relation is called a (signed or unsigned) **circular permutation**, and may be denoted using any of its elements.

In order to avoid confusion between linear and circular permutations, we use the notation $[\pi_1\ \pi_2\ \cdots\ \pi_n]$ to denote circular permutations; therefore, we have

$$[\pi_1\ \pi_2\ \cdots\ \pi_n] = \{(\pi_1\ \pi_2\ \cdots\ \pi_n), (\pi_2\ \pi_3\ \cdots\ \pi_n\ \pi_1), \ldots, (\pi_n\ \pi_1\ \cdots\ \pi_{n-1})\}.$$

We will also use the $^\circ$ symbol to explicitly denote circular permutations without expanding their content. For instance, π is a linear permutation, whereas π° is a circular one. If $\pi = (\pi_1\ \cdots\ \pi_n)$ is a linear (signed or unsigned) permutation, we refer to the circular permutation $\pi^{e\circ} = [\pi_1\ \cdots\ \pi_n\ n+1]$ as its corresponding **circular extension**. Any linear permutation in the equivalence class of a circular permutation π° is called a **linearization** of π°.

2.5.2 Genomic Circular Permutations

The following definition corresponds to how circular permutations are often described in combinatorics of genome rearrangement literature (see, e.g., Meidanis et al. [265], Solomon et al. [342]). Let $\equiv^\circ$ be the relation between signed or unsigned linear permutations defined as follows: we write $\pi \equiv^\circ \sigma$ if

- either there exists an integer k such that for all i in $\{1, 2, \ldots, n\}$, $\pi_i = \sigma_{(i+k \bmod n)+1}$
- or $(\pi_1, \ldots, \pi_n) = (\sigma_n, \ldots, \sigma_1)$ if π and σ are unsigned, and $(\pi_1, \ldots, \pi_n) = (-\sigma_n, \ldots, -\sigma_1)$ if π and σ are signed.

In other words, π and σ are in relation if π can be obtained by rotating the elements of σ or by reversing the whole set of elements, flipping the signs for signed permutations. The transitive closure of $\equiv^\circ$ is an equivalence relation, and an equivalence class under this equivalence relation is called a (signed or unsigned) **genomic circular permutation**, and may be denoted using any of its elements.

For each problem involving circular permutations, we will mention whether it is solved for classical or genomic circular permutations. Note that given one genomic signed circular permutation such that one linearization π contains only elements with a + sign, the subclass containing all permutations in which all elements have a + sign corresponds to the classical unsigned circular permutation of π. This observation will lower the number of different variants to consider.

2.6 First Measures of Similarity between Permutations

Some measures of similarity and dissimilarity between permutations (they are numerous in the literature; see, for example, the survey by Estivill-Castro and Wood [166]) are used in computational biology, even if they do not explicitly correspond to rearrangement events. They are nevertheless useful, and we present here two examples based on concepts that will be used by other distances we will introduce later.

2.6.1 Breakpoints

- Introduced by Sankoff and Blanchette [317].
- Complexity: polynomial. Computing the distance and diameter is a trivial problem, contained in the definition.

One of the first intuitions inspired by the parsimony criterion is that if a group of genes appears consecutively in several species, then they must have been present in the same order in the ancestral species, and were not separated during evolution.

This translates mathematically into the notions of adjacencies and breakpoints, which, despite their simplicity, deserve a clarification because their definition in the literature varies according to the models and operations under consideration (see also a discussion on multichromosomal genomes in section 10.2).

Definition 2.13 The **linear extension** of a (signed or unsigned) permutation π of $\{1, 2, \ldots, n\}$ is the permutation of $\{0, 1, \ldots, n+1\}$ defined by $\pi^l = (0\ \pi_1\ \cdots\ \pi_n\ n+1)$.

For the linear extension of signed permutations, it is a convention that 0 has a positive sign.

Definition 2.14 Let π^l be the linear extension of a (signed or unsigned) permutation π on $\{1, 2, \ldots, n\}$. A **point** of π is an ordered pair (π_i^l, π_{i+1}^l) for $0 \leq i \leq n$. Moreover:

- If $\pi_{i+1}^l = \pi_i^l + 1$, it is called an **adjacency**;
- If $\pi_{i+1}^l = \pi_i^l - 1$, it is called a **reverse adjacency**;
- If it is not an adjacency, it is called a **breakpoint**;
- If it is neither an adjacency nor a reverse adjacency, it is called a **strong breakpoint**.

For a circular permutation π°, its **canonical linearization** is the linearization of π° placing the last element n at the last position. Note that a genomic unsigned permutation has two canonical linearizations, and a classical signed permutation may have none. In this case, define the canonical linearization as the one of the corresponding genomic permutation. We say that the points, adjacencies, reverse adjacencies, and breakpoints are the points, adjacencies, reverse adjacencies, and breakpoints of the linear permutation $(\pi_1\ \pi_2 \cdots \pi_{n-1})$, where π is a canonical linearization. Note that in genomic unsigned permutations, adjacencies correspond to reverse adjacencies, and breakpoints to strong breakpoints.

We illustrate those concepts by the following example: points are indicated by •s in the permutation $(0 \bullet 4 \bullet 8 \bullet 9 \bullet 7 \bullet 6 \bullet 5 \bullet 1 \bullet 3 \bullet 2 \bullet 10)$. Of all those points, (8, 9) is an adjacency; (7, 6), (6, 5), and (3, 2) are reverse adjacencies; and all other points are breakpoints.

The number of points and breakpoints of a permutation π (be it signed or unsigned, linear or circular) are denoted by $p(\pi)$ and $bp(\pi)$, respectively. Moreover, we use the notation $sb(\pi)$ for the number of strong breakpoints. Note that if π° is a circular permutation of $\{1, 2, \ldots, n\}$, then $p(\pi^\circ) = n$, and if π is a linear permutation of $\{1, 2, \ldots, n\}$, then $p(\pi) = n + 1$.

Breakpoints yield a first simple example of an evolutionary distance between genomes: the **breakpoint distance** between two permutations π and σ is defined as $bp(\pi, \sigma) = bp(\sigma^{-1} \circ \pi)$. This distance is trivial to compute, and some authors have argued that it is not less realistic than others (see, e.g., Sankoff and Blanchette [317]). It corresponds to the number of adjacencies in one permutation that are not adjacencies in the other. It is extended to circular permutations by applying the same formula, removing element n from the linearization where n is the last element.

2.6.2 Common Intervals and Semipartitive Families

Common intervals are an extension of adjacencies that model the fact that groups of genes stay together in a genome, but not necessarily in the same order or with the same direction.

Definition 2.15 An **interval** (or **segment**) of a permutation π is a set $\{|\pi_a|, |\pi_{a+1}|, \ldots, |\pi_{b-1}|, |\pi_b|\}$, with $1 \le a \le b \le n$.

The elements π_a and π_b are the **extremities** of the interval.

Definition 2.16 A set I is said to be a **common interval** of permutations π and σ if it is an interval of both π and σ.

In the particular case where $\sigma = \iota$, an interval $I = \{|\pi_a|, \ldots, |\pi_b|\}$ of π is a common interval (common to π and ι) if, given $m = \min_{i \in [a,b]} |\pi_i|$ and $M = \max_{i \in [a,b]} |\pi_i|$, I

contains all integers in the range $[m, M]$, which is equivalent to requiring that $M - m = b - a$.

As an example, let us consider the permutation (8 9 7 6 5 3 1 4 2 10); its nontrivial (i.e., we do not list singletons nor the whole permutation) common intervals with respect to ι are indicated as line segments below.

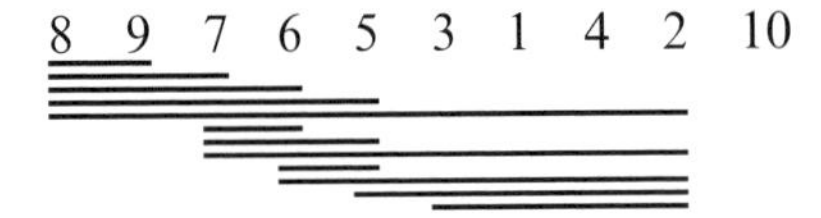

Computing the set of common intervals of two permutations is both easy and very interesting, because it raises a modular structure of differences between permutations that can be useful in many rearrangements studies. (For a method of construction, see, for example, Bui et al. [91]).

Definition 2.17 Two common intervals of permutations π and σ are said to **overlap** if they intersect and neither of them is contained in the other. A common interval is called **strong** if it does not overlap any other common interval.

Definition 2.18 The **strong interval tree** of permutations π and σ is the graph whose vertex set is the set of strong intervals of π and σ, and which contains an edge between two strong intervals I_1 and I_2 if $I_1 \subset I_2$ and there is no other strong interval I_3 such that $I_1 \subset I_3 \subseteq I_2$. This graph is a tree rooted at the interval $\{1, 2, \ldots, n\}$, so that each node has one parent and possibly several children, ordered according to their position in both permutations.

Lemma 2.1 (See, for example, Bui et al. [91].) Let I be a strong interval of two permutations π and σ. One of the following is true:

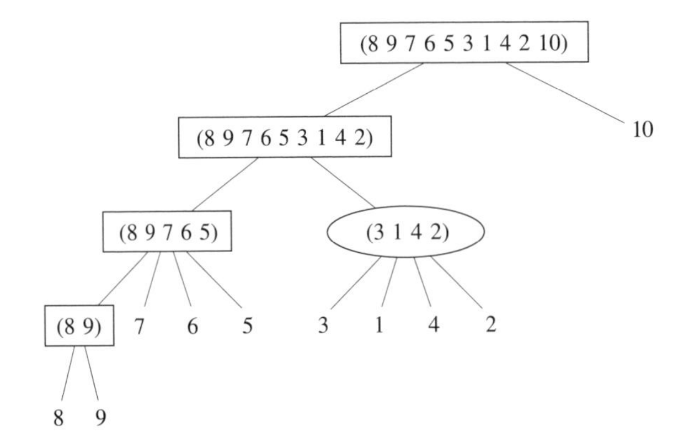

Figure 2.3
Example of a PQ-tree of common intervals for the permutation (8 9 7 6 5 3 1 4 2 10). Linear nodes are drawn as rectangles, and prime nodes are represented as ellipses

- No union of children of I in the tree of strong intervals is a common interval of π and σ; in this case I is called **prime**.
- All the subsets that are the union of consecutive children are common intervals of π and σ; in that case I is called **linear**.

Definition 2.19 The strong interval tree is called a **PQ-tree**, and the prime nodes are identified as **P-nodes**; linear nodes are identified as **Q-nodes**.

All these notions are immediately useful for common intervals of permutations, since the family of common intervals of two permutations is weakly partitive and linear. Figure 2.3 shows an example of a PQ-tree of a family of common intervals.

Bergeron et al. [49] use the structure of common intervals to infer a metric between permutations, whereas Bernt et al. [65] use the structure of common intervals to give a hint on the type of rearrangements that have occurred for specified instances.

3 Distances between Unsigned Permutations

Unsigned permutations were the first combinatorial model of genomes for the study of rearrangements (see Watterson et al. [369]), and are still used when the orientation of the markers is not known, as is the case, for example, when the data come from *in situ hybridization* (see Wienberg [371]). All permutations in this chapter are unsigned, unless explicitly stated otherwise.

3.1 Transposition Distance

We begin our survey of permutation-related genome rearrangement problems with the operation of *transposition* (the term "transposition" comes from biology and refers to *transposons*, which are sequences of DNA that can be displaced in a genome. They therefore have nothing to do with *algebraic transpositions* to which mathematicians are used). Transpositions consist in displacing an interval of the permutation or, equivalently, in exchanging two contiguous intervals of the permutation. However simple the problem of sorting by transpositions might seem, a lot of questions remain open, the most notable being the complexity of the sorting problem and computing the associated distance, as well as the diameter of the symmetric group under this operation.

- Introduced by Bafna and Pevzner [30].
- Complexity: unknown.
- Best approximation ratio: $\frac{11}{8}$ by Elias and Hartman [160].
- Diameter: unknown. Lies between $\lfloor\frac{n+1}{2}\rfloor$ (see Bafna and Pevzner [30]) and $\frac{2n}{3}$ (see Eriksson et al. [164]).

Definition 3.1 For any permutation π in S_n, the **transposition** $\tau(i,j,k)$ with $1 \leq i < j < k \leq n+1$ applied to π exchanges the intervals determined respectively by i and $j-1$ and by j and $k-1$, transforming π into $\pi \circ \tau(i,j,k)$. Therefore, $\tau(i,j,k)$ is the following permutation:

Table 3.1
The number of permutations π in S_n with $td(\pi) = k$

$n \backslash k$	0	1	2	3	4	5	6	7
1	1	0	0	0	0	0	0	0
2	1	1	0	0	0	0	0	0
3	1	4	1	0	0	0	0	0
4	1	10	12	1	0	0	0	0
5	1	20	68	31	0	0	0	0
6	1	35	259	380	45	0	0	0
7	1	56	770	2,700	1,513	0	0	0
8	1	84	1,932	13,467	22,000	2,836	0	0
9	1	120	4,284	52,512	191,636	114,327	0	0
10	1	165	8,646	170,907	1,183,457	2,010,571	255,053	0

$$\begin{pmatrix} 1 \cdots i-1 & \boxed{i\; i+1 \cdots j-2\; j-1} & \boxed{j\; j+1 \cdots k-1} & k \cdots n \\ 1 \cdots i-1 & \boxed{j\; j+1 \cdots k-1} & \boxed{i\; i+1 \cdots j-2\; j-1} & k \cdots n \end{pmatrix}.$$

The **transposition distance** of a permutation π will be denoted by $td(\pi)$. Table 3.1 shows the distribution of the transposition distance for $1 \leq n \leq 10$.

For a (classical) circular permutation π°, the action of a transposition can be modeled on the intervals of any of its linearizations. The problem of sorting a circular permutation by transpositions is equivalent to the problem of sorting linear permutations by transpositions, as proved, for example, by Hartman [201]. Indeed, it is easy to see that any transposition on an arbitrary linearization has the same result as another transposition on a canonical linearization. No study of the transposition distance of genomic circular permutations has been reported.

3.1.1 Lower Bounds on the Transposition Distance

3.1.1.1 Lower Bounds Based on Breakpoints It can easily be seen that one transposition can decrease the number of breakpoints by at most three, as illustrated by the following example:

$(5\ 1\ \boxed{3}\,\boxed{2}\ 4) \rightarrow (5\ 1\ 2\ 3\ 4).$

This yields a first lower bound on the transposition distance.

Theorem 3.1 [30] For all π in S_n, we have $td(\pi) \geq \frac{bp(\pi)}{3}$.

Breakpoints can be categorized into two different classes, which are well known in the literature of sorting procedures: breakpoints such that $\pi_{i+1} > \pi_i$ are called

ascents, and breakpoints such that $\pi_{i+1} < \pi_i$ are called **descents**. We use $des(\pi)$ to denote the number of descents in a permutation π.

It can be easily checked that whereas the number of breakpoints may decrease by three through the application of a transposition, the number of descents can decrease only by two (see our above example in the case of breakpoints). Since ι has no descent, we immediately have a lower bound of $\frac{des(\pi)}{2}$. However, it should be noted that, contrary to the number of breakpoints and to the transposition distance, the number of descents in a permutation may differ from the number of descents in its inverse. This yields a second lower bound, mentioned by Eriksson et al. [164].

Theorem 3.2 [164] For all π in S_n,

$$td(\pi) \geq \max\left(\frac{des(\pi)}{2}, \frac{des(\pi^{-1})}{2}\right).$$

An observation related to breakpoints allows us to restrict our study of sorting by transpositions to a particular class of permutations characterized by the following idea: since the identity permutation is the only permutation with no breakpoint, a first intuitive sorting strategy would be to preserve adjacencies and "repair" breakpoints. A lemma by Christie [115] confirms that the intuition of preserving adjacencies always leads to an optimal solution.

Lemma 3.1 [115] For a permutation π, there exists an optimal sorting sequence of transpositions that never breaks the adjacencies of π.

Every permutation π can be uniquely transformed into a permutation with no adjacency without affecting its distance, by partitioning π into strips.

Definition 3.2 A **strip** is a maximal interval of π containing no breakpoint.

The **reduced permutation** corresponding to π is obtained by discarding the leftmost (resp. rightmost) strip if it begins with 1 (resp. if it ends with n), then keeping the minimal element of each strip in π, and finally renumbering the remaining elements appropriately. For example, the reduced permutation corresponding to $(\underline{4\ 5\ 6}\ \underline{3}\ \underline{1}\ \underline{7\ 8}\ \underline{2})$, in which strips are underlined, is (4 3 1 5 2), through replacing strips 4 5 6 and 7 8. Lemma 3.1 implies that the transposition distance of the reduced permutation is the same as the transposition distance of π, and we can therefore restrict our attention to sorting reduced permutations.

3.1.1.2 Lower Bounds Based on the Cycle Graph

The cycle graph of a permutation, introduced by Bafna and Pevzner [30], is one of the many variants of a pervasive structure that has proved most useful in obtaining theoretical results on genome

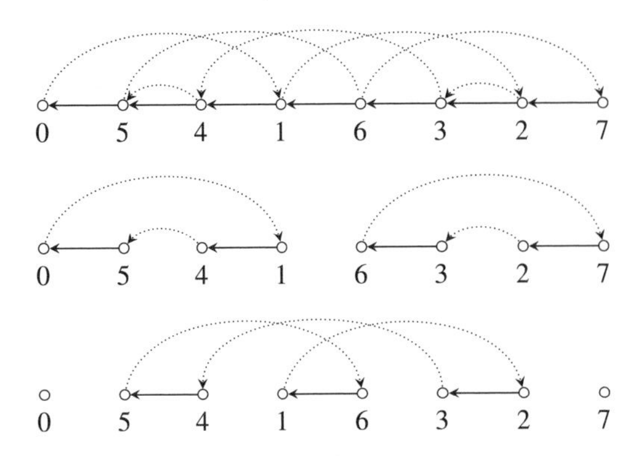

Figure 3.1
Cycle graph of the permutation (5 4 1 6 3 2) and its decomposition into three cycles

rearrangement problems. Indeed, most important results in the field (e.g., bounds, approximation algorithms, or formulas for computing rearrangement distances) make an extensive use of parameters based on the cycle graph or on a close structure.

Definition 3.3 The **cycle graph** of a permutation π of $\{1, 2, \ldots, n\}$ is the directed graph $G(\pi)$ with vertex set $\{0, 1, \ldots, n, n+1\}$ and whose arcs consist in

- **black** (or **reality**) arcs $(\pi_i^l, \pi_{(i-1)}^l)$ for $1 \leq i \leq n+1$,
- **gray** (or **desire**) arcs $(i, (i+1))$ for $0 \leq i \leq n$,

where π^l is the linear extension of π (see definition 2.13).

Reality arcs represent what we have (the permutation π), whereas desire arcs indicate what we want to obtain (the permutation ι). Each vertex of the cycle graph has one incoming arc of each color and one outgoing arc of each color. As a straightforward consequence, the cycle graph decomposes in a single way into **alternating cycles** (i.e., cycles that alternate black and gray arcs). The number of alternating cycles in $G(\pi)$ is denoted by $c(G(\pi))$.

Figure 3.1 shows the cycle graph of a linear permutation. It is straightforward to define the cycle graph using the circular extension instead of the linear extension: identify 0 and $n+1$ in definition 3.3. This gives the circular layout of the cycle graph used in figure 3.2, which is frequently encountered in the literature. Since sorting a permutation, its linear extension or its circular extension is equivalent, any model can be preferred for devising bounds or algorithms.

Since the identity permutation ι is the only permutation whose cycle graph contains $n+1$ cycles, which is the maximum possible number, sorting a permutation π by transpositions comes down to, from a graph-theoretic point of view, increasing

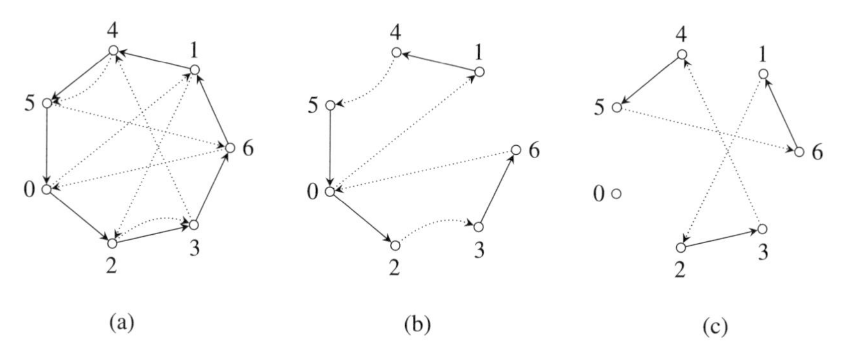

Figure 3.2
(*a*) The cycle graph of [0 5 4 1 6 3 2], and its unique decomposition into three alternating cycles ((*b*) two 2-cycles, and (*c*) a 3-cycle)

the number of cycles in $G(\pi)$ in as few steps as possible. The remark of Bafna and Pevzner [30] that a transposition may increase the number of cycles by at most two yields a lower bound on the transposition distance.

Theorem 3.3 [30] For every permutation π, we have $td(\pi) \geq \frac{p(\pi)-c(G(\pi))}{2}$.

This bound can be sharpened by taking the parity of cycles into account. An alternating cycle is said to be **odd** if it has an odd number of gray (or black) arcs, and **even** otherwise. The number of odd (resp. even) cycles in $G(\pi)$ is denoted by $c_{odd}(G(\pi))$ (resp. $c_{even}(G(\pi))$).

Theorem 3.4 [30] For every permutation π, we have $td(\pi) \geq \frac{p(\pi)-c_{odd}(G(\pi))}{2}$.

3.1.2 Upper Bounds

3.1.2.1 Upper Bounds Based on Breakpoints The first upper bound based on the number of breakpoints was computed by Bafna and Pevzner [30], and stated that for all π in S_n, $td(\pi) \leq \frac{3}{4}bp(\pi)$. It is deduced from a $\frac{3}{2}$-approximation algorithm (see section 3.1.5) and the lower bound of theorem 3.4. It has been outperformed by Eriksson et al. [164].

Theorem 3.5 [164] For every permutation π,

$$td(\pi) \leq \begin{cases} \left\lceil \frac{2}{3}bp(\pi) \right\rceil & \text{if } n < 9; \\ \left\lfloor \frac{2bp(\pi)-2}{3} \right\rfloor & \text{if } n \geq 9. \end{cases}$$

This bound is obtained by proving that two operations, either on the permutation *or on its inverse*, are sufficient to create three adjacencies.

3.1.2.2 Upper Bounds Based on the Cycle Graph A first straightforward upper bound is deduced from the possibility to increase the number of cycles of $G(\pi)$ by at least one at each step, which corresponds to creating one adjacency at a time.

Theorem 3.6 [30] For all π in S_n,

$$td(\pi) \leq p(\pi) - c(G(\pi)).$$

The following upper bound is based on a $\frac{3}{2}$-approximation algorithm and the lower bound of theorem 3.4.

Theorem 3.7 [30] For all π in S_n,

$$td(\pi) \leq \frac{3(p(\pi) - c_{odd}(G(\pi)))}{4}.$$

3.1.2.3 Upper Bounds Based on the Γ-Graph Labarre [240] introduced a slight variant of the graph of a permutation that proved useful for the problem studied in this section. It is essentially the same graph, except that vertices are ordered by position.

Definition 3.4 The **Γ-graph** of a permutation π in S_n is the directed graph $\Gamma(\pi)$ with ordered vertex set $(\pi_1, \ldots, \pi_n)$ and arc set $\{(\pi_i, \pi_j) \mid \pi_i = j\}$.

Figure 3.3 shows an example of a Γ-graph. If $C = (i_1, i_2, \ldots, i_k)$ is a cycle of π, we obtain a cycle $(\pi_{i_1}, \pi_{i_2}, \ldots, \pi_{i_k})$, which we also denote as C, in $\Gamma(\pi)$, and call it a k-**cycle**. The **length** of a cycle in $\Gamma(\pi)$ is therefore k.

In a fashion quite similar to the parity of cycles defined in the context of $G(\pi)$, a k-cycle in $\Gamma(\pi)$ is **odd** (resp. **even**) if k is odd (resp. even). Likewise, $c(\Gamma(\pi))$ denotes the number of cycles in $\Gamma(\pi)$, and $c_{odd}(\Gamma(\pi))$ (resp. $c_{even}(\Gamma(\pi))$ denotes the number of odd (resp. even) cycles in $\Gamma(\pi)$. Labarre [240] derived an upper bound based on these cycles.

Theorem 3.8 [240] For all π in S_n,

$$td(\pi) \leq n - c_{odd}(\Gamma(\pi)).$$

3.1.2.4 Other Upper Bounds An upper bound based on the length of a longest increasing subsequence of a linear permutation was deduced by Guyer et al. [192].

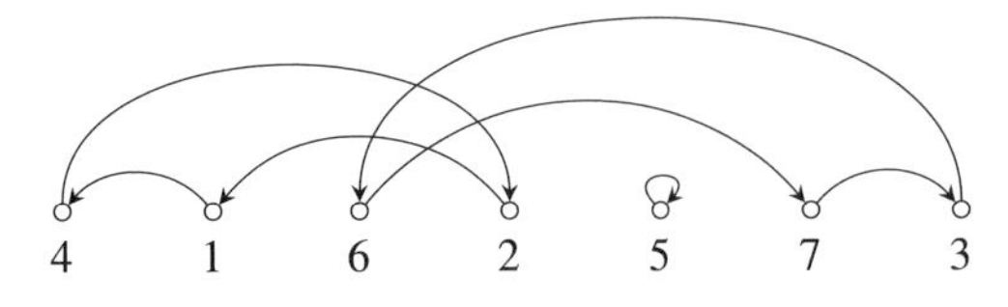

Figure 3.3
The Γ-graph of the permutation (4 1 6 2 5 7 3)

Definition 3.5 Given a permutation π, a **subsequence** of π is a subset $\{\pi_{i_1}, \ldots, \pi_{i_k}\}$ of nonnecessarily contiguous elements of π, with $i_1 < i_2 < \cdots < i_k$. The subsequence is **increasing** if $\pi_{i_1} < \cdots < \pi_{i_k}$, and it is a **longest increasing subsequence** if there is no other increasing subsequence in π with more elements.

Guyer et al. [192] observed that a permutation could be sorted by transpositions by "growing" its longest increasing subsequence. That subsequence can always be increased by at least 1 at each step, which yields the following upper bound.

Observation 3.1 [192] For all π in S_n,

$$td(\pi) \leq n - |LIS(\pi)|,$$

where $|LIS(\pi)|$ is the length of a longest increasing subsequence of π.

The right-hand side of the above inequality is also known as *Ulam's distance* (see Diaconis [135]), which we denote as $ulam(\pi)$.

Benoît-Gagné and Hamel [40] proposed another view of sorting by transpositions, in order to avoid using the cycle graph.

Definition 3.6 For any π in S_n, the **left code** of an element π_i is

$$lc(\pi_i) = |\{\pi_j \mid \pi_j > \pi_i \text{ and } 1 \leq j \leq i-1\}|,$$

and the **right code** of an element π_i is

$$rc(\pi_i) = |\{\pi_j \mid \pi_j < \pi_i \text{ and } i+1 \leq j \leq n\}|.$$

The left (resp. right) code of a permutation is the sequence formed by the left (resp. right) codes of its elements.

The left code of an element π_i simply counts the number of elements that are larger than π_i and precede it in π, and the right code of an element π_i counts the number of elements that are smaller than π_i and follow it in π. For example, the permutation $\pi = (5\ 4\ 1\ 6\ 3\ 2)$ has the left code $lc(\pi) = (0, 1, 2, 0, 3, 4)$ and the right code $rc(\pi) = (4, 3, 0, 2, 1, 0)$; it can easily be seen that only the identity permutation has $(0, 0, \ldots, 0)$ as both left and right codes, so the goal is to increase the number of 0s in the left or right code of a permutation.

Definition 3.7 A **plateau** in a sequence S is a subsequence of contiguous elements that have the same nonzero value. We denote as $plat(S)$ the number of plateaus in S.

Lemma 3.2 [40] For any π in S_n, we have

$$td(\pi) \leq \min\{plat(lc(\pi)), plat(rc(\pi))\}.$$

3.1.2.5 Upper Bounds for Special Classes of Permutations Elias and Hartman [160] proved upper bounds on the distance of three special classes of permutations.

Definition 3.8 A permutation π in S_n is **simple** if $G(\pi)$ contains no cycle with more than three black arcs (the total number of arcs is at most six).

Definition 3.9 A permutation π in S_n is a 2-**permutation** (resp. 3-**permutation**) if all cycles in $G(\pi)$ have two (resp. three) black arcs.

Note that a 2-permutation (resp. 3-permutation) exists only if $p(\pi)$ can be divided by 4 (resp. 3). Simple permutations were initially introduced because they are easier to study than arbitrary permutations, due to the simpler structure of their cycle graphs. Any permutation can be transformed into a simple permutation in linear time (see Gog and Bader [186]), and the transformation preserves the lower bound of theorem 3.4 on the transposition distance.

Theorem 3.9 [160] For every simple permutation π that is not a 3-permutation,

$$d(\pi) \leq \left\lfloor \frac{p(\pi)}{2} \right\rfloor.$$

Theorem 3.10 [160] For every 3-permutation π,

$$d(\pi) \leq 11 \left\lfloor \frac{p(\pi)}{24} \right\rfloor + \left\lfloor \frac{3\left(\frac{p(\pi)}{3} \bmod 8\right)}{2} \right\rfloor + 1.$$

3.1.3 Improving Bounds Using Toric Permutations

Eriksson et al. [164] introduced a useful equivalence relation on S_n whose equivalence classes are called *toric permutations*. As in the work of Hultman [214], for x in $\{0, 1, 2, \ldots, n\}$, let $\bar{x}^m = (x + m) \pmod{n+1}$, and define the following operation on extended circular permutations:

$$m + \pi^{eo} = [\bar{0}^m \; \overline{\pi_1}^m \; \overline{\pi_2}^m \; \cdots \; \overline{\pi_n}^m].$$

Two permutations π, σ in S_n are said to be **torically equivalent** if there exists m $(0 \leq m \leq n)$ such that $\sigma^{eo} = m + \pi^{eo}$. An equivalence class for this relation is called a **toric permutation**, and is denoted by $\pi^{\circ}_{\circ}$ when it contains the permutation π. We illustrate those concepts with the following example: let $\pi = (3\ 1\ 5\ 2\ 4\ 6)$; then $\pi^{eo} = [0\ 3\ 1\ 5\ 2\ 4\ 6]$, and

$$0 + \pi^{eo} = [0\ 3\ 1\ 5\ 2\ 4\ 6],$$

$$1 + \pi^{eo} = [1\ 4\ 2\ 6\ 3\ 5\ 0],$$

$2 + \pi^{e\circ} = [2\ 5\ 3\ 0\ 4\ 6\ 1]$,

$3 + \pi^{e\circ} = [3\ 6\ 4\ 1\ 5\ 0\ 2]$,

$4 + \pi^{e\circ} = [4\ 0\ 5\ 2\ 6\ 1\ 3]$,

$5 + \pi^{e\circ} = [5\ 1\ 6\ 3\ 0\ 2\ 4]$,

$6 + \pi^{e\circ} = [6\ 2\ 0\ 4\ 1\ 3\ 5]$,

which yields $\pi_\circ^\circ = \{(3\ 1\ 5\ 2\ 4\ 6),\ (1\ 4\ 2\ 6\ 3\ 5),\ (4\ 6\ 1\ 2\ 5\ 3),\ (2\ 3\ 6\ 4\ 1\ 5),\ (5\ 2\ 6\ 1\ 3\ 4),\ (2\ 4\ 5\ 1\ 6\ 3),\ (4\ 1\ 3\ 5\ 6\ 2)\}$, and all permutations in that set are torically equivalent.

It is easily seen that any two permutations that are torically equivalent have the same number of breakpoints; but they have more in common, as shown by the following property, which is actually the main reason why toric permutations were introduced.

Lemma 3.3 [164] Any two permutations that are torically equivalent have the same transposition distance.

This implies that any bound that is valid for a given permutation remains valid for all permutations to which it is torically equivalent. This does not improve any bound that is based on parameters preserved by toric permutations, such as the number of breakpoints or the number of cycles in the cycle graph, as shown by Hultman [214]. However, any bound based on parameters that are *not* preserved by the toric equivalence relation can be improved using that relation. This is, for example, the case for the cycles of the Γ-graph, which helps sharpen the bound of theorem 3.11.

Theorem 3.11 [240] For all $\pi \neq \iota$ in S_n,

$$td(\pi) \leq n - \max_{\sigma \in \pi_\circ^\circ} c_{odd}(\Gamma(\sigma)).$$

3.1.4 Easy Cases

This section lists some linear permutations whose transposition distance can be computed in polynomial time.

- The **reversed permutation** $\chi = (n\ n-1\ \cdots\ 2\ 1)$ has transposition distance $\lfloor \frac{n}{2} \rfloor + 1$ (see Christie [115] and a simple proof by Eriksson et al. [164]).
- 2-permutations have a transposition distance equal to $n/2$ (see Christie [115]).
- Permutations of the form $(2\ 4\ 6\ \cdots\ n\ 1\ 3\ 5\ \cdots\ n-1)$ with n even have transposition distance $n/2$ (see Christie [115]).

• The permutation κ_k, defined by

$$\begin{cases} \kappa_1 = (4\ 3\ 2\ 1) \\ \kappa_2 = (4\ 3\ 2\ 1\ 5\ 9\ 8\ 7\ 6) \\ \vdots \\ \kappa_k = (4\ 3\ 2\ 1\ 5\ 9\ 8\ 7\ 6\ 10 \cdots 5k\ k+3\ k+2\ k+1\ k), \end{cases}$$

has transposition distance $2k + \lceil \frac{k}{2} \rceil$ (see Christie [115]).

• The permutation $\pi = (4\ 3\ 2\ 1\ 5\ 13\ 12\ 11\ 10\ 9\ 8\ 7\ 6\ \sigma)$, where σ is any 2-permutation constructed using numbers starting from 14, has distance $\lfloor \frac{n+1}{2} \rfloor + 1$, where n is the size of the permutation. It is currently the permutation with the largest known value for the transposition distance (see Elias and Hartman [160]).

• γ-**permutations** are linear reduced permutations that fix all even elements (thus n must be odd). An example of a γ-permutation is $(3\ 2\ 1\ 4\ 7\ 6\ 9\ 8\ 5)$. Any permutation π that reduces to a γ-permutation has distance $td(\pi) = n - c_{odd}(\Gamma(\pi))$ (see Labarre [240]).

Other constructions that are solvable in polynomial time were considered by Christie [115]. Special cases of γ-permutations were used by Labarre [240] to deduce a variety of other classes of permutations whose distance can be computed in polynomial time.

3.1.5 Approximation Algorithms

The first polynomial-time approximation algorithm for sorting by transpositions was given by Bafna and Pevzner [30], and had an approximation ratio of 3/2, while running in quadratic time, in the size of the permutation. It is based on the two theorems that follow definition 3.10.

Definition 3.10 For a permutation π, a k-**transposition** is a transposition τ such that $c(G(\pi \circ \tau)) = c(G(\pi)) + k$.

Theorem 3.12 [30] For all π in S_n, if $G(\pi)$ contains a cycle with at least three black arcs, then it is possible to apply

• either a 2-transposition or

• a 0-transposition followed by two consecutive 2-transpositions.

Theorem 3.13 [30] For all π in S_n, if $G(\pi)$ contains no cycle with more than two black arcs, then it is always possible to apply a transposition that creates two odd cycles in $G(\pi)$.

Bafna and Pevzner's algorithm consists in repeatedly using either 2-transpositions or 0-transpositions followed by two consecutive 2-transpositions, whereas $G(\pi)$ contains cycles with more than two black arcs, then using 2-transpositions as long as π is not sorted. Since theorem 3.12 guarantees that we can always create at least four odd cycles using three transpositions, the approximation ratio of 3/2 follows from theorem 3.4.

Following Bafna and Pevzner [30], Christie [115], Gu et al. [190], and Hartman [201] devised simpler and faster 3/2-approximation algorithms based on the same principles. The 3/2-approximations of Gu et al. [190] and Hartman [201] use a transformation of the initial permutation into a simple one for which the same lower bound holds.

The approximation ratio of 3/2 was outperformed by Elias and Hartman [160], who proposed an 11/8-approximation algorithm with time complexity $O(n^2)$. It is based on the same principle as the other algorithms described above (namely, the lower bound of theorem 3.4), and consists in applying series of transpositions, a certain proportion of which are 2-transpositions. If the series have length 3, and at least two transpositions in each series are 2-transpositions, then this strategy yields a 3/2-approximation. Elias and Hartman [160] proved that for simple permutations of n elements, where n is sufficiently large there always exists a series of 11 transpositions, at least eight of which are 2-transpositions. The proof of correctness of that algorithm is heavily computer-driven and based on a huge case analysis requiring the verification of more than 80,000 configurations.

All algorithms discussed above are based on the cycle graph. Another structure, called the *breakpoint diagram*, was introduced by Dias et al. [140] and used to derive a 2.25-approximation running in time $O(n^2)$. The authors claim that although their algorithm has a poorer ratio, it has the advantage of being much easier to implement. To date, it does not seem to have been used by other authors. Benoît-Gagné and Hamel [40] also tried to bypass the cycle graph, and obtained a 3-approximation; they noted that their algorithm and the heuristics they proposed actually perform much better in practice, but the theoretical bound has not been improved.

Feng and Zhu [171] designed a variant of a balanced binary tree to encode permutations, which they call a *permutation tree*. This structure allows improvement of the running time of algorithms that use it; the authors illustrate this by showing that they can implement Hartman and Shamir's algorithm [202] in such a way that its running time becomes $O(n \log n)$.

3.1.6 Conjectures and Open Problems

The main open problems regarding transpositions are obviously to determine the complexity of the sorting problem and of the distance computation problem. No exact polynomial-time algorithm is yet known, but Dias and Carvalho de Souza [136] proposed an integer linear programming model with $O(n^4)$ variables and

Table 3.2
Known values of the diameter of S_n under transpositions

n	1	2	3	4	5	6	7	8	9	10	11	12	13	14	15
diameter	0	1	2	3	3	4	4	5	5	6	6	7	8	8	9

Source: Eriksson et al. [164]

$O(n^6)$ constraints; however, it seems that even though it solves the sorting problem to optimality, it is exceedingly time-consuming even for permutations of size 8. On a related matter, Radcliffe et al. [309] have proved that sorting binary strings by transpositions is **NP**-hard (see part II).

Another problem is to determine the diameter of S_n under transpositions. On the one hand, an upper bound can be derived from theorem 3.5: it is less than $\frac{2}{3}n$. On the other hand, the best lower bound has been shown by Elias and Hartman [160], disproving a conjecture by Eriksson et al. [164]: as we have seen in section 3.1.4, some permutations need $\lfloor \frac{n+1}{2} \rfloor + 1$ transpositions to be sorted. The known values of the diameter of S_n under transpositions are shown in table 3.2.

As observed by Bafna and Pevzner [30], each transposition can change the number of cycles in the cycle graph by 0, 2, or -2. Since sorting by transpositions can be formulated as "creating odd alternating cycles as fast as possible," it is intuitively foolish to use -2-transpositions in a sorting algorithm. All approximation algorithms based on the cycle graph make the assumption that such transpositions can always be disregarded, but they still have not been proved to be useless.

Conjecture 3.1 [115] For any permutation, there always exists an optimal sorting sequence of transpositions that contains no -2-transposition.

Christie [115] asked whether it was possible to decide in polynomial time whether a permutation π is tight with respect to the lower bound of theorem 3.1, a question that is still unanswered today, even for 3-permutations. The same "tightness question" concerning the other lower bounds does not seem to have received much attention.

The question of sorting circular genomic permutations (see page 18) has never been handled.

3.2 Prefix Transposition Distance

Since the problem of sorting by transpositions seems very challenging, some researchers have tried to study variants of this problem, in the hope that they would shed some light on the original problem. In this section, we discuss a restricted version of sorting by transpositions in which only the "beginning" of the permutation may be moved. Due to the definition, it is studied only on linear permutations.

Table 3.3
The number of permutations π in S_n with $ptd(\pi) = k$; $1 \leq n \leq 10$

n \ k	0	1	2	3	4	5	6	7	8
1	1	0	0	0	0	0	0	0	0
2	1	1	0	0	0	0	0	0	0
3	1	3	2	0	0	0	0	0	0
4	1	6	14	3	0	0	0	0	0
5	1	10	50	55	4	0	0	0	0
6	1	15	130	375	194	5	0	0	0
7	1	21	280	1,575	2,598	562	3	0	0
8	1	28	532	4,970	18,096	15,532	1,161	0	0
9	1	36	924	12,978	85,128	188,386	74,183	1,244	0
10	1	45	1,500	29,610	308,988	1,364,710	1,679,189	244,430	327

- Introduced by Dias and Meidanis [138].
- Complexity: unknown.
- Best approximation ratio: 2, by Dias and Meidanis [138].
- Diameter: unknown. Lies between $\lfloor \frac{3n+1}{4} \rfloor$ (Labarre [241]) and $n - \log_8 n$ (Chitturi and Sudborough [110]).

Definition 3.11 A **prefix transposition** is a transposition (definition 3.1) $\tau(i, j, k)$ with $i = 1$.

The **prefix transposition distance** of a permutation π will be denoted by $ptd(\pi)$. Table 3.3 shows the distribution of the prefix transposition distance for $1 \leq n \leq 10$.

3.2.1 Lower Bounds

It is a trivial fact that the prefix transposition distance is always at least as large as the transposition distance; therefore, any lower bound on $td(\pi)$ is also a lower bound on $ptd(\pi)$. Dias and Meidanis [138] obtained another lower bound, based on the following concepts.

Definition 3.12 A **prefix transposition breakpoint** in a permutation π in S_n is a breakpoint of π, except that $(0, \pi_1)$ is always a prefix transposition breakpoint. The number of prefix transposition breakpoints in a permutation π is denoted by $ptb(\pi)$.

Definition 3.13 For a permutation π, a k-**prefix transposition** is a transposition τ such that $ptb(\pi \circ \tau) = ptb(\pi) + k$.

As in the case of transpositions, the set of possible values for this parameter is bounded.

Lemma 3.4 [138] For any k-prefix transposition τ, we have $k \in \{-2, -1, 0, 1, 2\}$.

This immediately yields the following lower bound.

Lemma 3.5 [138] For all π in S_n, we have

$$ptd(\pi) \geq \left\lceil \frac{ptb(\pi) - 1}{2} \right\rceil.$$

Chitturi and Sudborough [110] then obtained another lower bound, using the following concept based on permutations of $\{0, 1, 2, \ldots, n-1\}$ rather than $\{1, 2, \ldots, n\}$: a **clan** is a maximal interval of π that contains only reverse adjacencies.

Lemma 3.6 [110] For any π in S_n, let $\Upsilon(\pi)$ denote the set of all clans of π of length at least 3, and $s(\pi)$ denote the number of strips of π. Then

$$ptd(\pi) \geq \frac{s(\pi) + \frac{\sum_{C \in \Upsilon(\pi)}(|C|-2)}{3}}{2}.$$

Finally, Labarre [241] proved another lower bound and showed that its value is always at least as large as that given by lemma 3.5.

Theorem 3.14 [241] For any π in S_n, we have

$$ptd(\pi) \geq \frac{n + 1 + c(G(\pi))}{2} - c_1(G(\pi)) - \begin{cases} 0 & \text{if } \pi_1 = 1 \\ 1 & \text{otherwise} \end{cases},$$

where $c_1(G(\pi))$ is the number of cycles of length 1 in $G(\pi)$.

3.2.2 Upper Bounds

The observation that in the worst case, only one prefix transposition breakpoint can be removed at each step, and the fact that we can never remove $(0, \pi_1)$, yield the following upper bound.

Lemma 3.7 [138] For all π in S_n, we have $ptd(\pi) \leq ptb(\pi) - 2$.

Chitturi and Sudborough [110] proved another upper bound.

Theorem 3.15 [110] For all π in S_n, we have $ptd(\pi) \leq n - \log_8 n$.

3.2.3 Diameter

The lower bound proved by Dias and Meidanis [138] immediately yields a lower bound of $n/2$ on the prefix transposition diameter, for instance, when π is the reversed permutation $(n\ n-1\ n-2 \cdots 2\ 1)$. Chitturi and Sudborough [110] used the same permutation and their lower bound (lemma 3.6) to improve that result to

$2n/3$. Finally, Labarre [241] improved that lower bound, using theorem 3.14, to $\lfloor \frac{3n+1}{4} \rfloor$. That value is reached by any 2-permutation (recall definition 3.9), and other such instances can easily be constructed for values of n for which 2-permutations do not exist.

As far as upper bounds on the diameter are concerned, Dias and Meidanis's [138] upper bound (lemma 3.7) yields a value of $n - 1$, for instance, when π is the reversed permutation. Chitturi and Sudborough [110] improved this result to $n - \log_8 n$. However, Dias and Meidanis [138] were able to sort the reversed permutation using $n - \lfloor \frac{n}{4} \rfloor$ prefix transpositions (but did not prove that this is optimal), and conjectured that the prefix transposition diameter is $n - \lfloor \frac{n}{4} \rfloor$, under the assumptions that there is no "harder" permutation than the reversed permutation and that $n - \lfloor \frac{n}{4} \rfloor$ is indeed its actual prefix transposition distance.

3.2.4 Easy Cases

Dias and Meidanis [138] show that a particular family of permutations can be sorted by prefix transpositions in linear time; those are the permutations defined by

$$(k+1\ k\ k+2\ k-1\ k+3\ k-2\cdots 2k-1\ 2\ 2k\ 1).$$

The prefix transposition distance of those permutations is k.

Dias and Meidanis [138] also provide an $O(n^2)$ algorithm for determining whether a given permutation has a prefix transposition distance equal to $\frac{ptb(\pi)-1}{2}$, based on the following lemma.

Lemma 3.8 [138] For all π in S_n, there exists at most one prefix transposition τ such that $ptb(\pi \circ \tau) = ptb(\pi) - 2$.

Therefore, determining whether π is tight with respect to the lower bound of lemma 3.5 is done by repeatedly applying a prefix transposition of the kind described in lemma 3.8. If at any stage there is no such operation, then the permutation is not tight; otherwise, it is.

Theorem 3.16 [177] If a permutation π is tight with respect to the lower bound of lemma 3.5, then $ptd(\pi) = td(\pi)$.

3.2.5 Approximation Algorithms

Dias and Meidanis [138] have shown that every transposition can be simulated by at most two prefix transpositions, and therefore any k-approximation algorithm for computing $td(\pi)$ is a $2k$-approximation algorithm for computing $ptd(\pi)$. Therefore, Elias and Hartman's algorithm [160] can be converted into a 2.75-approximation algorithm for sorting by prefix transpositions. However, a straightforward 2-approximation follows from the upper bound based on prefix transposition breakpoints.

3.2.6 Variant: Insertion of the Leading Element

In this further restricted variant, only transpositions of the form $\tau(1,2,k)$ are considered. It has been investigated by Aigner and West [3]. The number of operations required to sort a permutation is $n-k$, where n is the size of the permutation and k is the largest integer such that the last k entries of the permutation form an increasing sequence.

3.3 Reversal Distance

We will now examine one of the most studied problems in the field of genome rearrangements, the problem of sorting by reversals. This is the first combinatorially studied problem in the field, and it has spawned the theory based on the breakpoint graph introduced by Bafna and Pevzner [29].

- Introduced by Watterson et al. [369] on circular permutations.
- Complexity: **NP**-hard (see Caprara [93]) and not approximable within 1.0008 (see Berman and Karpinski [57]).
- Best approximation ratio: $\frac{11}{8}$ (see Berman et al. [58]).
- Diameter: $n-1$ (see Bafna and Pevzner [29]). The diameter is reached by only two permutations: the **Gollan permutation**, γ_n, and its inverse, defined by

$$\gamma_n = \begin{cases} (1,3,5,7,\ldots,n-1,n,\ldots,8,6,4,2) & \text{if } n \text{ is even} \\ (1,3,5,7,\ldots,n,n-1,\ldots,8,6,4,2) & \text{if } n \text{ is odd.} \end{cases}$$

Definition 3.14 For any permutation π in S_n, the **reversal** $\rho(i,j)$ with $1 \leq i < j \leq n$ applied to π reverses the closed interval determined by i and j, transforming π into $\pi \circ \rho(i,j)$. Therefore $\rho(i,j)$ is the following permutation:

$$\begin{pmatrix} 1 & \cdots & i-1 & \underline{i \; i+1 \; \cdots \; j-1 \; j} & j+1 & \cdots & n \\ 1 & \cdots & i-1 & j \; j-1 \; \cdots \; i+1 \; i & j+1 & \cdots & n \end{pmatrix}.$$

The **reversal distance** of a permutation π will be denoted by $rd(\pi)$. Table 3.4 shows the distribution of the reversal distance, for $1 \leq n \leq 10$.

3.3.1 Lower Bounds

3.3.1.1 Lower Bound Based on Breakpoints Since a reversal can reduce the number of strong breakpoints by at most two (e.g., (1 $\underline{3\ 2}$ 4) $\rightarrow$ (1 2 3 4)), we get the following first bound:

Theorem 3.17 [29] For all π in S_n, we have $rd(\pi) \geq sb(\pi)/2$.

Table 3.4
The number of permutations π in S_n with $rd(\pi) = k$; $1 \leq n \leq 10$

$n \backslash k$	0	1	2	3	4	5	6	7	8	9
1	1	0	0	0	0	0	0	0	0	0
2	1	1	0	0	0	0	0	0	0	0
3	1	3	2	0	0	0	0	0	0	0
4	1	6	15	2	0	0	0	0	0	0
5	1	10	52	55	2	0	0	0	0	0
6	1	15	129	389	184	2	0	0	0	0
7	1	21	266	1,563	2,539	648	2	0	0	0
8	1	28	487	4,642	16,445	16,604	2,111	2	0	0
9	1	36	820	11,407	69,863	169,034	105,365	6,352	2	0
10	1	45	1,297	24,600	228,613	1,016,341	1,686,534	654,030	17,337	2

3.3.1.2 Lower Bound Based on Matchings Kececioglu and Sankoff [232] constructed a graph $G = (V, E)$ based on π, where V is the set of strong breakpoints of π and whose edges connect two vertices such that the corresponding strong breakpoints can be eliminated using a single reversal. Let m be the number of vertices in a maximum cardinality matching of G; then

$$rd(\pi) \geq \left\lceil \frac{m}{2} + \frac{2(sb(\pi) - m)}{3} \right\rceil.$$

This formalizes the idea that, in the best case, two consecutive reversals can remove at most three strong breakpoints. Improved lower bounds can be obtained by considering k-tuples of strong breakpoints instead of pairs, but this results in an increased time complexity. Kececioglu and Sankoff [232] propose a linear programming model formulation for that purpose.

3.3.1.3 Lower Bounds Based on the Breakpoint Graph Bafna and Pevzner [29] proposed the following undirected version of the cycle graph (definition 3.3). Just like the cycle graph, it can be defined equivalently for circular and linear permutations, and can be represented using a circular or a linear layout.

Definition 3.15 The **breakpoint graph** of a permutation π is the undirected graph $BG(\pi)$, whose vertex set is the vertex set of the cycle graph of π, and whose edges are the arcs of the cycle graph of π taken without their orientation.

Figure 3.4 shows an example of a breakpoint graph of a linear permutation, using the linear layout. Since each vertex has the same number of incident gray and black edges (at least in the circular layout), the breakpoint graph decomposes into edge-disjoint

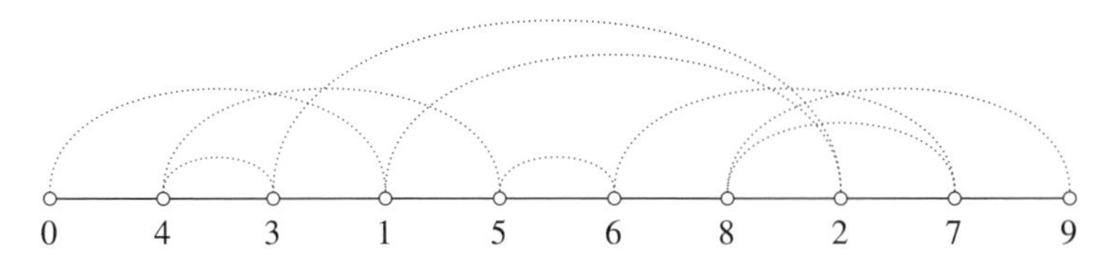

Figure 3.4
The breakpoint graph of the permutation (4 3 1 5 6 8 2 7)

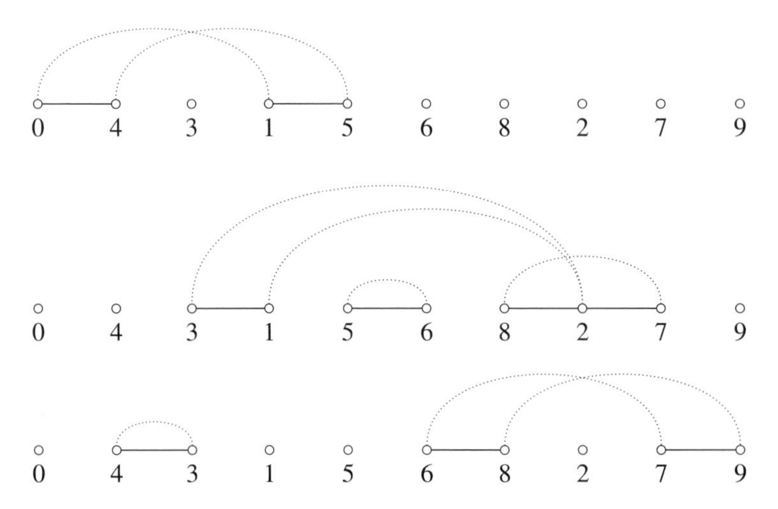

Figure 3.5
A maximal alternating cycle decomposition of the breakpoint graph of figure 3.4 into five cycles

alternating cycles. However, by contrast with the cycle graph introduced in the context of sorting by transpositions, this decomposition is not unique (this is actually the main reason why sorting by reversals is **NP**-hard; see Caprara [93]). Since the breakpoint graph of ι has the largest possible number of cycles, one would want to start with a configuration that is as close as possible to that of ι, and get a decomposition in alternating cycles which is *maximal* in the sense that it should contain the largest number of alternating cycles. A maximal alternating cycle decomposition of the breakpoint graph of figure 3.4 is shown in figure 3.5.

We denote $c^*(BG(\pi))$ as the number of cycles in a maximal alternating cycle decomposition of $BG(\pi)$. The same kind of lower bound that was shown in the case of transpositions (theorems 3.3 and 3.4) is obtained for reversals, except that the parameter $c^*(BG(\pi))$ on which it relies is not easy to compute.

Theorem 3.18 [29] For all π in S_n, $rd(\pi) \geq p(\pi) - c^*(BG(\pi))$.

Caprara [94] studied the tightness of the lower bound of theorem 3.18, and concluded that the probability that it is not tight for a random permutation of n elements is asymptotically $O(1/n^5)$.

3.3.2 Upper Bounds

The only upper bound on the unsigned reversal distance we are aware of is the following trivial bound presented, for example, in Kececioglu and Sankoff [232].

Lemma 3.9 For all π in S_n, we have $rd(\pi) \leq sb(\pi) - 1$.

Other upper bounds can be obtained using a connection with the *signed* version of sorting by reversals, as we explain in section 3.3.3.1.

3.3.3 Easy Cases

3.3.3.1 A Link with the Signed Version There exists a version of the problem of sorting by reversals for signed permutations, which will be discussed in detail in section 4.2 and in which a reversal reverses not only the order but also the signs of the elements of the target interval. The corresponding sorting and distance computation problems are, surprisingly, solvable in polynomial time, and turn out to be particular cases of the unsigned version.

Definition 3.16 A **spin** of a permutation π in S_n is a signed permutation $\vec{\pi}$ in $S_n^{\pm}$ such that $|\vec{\pi}_i| = \pi_i$ for all $1 \leq i \leq n$.

Lemma 3.10 [198] For all π in S_n, denote $\vec{\Pi}$ as the set of all spins of π; we have

$$rd(\pi) = \min_{\vec{\Pi} \in \vec{\Pi}} srd(\vec{\Pi}),$$

where $srd(\vec{\Pi})$ denotes the signed reversal distance of the spin $\vec{\Pi}$.

Sorting by reversals can therefore be formulated as finding an optimal spin of the given permutation. In some cases, characterized using the concepts below, such a spin can be found in polynomial time.

A **strong strip** is a maximal interval containing no strong breakpoint, and it is called **long** if its length is at least 3. A **singleton** is a strong strip of length 1. The **canonical sign** of an element in a strip is positive if the strong strip is **increasing** (i.e., it consists of adjacencies) and negative if the strong strip is **decreasing** (i.e., it consists of reverse adjacencies). Its **anticanonical sign** is the opposite sign of the canonical sign.

Hannenhalli and Pevzner [198] proved that signing long strong strips with canonical signs always leads to an optimal spin. For strong strips of size 2, the same procedure works, except in a particular case where $\vec{\pi}$ contains "unoriented components" (see section 4.2 for more details and definitions). In that case, one strip of size 2 per unoriented component is given the anticanonical sign. This provides a polynomial-time algorithm in every case.

Lemma 3.11 [198] For all π in S_n, there exists an optimal sorting sequence of reversals that never cuts long strips.

Theorem 3.19 [198] For all π in S_n, there exists an optimal sorting sequence of reversals that never increases the number of strong breakpoints.

3.3.3.2 Tightness of the Breakpoint Lower Bound

Tran [359] and Christie [115] independently proved that it is possible to decide in polynomial time whether a permutation reaches the lower bound of theorem 3.17.

Definition 3.17 Given a permutation π, two strong breakpoints (π_i, π_{i+1}) and (π_j, π_{j+1}) define an **active interval** $[i, j]$ if $|\pi_i - \pi_j| = 1$ and $|\pi_{i+1} - \pi_{j+1}| = 1$. Similarly, they define a **passive interval** $|\pi_i - \pi_{j+1}| = 1$ and $|\pi_{i+1} - \pi_j| = 1$.

Tran [359] defines B_π as the graph whose vertices are strong breakpoints in π and whose edges connect those pairs of strong breakpoints that define active or passive intervals. If B_π has a perfect matching M, let I_M be the graph whose vertices are the intervals defined by the edges of M, and whose edges connect intersecting intervals (i.e., intervals $[i, j]$ and $[k, l]$ with either $i < k < j < l$ or $k < i < l < j$).

Theorem 3.20 [359] For any π in S_n, we have $rd(\pi) = sb(\pi)/2$ if and only if there exists a perfect matching M of B_π such that each connected component of the graph I_M includes at least one active interval of π.

3.3.4 Computational Complexity

Caprara [97] proved that sorting by reversals is **NP**-hard, through a reduction of ALTERNATING CYCLE DECOMPOSITION, the latter being proved to be **NP**-hard by a reduction from EULERIAN CYCLE DECOMPOSITION. Berman and Karpinski [57] strengthened this result by proving that sorting by reversals is not approximable within ratio 1.0008, unless **P** = **NP**. This proves in particular that the problem is **APX**-hard, and justifies the numerous algorithmic studies on approximations, exact algorithms, and heuristics.

Sorting genomic circular permutations (see page 18) by reversals is also **NP**-hard, as proved by Solomon et al. [342], thanks to the **APX**-hardness result of Berman and Karpinski [57]. The result is not immediately deducible from the same result for linear permutations, because of a gap of at most three reversals between the solutions for a circular permutation and one of its linearizations.

3.3.5 Approximation Algorithms

A trivial approximation algorithm has been suggested by Watterson et al. [369]. It consists simply in repeatedly applying reversals that repair at least one strong breakpoint at each step, thereby guaranteeing that any permutation can be sorted using $sb(\pi) - 1$ reversals where n is the size of permutation π. The lower bound of theorem 3.17 implies that this algorithm is a 2-approximation.

Kececioglu and Sankoff [232] proposed a greedy trick to improve this algorithm, which consists in applying at each step a reversal that removes the largest possible number of strong breakpoints in the permutation, favoring reversals that leave decreasing strips. The approximation ratio is, however, still 2, and it runs in time $O(n^2)$.

Bafna and Pevzner [29] introduced the breakpoint graph and proposed an approximation algorithm based on that structure which finds a decomposition of the breakpoint graph that is not necessarily maximal, but that contains a large number of 4-cycles. The analysis of the algorithm is quite involved, but the idea is to use 2-reversals while it is possible, then use an approximation of INDEPENDENT SET on the 4-*cycle graph* of the resulting permutation, denoted by σ, that guarantees a decomposition of $BG(\sigma)$ with at least $\frac{2}{5}c_4(\sigma)$ 4-cycles, where $c_4(\sigma)$ denotes the number of 4-cycles in a maximal cycle decomposition of $BG(\sigma)$. The performances of the algorithm are improved by modifying the way in which 2-reversals are applied in the first step, in order to obtain a permutation σ with a bipartite 4-cycle graph. This allows finding a decomposition of $BG(\sigma)$ that contains a *maximum* number of 4-cycles, still using the INDEPENDENT SET approach. The last step of the algorithm remains the same, and this yields a 7/4-approximation, running in $O(n^2)$ time.

Christie [114] then improved the approximation ratio down to 3/2, using a novel structure called the *reversal graph* as an auxiliary tool for finding, given a cycle decomposition of the breakpoint graph of a permutation, a sequence of reversals that sorts the permutation. Another graph, called the *matching graph*, is used to find a "good enough" alternating cycle decomposition of a given breakpoint graph. Christie's algorithm has time complexity $O(n^4)$, but he notes that it can be reduced to $O(n^2)$, using techniques proposed by Kaplan et al. [228] in the context of sorting signed permutations by reversals.

Caprara and Rizzi [98], and then Lin and Jiang [251], proposed improved approximation algorithms for finding a large cardinality decomposition of the breakpoint graph. The former improved the 3/2-approximation ratio to $33/23 + \varepsilon$, which is about $1.4348 + \varepsilon$; the latter further improved it to $\frac{5073 - 15\sqrt{1201}}{3208} + \varepsilon$, which is about $1.4193 + \varepsilon$, for any positive ε. Both approximations use variants of INDEPENDENT SET and SET PACKING problems.

Berman et al. [58] then came up with an 11/8-approximation, the best ratio that has been reached so far. Their algorithm is based on a better method to solve

ALTERNATING CYCLE DECOMPOSITION, which again gives a spin of the permutation, which can then be sorted optimally, using a polynomial algorithm. They prove that ALTERNATING CYCLE DECOMPOSITION is $11/8$-approximable for instances with more than 48 strong breakpoints.

There is also a heuristic proposed by Auyeung and Abraham [23], which is a genetic algorithm with time complexity $O(n^4)$. It uses Boolean vectors to represent solutions that correspond to spins of the original permutation and whose fitness is the reversal distance of the corresponding spin, computable in linear time (see section 4.2). Mo and Zeng [272] have proposed an improved genetic algorithm, which they claim outperforms both Auyeung and Abraham's algorithm and Christie's 3/2-approximation.

3.3.6 Exact Algorithms

A branch-and-bound algorithm was proposed by Kececioglu and Sankoff [232], who also presented bounding techniques in order to reduce the search space (see section 3.3.1.2). Hannenhalli and Pevzner [198] proposed another algorithm, based on concepts explained in section 3.3.3.1. It runs in polynomial time on singleton-free permutations and on permutations with $O(\log n)$ singletons, and consists in exploring all **canonical spins** of the permutation (i.e., all spins in which elements in each increasing (resp. decreasing) strip are signed positively (resp. negatively)); for each such configuration, it then finds in polynomial time how to modify that spin so as to obtain the lowest possible distance; and finally, it computes in polynomial time the signed reversal distance of that spin, keeping, in the end, the spin with the lowest distance. An optimal sorting sequence of signed reversals is then computed in polynomial time for that spin, which is eventually mimicked on the original, unsigned permutation.

Hannenhalli and Pevzner [198] claimed that their algorithm had a running time of $O(2^k n^3 + n^4)$, where k is the number of singletons in the permutation. However, advances on the problem of sorting signed permutations by reversals allow an improvement of this running time (references and details are given in section 4.2). Indeed, the n^4 part was due to the sorting procedure, which can now be performed in time $O(n^{3/2})$, and the n^3 part was due to the computation of the signed reversal distance, which can now be done in time $O(n)$. Therefore, the time complexity of this algorithm becomes $O(2^k n + n^{3/2})$.

A remarkable idea, proposed by Caprara et al. [99], is a linear programming formulation of sorting by unsigned reversals. This approach allowed them to efficiently and optimally solve very large instances of the problem. The computational limit of these exact algorithms seems to be random permutations of size around $n = 200$, for a few minutes of computation. Dias and Carvalho de Souza [136] later proposed another integer linear programming model, which is polynomial-sized. It uses $O(n^3)$

variables and $O(n^5)$ constraints, whereas the model proposed by Caprara et al. [99] uses an exponential number of variables and constraints.

3.4 Prefix Reversal Distance (Pancake-Flipping)

Just as in the case of transpositions, there exists a variant of sorting by reversals where valid reversals may act only on the "beginning" of the permutation.

- Introduced by Harry Dweighter [151] (a nickname obviously a pun on "harried waiter"; the author's real name is Jacob E. Goodman).
- Complexity: unknown.
- Best approximation ratio: 2, by Fischer and Ginzinger [176].
- Diameter: unknown. Lies between $\frac{15n}{14}$ (Heydari and Sudborough [212]) and $\frac{18}{11}n + O(1)$ (see Chitturi et al. [111]).

Definition 3.18 A **prefix reversal** is a reversal (definition 3.14) $\rho(i, j)$ with $i = 1$.

The prefix reversal distance of a permutation π will be denoted by $prd(\pi)$. Table 3.5 shows the distribution of the prefix reversal distance, for $1 \leq n \leq 10$.

3.4.1 Lower Bounds

Since prefix reversals are restricted reversals, any lower bound on the reversal distance will be a lower bound on the prefix reversal distance. Fischer and Ginzinger [176] prove the following lower bound.

Lemma 3.12 [176] For all π in S_n, we have $prd(\pi) \geq sb(\pi^{eo})$.

Table 3.5
The number of permutations π in S_n with $prd(\pi) = k$; $1 \leq n \leq 10$

n \ k	0	1	2	3	4	5	6	7	8	9	10	11
1	1	0	0	0	0	0	0	0	0	0	0	0
2	1	1	0	0	0	0	0	0	0	0	0	0
3	1	2	2	1	0	0	0	0	0	0	0	0
4	1	3	6	11	3	0	0	0	0	0	0	0
5	1	4	12	35	48	20	0	0	0	0	0	0
6	1	5	20	79	199	281	133	2	0	0	0	0
7	1	6	30	149	543	1,357	1,903	1,016	35	0	0	0
8	1	7	42	251	1,191	4,281	10,561	15,011	8,520	455	0	0
9	1	8	56	391	2,278	10,666	38,015	93,585	132,697	79,379	5,804	0
10	1	9	72	575	3,963	22,825	106,461	377,863	919,365	1,309,756	814,678	73,232

3.4.2 History

Oppositely to transpositions, the variant discussed in this section has not been introduced in order to get more insight on the general problem. In fact, the study of sorting by prefix reversals *precedes* that of sorting by reversals, and began in 1978 and 1979 with two independent papers, one by Györi and Turán [193] and the other by Gates and Papadimitriou [183]. The study of sorting by prefix reversals is not at all motivated by genome rearrangements, though Pevzner and Waterman [298] reinterpret it as a genome rearrangement problem with a hot point of mutation, but by a problem stated as a little story by Dweighter [151], which we reproduce below:

> The chef in our place is sloppy, and when he prepares pancakes they come out all in different sizes. Therefore, when I deliver them to a customer, on the way to the table I rearrange them (so that the smallest winds up on top, and so on, down to the largest at the bottom) by grabbing several from the top and flipping them over, repeating this (varying the number I flip) as many times as necessary. If there are n pancakes, what is the maximum number of flips (as a function of n) that I shall ever have to use to rearrange them?

This little story gave the problem of sorting by prefix reversals the name "pancake-flipping." As indicated by the question above, people studying this problem were primarily interested in the diameter rather than in the sorting problem. Pancake-flipping has since found applications in parallel computing, primarily in the design of symmetric interconnection networks—the so-called *pancake network*, which is the Cayley graph of the symmetric group under prefix reversals, seems to be a good model for processor interconnections (see, e.g., Akers and Krishnamurthy [5] or Qiu et al. [308]).

The study of interconnection networks has spawned a large amount of literature that is far beyond the scope of this book. However, it is interesting to note that much attention in that field has been devoted to interconnection networks based on Cayley graphs of permutation groups, starting with the seminal paper by Akers and Krishnamurthy [5]. Moreover, problems of interest in that field also include determining the diameter of a given network, and sorting permutations using a given set of operations, although the approaches used to tackle these problems differ completely from those used in genome rearrangement problems. For more information on that field, a good starting point is the survey by Lakshmivarahan et al. [242].

3.4.3 Variants

The exploration of the properties of the pancake network has motivated several variants. For example, Bass and Sudborough [34] remark that even though prefix reversals constitute a set of generators of the symmetric group, it is not a minimal one, and they examine the possibility of restricting the possibilities of prefix pancakes to a constant of logarithmic number related to the number of elements in a permutation. A number of other operations have been introduced, motivated by the properties of the associated network, such as an operation of shuffle exchange (see Bass and Sudborough [33]), which consists in exchanging the first two elements or shifting the permutation (transforming $(\pi_1\ \pi_2\ \cdots\ \pi_n)$ into $(\pi_2\ \cdots\ \pi_n\ \pi_1)$ or $(\pi_n\ \pi_1\ \cdots\ \pi_{n-1})$). Bass and Sudborough [33] prove an upper bound of $\frac{5}{8}n^2$ on the diameter of this variant.

3.5 Variants

3.5.1 Block Interchange Distance

Block interchanges can be viewed as a generalization of transpositions, in that they exchange nonnecessarily contiguous intervals, whereas transpositions exchange only contiguous intervals.

- Introduced by Christie [113].
- Complexity: polynomial.
- Best algorithm: the block-interchange distance can be computed in $O(n)$ time (Christie [113]). If this distance equals δ for a given permutation, then sorting that permutation by block interchanges can be done in $O(\delta n)$ time (Lin et al. [254]). Alternatively, the sorting problem can be solved in $O(n \log n)$ time, using a data structure by Feng and Zhu [171].
- Diameter: $n/2$ (Christie [113]).

Definition 3.19 For any π in S_n, the **block interchange** $\beta(i,j,k,l)$ with $1 \leq i < j \leq k < l \leq n+1$ applied to π, exchanges the closed intervals determined respectively by i and $j-1$ and by k and $l-1$, transforming π into $\pi \circ \beta(i,j,k,l)$. Therefore, $\beta(i,j,k,l)$ is the following permutation:

$$\begin{pmatrix} 1 \cdots i-1 & \boxed{i \cdots j-1} & j\ j+1 \cdots k-1 & \boxed{k \cdots l-1} & l\ l+1 \cdots n \\ 1 \cdots i-1 & \boxed{k \cdots l-1} & j\ j+1 \cdots k-1 & \boxed{i \cdots j-1} & l\ l+1 \cdots n \end{pmatrix}.$$

The block-interchange distance of a permutation π is denoted by $bid(\pi)$. Table 3.6 shows the distribution of the block-interchange distance for $1 \leq n \leq 10$. This distribution follows from the work of Doignon and Labarre [147], who enumerate permutations whose cycle graph belongs to a given conjugacy class. As a corollary, the number of permutations whose cycle graph has k cycles is characterized, and the distribution of the block-interchange distance follows from that result and theorem 3.21 (below).

Using the cycle graph introduced in the context of sorting by transpositions (see page 27), Christie [113] proved a formula for computing the block-interchange distance.

Table 3.6
The number of permutations π in S_n with $bid(\pi) = k$; $1 \leq n \leq 10$

n \ k	0	1	2	3	4	5
1	1	0	0	0	0	0
2	1	1	0	0	0	0
3	1	5	0	0	0	0
4	1	15	8	0	0	0
5	1	35	84	0	0	0
6	1	70	469	180	0	0
7	1	126	1,869	3,044	0	0
8	1	210	5,985	26,060	8,064	0
9	1	330	16,401	152,900	193,248	0
10	1	495	39,963	696,905	2,286,636	604,800

Theorem 3.21 [113] For all π in S_n, we have

$$bid(\pi) = \frac{n + 1 - c(G(\pi))}{2}.$$

Computing $bid(\pi)$ is straightforwardly done by building its cycle graph, then counting its cycles, and both tasks are achievable in $O(n)$ time. Finding an optimal sequence is done by repeatedly applying block interchanges that create two new cycles in $G(\pi)$; such moves can be identified in $O(n)$ time, yielding a time complexity of $O(n^2)$ for sorting by block interchanges. Lin et al. [254] obtained the same results as Christie, using the "algebraic formulation" of Meidanis and Dias [264] (see chapter 11). Their approach yields a faster algorithm: if $bid(\pi) = \delta$, then π can be sorted by block interchanges in $O(\delta n)$ time. Using the structure introduced by Feng and Zhu [171], which we discussed in section 3.1.5, the running time of Christie's algorithm can be improved to $O(n \log n)$. The block-interchange operation is of special interest when combined with signed reversals, yielding the "double cut-and-join" operation (see section 4.4).

3.5.2 Element Interchange Distances

3.5.2.1 Exchanges A particular case of block interchanges is the case where the size of the blocks to be exchanged is exactly 1: in other words, permutations are to be sorted using only exchanges of two (nonnecessarily adjacent) elements. If π is a linear permutation of n elements, then the problem is easily solved and the associated distance, called the **Cayley distance**, is exactly $exc(\pi) = n - c(\Gamma(\pi))$ (see, for instance, Diaconis [135]). Table 3.7 shows the distribution of the Cayley distance for $1 \leq n \leq 10$. It is straightforward from the distance formula that the number of per-

Table 3.7
The number of permutations π in S_n with $exc(\pi) = k$; $1 \leq n \leq 10$

$n \backslash k$	0	1	2	3	4	5	6	7	8	9
1	1	0	0	0	0	0	0	0	0	0
2	1	1	0	0	0	0	0	0	0	0
3	1	3	2	0	0	0	0	0	0	0
4	1	6	11	6	0	0	0	0	0	0
5	1	10	35	50	24	0	0	0	0	0
6	1	15	85	225	274	120	0	0	0	0
7	1	21	175	735	1,624	1,764	720	0	0	0
8	1	28	322	1,960	6,769	13,132	13,068	5,040	0	0
9	1	36	546	4,536	22,449	67,284	118,124	109,584	40,320	0
10	1	45	870	9,450	63,273	269,325	723,680	1,172,700	1,026,576	362,880

mutations of n elements with exchange distance equal to $n - k$ is exactly the **Stirling number of the first kind** $\left[{n \atop k} \right]$, which counts the number of permutations in S_n that decompose into k disjoint cycles.

3.5.2.2 Adjacent Exchanges One can further restrict those valid exchanges to adjacent elements, which could also be seen as restricting valid moves to reversals of length 2. Again, the sorting problem is trivially solved in polynomial time on linear permutations by a well-known sorting algorithm called "bubble sort" (see, e.g., Knuth [237]), and we denote the associated distance by $inv_2(\pi)$. This measure is usually referred to as **Kendall's tau** in statistics (see Diaconis [135]). Both the distance computation and the sorting problems are easily solved in $O(n^2)$ time, and the corresponding diameter is $\binom{n}{2}$ (see, e.g., Knuth [237]).

Designing an algorithm for sorting *circular* permutations by reversals of length 2 is mentioned as an open problem by Pevzner [296] (problem 12.105, page 268). However, Jerrum [221] proposed an integer programming formulation that can be used to compute the associated distance in $O(n^2)$ time, and mentions that the techniques used in the proofs of that result imply polynomial-time algorithms for solving the sorting problem as well. Bafna et al. [31] proved that the corresponding diameter is at most $\lceil \frac{n-1}{2} \rceil \lfloor \frac{n-1}{2} \rfloor$.

3.5.2.3 Prefix Exchanges Finally, the problem of sorting by prefix exchanges has also been considered: the operation consists in swapping the first element of the permutation with any other element of the permutation. This problem is not motivated by genome rearrangements, but has applications in the design of interconnection networks, just like the pancake-flipping problem (see section 3.4).

- Introduced by Akers et al. [6].
- Complexity: polynomial.
- Best algorithm: sorting can be done in $O(n^2)$ time, whereas merely computing the distance can be done in $O(n)$ time (see Akers et al. [6]).
- Diameter: $\left\lfloor \frac{3(n-1)}{2} \right\rfloor$ (see Akers et al. [6]).

The **prefix exchange distance** of a permutation π will be denoted by $pexc(\pi)$. Table 3.8 shows the distribution of the prefix exchange distance, for $1 \leq n \leq 10$. This distribution was fully characterized by Portier and Vaughan [303].

Akers et al. [6] prove a formula for computing the prefix exchange distance.

Theorem 3.22 [6] The prefix exchange distance of π in S_n is equal to

$$pexc(\pi) = n + c(\Gamma(\pi)) - 2c_1(\Gamma(\pi)) - \begin{cases} 0 & \text{if } \pi_1 = 1 \\ 2 & \text{otherwise} \end{cases},$$

where $c_1(\Gamma(\pi))$ is the number of cycles of length 1 in $\Gamma(\pi)$, that is, the number of fixed points of π.

3.5.3 Weighted Reversals

Assigning a different weight to each reversal is motivated by the hypothesis that some operations are more likely to happen in some genomes than in others, and particularly short reversals. Therefore, a model that would assign a larger weight to long reversals would be more realistic, as argued, for example, by Sankoff et al. [329]. In those weighted variants, the goal is no longer to sort using as few operations as possible, but rather to select operations so as to minimize the value of a cost function that assigns a value to each operation used in a sorting sequence.

The weight function that has been most studied is the function $f(l) = l^\alpha$, where l is the *length* of a reversal and α is a positive real constant. The length of a reversal can be computed directly on the DNA molecule by counting how many nucleotides are involved in the inverted segment. However, such an accurate counting is hardly achievable because reversals do not frequently break the molecule so precisely, and hence counting the number of genes (or any measure at this scale) in the inverted segment is a more realistic approach.

- Introduced by Pinter and Skiena [299].
- Complexity: unknown.
- Best approximation ratio and diameter: see table 3.9.

Pinter and Skiena [299], Bender et al. [38], and Swidan et al. [346] have investigated this problem, and their results for different values of α are collected in table 3.9.

Table 3.8
The number of permutations π in S_n with $pexc(\pi) = k$; $1 \leq n \leq 10$

$n \backslash k$	0	1	2	3	4	5	6	7	8	9	10	11	12	13
1	1	0	0	0	0	0	0	0	0	0	0	0	0	0
2	1	1	0	0	0	0	0	0	0	0	0	0	0	0
3	1	2	2	1	0	0	0	0	0	0	0	0	0	0
4	1	3	6	9	5	0	0	0	0	0	0	0	0	0
5	1	4	12	30	44	26	3	0	0	0	0	0	0	0
6	1	5	20	70	170	250	169	0	0	0	0	0	0	0
7	1	6	30	135	460	1,110	1,689	1,254	340	15	0	0	0	0
8	1	7	42	231	1,015	3,430	8,379	13,083	10,048	3,409	315	0	0	0
9	1	8	56	364	1,960	8,540	28,994	71,512	114,064	96,116	36,260	4,900	105	0
10	1	9	72	540	3,444	18,396	80,262	273,546	680,448	1,106,460	978,696	411,984	71,477	3,465

Table 3.9
Bounds for the diameter and approximation ratios for sorting by weighted reversals, using the cost function that assigns weight l^α to a reversal of length l

Value of α	Lower bound on diameter	Upper bound on diameter	Approximation ratio
$0 \leq \alpha < 1$	$\Omega(n)$	$O(n \log n)$	
$\alpha = 1$	$\Omega(n \log n)$	$O(n \log^2 n)$	$O(\log n)$
$1 < \alpha < 2$	$\Omega(n^\alpha)$	$\Theta(n^\alpha)$	$O(\log n)$
$\alpha \geq 2$	$\Omega(n^2)$	$\Theta(n^2)$	2

Redrawn from [39].

Experimental studies on chloroplast genomes of plants suggest that the value of α in that context is around 0.5. Nguyen et al. [281] use the same function but forbid reversals that exceed a certain length. They provide bounds and approximation algorithms, and claim that their model is biologically the most meaningful.

3.5.4 Fixed-Length Reversals

Chen and Skiena [105] introduced a variant of sorting by reversals that consists in sorting permutations using only reversals of a prescribed fixed length; they give several characterizations of feasibility and optimization. This particular case is not motivated by genome rearrangements, but by a game called TopSpin™.

- Introduced by Chen and Skiena [105].
- Complexity: unknown.
- Diameter: unknown. Lies between $\Omega\left(\frac{n^2}{k^2} + n\right)$ and $O\left(\frac{n^2}{k} + kn\right)$ (see Chen and Skiena [105]).

Definition 3.20 A k-**reversal** $\rho(i, j)$ is a reversal (definition 3.14) with $j - i = k$.

Whereas reversals generate the symmetric group, and as a consequence all permutations can be sorted by reversals, the same is not necessarily true for k-reversals (although it is true in the easy particular case where $k = 2$, discussed in section 3.5.2). Therefore, the first problem is to determine which permutations k-reversals generate. The connected components of the k-reversal Cayley graph are investigated by Chen and Skiena [105], who give their number for all combinations of n, k.

3.5.5 Bounded Variants

3.5.5.1 Bounded Reversals Instead of allowing only k-reversals for a fixed k, some authors have considered allowing i-reversals, for all $i \leq k$, for a fixed k. If $k = 3$, the problem is known as sorting by *short swaps*.

- Introduced by Heath and Vergara [209].
- Complexity: unknown.
- Best approximation ratio: 2 (see Heath and Vergara [209]) for short swaps.
- Diameter: unknown. Lies between $\lceil \binom{n}{2}/3 \rceil$ (see Heath and Vergara [209]) and $\frac{3}{16}n^2 + O(n \log n)$ (see Feng et al. [173]) for short swaps.

Short swaps are also special cases of *block interchanges* (see section 3.5.1), or the factorization of permutations into 2-cycles.

If $v(\pi) = \sum_i |i - \pi(i)|$, Heath and Vergara [209] give an algorithm that sorts a permutation with 2-reversals and 3-reversals in at least $\frac{v(\pi)}{4}$ and at most $\frac{v(\pi)}{2}$ operations.

3.5.5.2 Bounded Transpositions This variant of sorting by transpositions was introduced by Heath and Vergara [207] under the name of *bounded block moves.*

Definition 3.21 A (p, q)-**transposition** $\tau(i, j, k)$ is a transposition (definition 3.1) with $j - i \leq p$ and $k - j \leq q$, or $j - i \leq q$ and $k - j \leq p$.

Heath and Vergara [207] note that the minimum number of $(1, n-1)$-transpositions required to sort a permutation π is exactly the length of a longest increasing subsequence of π. This yields an $O(n \log\log n)$ algorithm to solve the $(1, n-1)$-transposition distance. The $(1, 2)$-transposition problem is studied in a few papers as the *short block moves* problem. It is known that the diameter of the $(1, 2)$-transposition problem is $\lceil \binom{\frac{n}{2}}{2}/2 \rceil$ (see Heath and Vergara [207]), and there exists a $\frac{4}{3}$-approximation algorithm for the $(1, 2)$-transposition problem (see Heath and Vergara [208]).

Some classes of permutations for which the $(1, 2)$-transposition problem is solvable in polynomial time have been characterized (see Heath and Vergara [207, 208] or Mahajan et al. [262]), but no complexity result is known.

3.5.6 Cut-and-Paste

Cranston et al. [126] introduced the **cut-and-paste** operation, which consists in cutting a segment of the permutation and pasting it elsewhere, possibly reversed. In other words, the allowed operations are transpositions, and transpositions followed by a reversal of the transposed segment. This variant has been investigated in only a single paper, which focused on bounds on the diameter: more specifically, Cranston et al. [126] prove that the diameter lies between $\lceil n/2 \rceil$ and $\lceil 2n/3 \rceil$. No complexity result is known.

3.5.7 Strip Moves

Also referred to as a "block move" by authors who study this problem (we choose to be coherent with our own definitions, which are used by most authors in computational biology), a **strip move** is a transposition that displaces a strip of the permutation, that is, a segment without breakpoints. Since strip moves are restricted transpositions, any lower bound on the transposition distance is a lower bound on the strip move distance. This problem is not motivated by genome rearrangements, but has applications in computing vision.

- Introduced by Bein et al. [36].
- Complexity: **NP**-hard (see Bein et al. [36]).
- Best approximation ratio: 2 (see Bein et al. [37] and Mahajan et al. [261]).
- Diameter: n, attained by the reversed permutation (see Bein et al. [36]).

3.5.8 Stack-Sorting

Operations on stacks are not at all inspired by genome rearrangements, but it is worth mentioning that researchers from other fields have looked at sorting problems related to stacks. The term "stack-sorting" encompasses many variants, which have been surveyed by Bóna [75], sharing the common feature that all admissible operations are described through one or several stacks.

A **stack** is a last-in, first-out (LIFO) linear sequence accessed at one end called the **top**. Elements are added and removed from the top by means of **push** and **pop** operations, respectively. The **full-pop** operation pops the entire stack. A stack can be used to rearrange a permutation π as follows: the elements of π are pushed onto an initially empty stack, and an output permutation is formed by popping elements from the stack. The output permutation obviously depends on how the push and pop operations are interleaved.

The **stack-sorting** problem is to transform an input permutation π on $\{1, 2, \dots, n\}$ into $\iota = (1\ 2\ \cdots\ n)$ through a sequence of push, pop, and full-pop operations. Permutations that can be transformed into ι through a stack are said to be **stack-sortable**. As an illustration, permutation (2 1 4 3) is stack-sortable (see figure 3.6). Of course, not all permutations are stack-sortable; for example, the permutation (2 4 1 3) is not stack-sortable.

The stack-sorting problem generalizes to systems of stacks. Two special cases have attracted researchers over many years: stacks in parallel and stacks in series:

- If stacks $S_1, S_2, \dots, S_k$ are in parallel, at any point in the sorting process one may push the next input symbol onto one of the k stacks or may pop one of the k stacks and thereby create another symbol of the output permutation.
- If stacks $S_1, S_2, \dots, S_k$ are in series, one seeks to sort a permutation by pushing its symbols onto S_1, popping them off S_1 and onto S_2, transferring them from S_2 and pushing them onto S_3, and so on, until they emerge from S_k to become new output symbols.

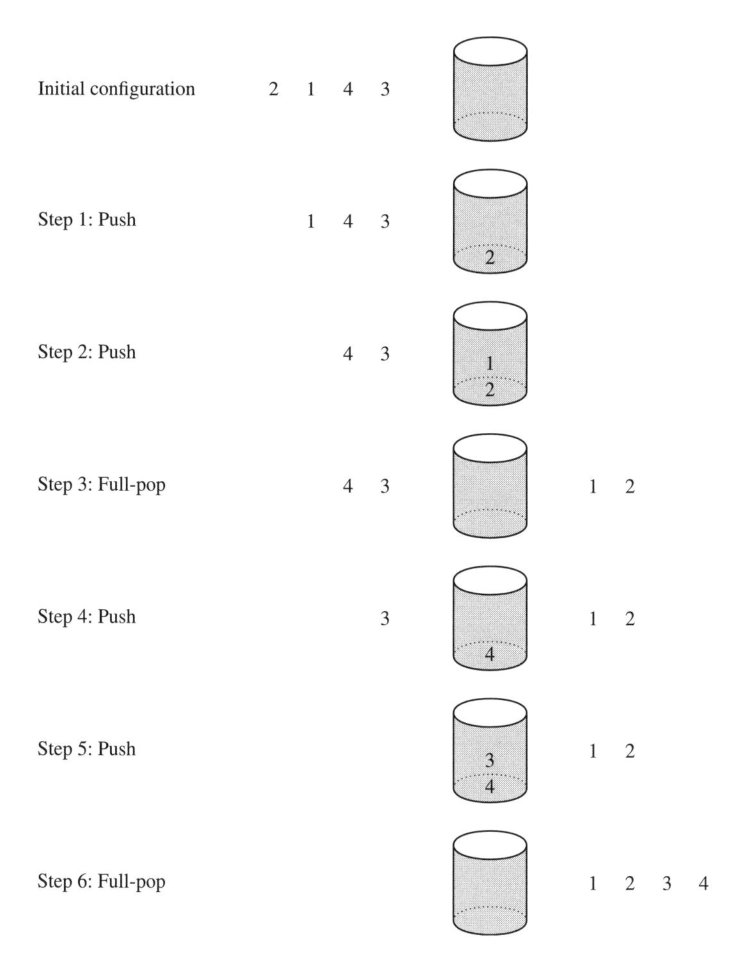

Figure 3.6
Sorting (2 1 4 3) with a stack

The stack-sorting problem was introduced in the 1960s (see Knuth [237]), and was motivated by the study of patterns in permutations.

Definition 3.22 For a permutation π in S_n, a permutation $t = (t_1 \cdots t_k)$, $k < n$, is a **pattern** of π if there is an injection σ of $\{1, 2, \ldots, k\}$ into $\{1, 2, \ldots, n\}$ such that for all $i, j \leq k$:

- $i < j$ if and only if $\sigma(i) < \sigma(j)$;
- $t_i < t_j$ if and only if $t_\sigma(i) < t_\sigma(j)$.

The general problem of deciding whether a given permutation contains a given pattern is **NP**-complete (Bose et al. [80]). Few polynomial-time solvable cases are known, such as separable permutations (see Bose et al. [80]). Of particular importance, Knuth [237] first observed that a permutation is stack-sortable if and only if it does not contain the pattern (2 3 1).

In fact, there are strong connections between stack-sorting and pattern-matching in permutations. Research in stack-sorting has led to the discovery of many results, among which we shall mention the one of Avis and Newborn [24], which states that a permutation can be sorted by a series of stacks using only the operations "push" and "full-pop" if and only if it does not contain the pattern (3 1 4 2) or (2 4 1 3). The full-pop operation reverses a segment of the permutation, and hence the problem corresponds to sorting permutations by reversals, and the number of reversals needed is the number of stacks in the series. This is why the result of Avis and Newborn [24] has been rediscovered and generalized to signed permutations in the scope of genome rearrangement problems, in a variant that we describe in section 4.3.1.

For a recent account on stack-sorting, see Bóna [75]. The study of patterns of permutations, and of classes of permutations avoiding certain patterns, is covered by Bóna [76]. A pattern avoidance database is maintained by Tenner [354].

3.5.9 Tandem Duplications and Random Losses

A tandem duplication (page 3) consists in inserting a copy of a given segment next to that segment (for instance, the permutation (1 5 4 2 6 3) undergoing a tandem duplication of the segment containing 5, 4, and 2 will be transformed into the string (1 5 4 2 5 4 2 6 3)). A **random loss** follows a duplication event and consists in the deletion of one copy of each of the duplicated genes. In that model, a tandem duplication followed by a random loss is counted as a single event, referred to as a "duplication loss." See figure 3.7 for an example.

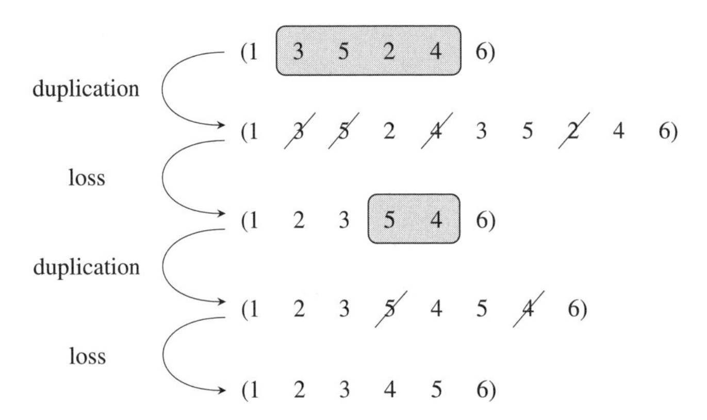

Figure 3.7
Sorting (1 3 5 2 4 6) by duplication loss operations

- Introduced by Chaudhuri et al. [102].
- Complexity: polynomial.
- Best algorithm: as explained below, the distance depends on a fixed integer parameter $\alpha \geq 1$; the distance can be computed in $O(n)$ time if $\alpha = 1$, and in $O(n \log n)$ time otherwise (see Chaudhuri et al. [102]).
- Diameter: $\lceil \log_2 n \rceil$ if $\alpha = 1$, reached only by the reversed permutation $(n\ n-1 \cdots 2\ 1)$, and $\binom{n}{2}$ otherwise (see Chaudhuri et al. [102]).

Chaudhuri et al. [102] study a model in which each tandem duplication is weighted in the following way: if the length of the duplicated segment is k, then the duplication event has weight α^k, where $\alpha \geq 1$ is a fixed integer parameter. Under that weight function, the **tandem-duplication random-loss distance** with natural parameter α of a linear, unsigned permutation π, denoted by $tdrld_\alpha(\pi)$, is the number of duplication-loss events in a minimum weight sequence transforming ι into π. We wish to stress that, unlike all other distances we have encountered so far, here we transform the identity permutation into π, and not the contrary. This is important, because the distance under consideration is asymmetric: transforming ι into π may not require as many steps as transforming π into ι. For example, one duplication-loss event of length 4 (e.g., the duplication of (2 3 4 5) followed by an adequate random loss) is enough to transform $\iota = (1\ 2\ 3\ 4\ 5\ 6)$ into $\pi = (1\ 3\ 5\ 2\ 4\ 6)$, whereas two duplication-loss events (e.g., one of length 4 and one of length 2; see figure 3.7) are needed to transform π into ι.

Theorem 3.23 [102] Given a natural $\alpha \geq 1$ and any π in S_n:

1. if $\alpha = 1$, then $tdrld_\alpha(\pi) = \lceil \log_2(des(\pi) + 1) \rceil$ and can be computed in $O(n)$ time;
2. if $\alpha \geq 2$, then $tdrld_\alpha(\pi) = inv_2(\pi)$ and can be computed in $O(n \log n)$ time.

Bouvel and Rossin [86] introduce another variant, in which each duplication-loss event has weight 1 if the duplicated segment has a length strictly less than K and infinite otherwise, for some fixed $K \geq 2$ in $\mathbb{N} \cup \{\infty\}$. They prove some results that connect this model and permutation patterns—more specifically, with pattern avoidance.

3.5.10 Combined Operations: Reversals and Transpositions

All distances we have examined so far are based on only one kind of operation. It is certainly unrealistic, from a biological point of view, to assume that genomes evolve by only one type of mutation, and a few researchers have therefore tried to take several kinds of mutations into account. However, most rearrangement problems combining different kinds of operations have been studied in the signed case (see section 4.4), and very few results have been obtained in the unsigned case, where it seems that the only two operations to have been considered together are reversals and transpositions.

- Introduced by Walter et al. [363].
- Complexity: unknown.
- Best algorithm: sorting by unsigned reversals and transpositions is approximable within $2.5909 + \delta$, for any $\delta > 0$ (see Rahman et al. [310]).
- Diameter: not studied; upper-bounded by the minimum of the transposition diameter and of the reversal diameter.

As far as lower bounds are concerned, Walter et al. [363] made the simple observation that the number of strong breakpoints decreases by at most two using a reversal, and by at most three using a transposition, which yields a lower bound of (strong breakpoints)/3, and since it is always possible to create a new adjacency, a 3-approximation follows.

No upper bound is known other than those on the transposition distance (section 3.1.2) and on the reversal distance (section 3.3.2), which trivially remain upper bounds on the reversal and transposition distance.

Rahman et al. [310] then gave a $(4 - \frac{2}{k})$-approximation, where k is the approximation ratio of the algorithm used to find a maximal decomposition of the breakpoint graph (as discussed in section 3.3.5). Since at the time of publication, the best approximation algorithm for achieving this task had a ratio of $1.4193 + \varepsilon$ for any $\varepsilon > 0$ (see Lin and Jiang [251]), they obtained an approximation algorithm with the ratio $2.5909 + \delta$, for any $\delta > 0$. Finally, we mention a polynomial-sized integer linear

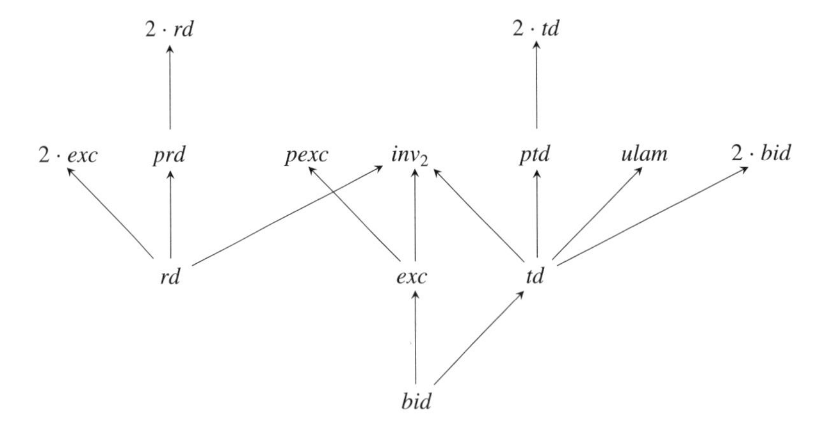

Figure 3.8
Some relations between distances on unsigned permutations; an arrow from distance d_1 to distance d_2 means that for all π in S_n, $d_1(\pi) \leq d_2(\pi)$

programming formulation of the problem by Dias and Carvalho de Souza [136] that requires $O(n^4)$ variables and $O(n^6)$ constraints. They observe that though these sizes are polynomial, they grow sufficiently fast not to allow the exact computation of the distance for values of n larger than 7.

3.6 Relations between Distances on Unsigned Permutations

We conclude this chapter with a summary of some relations between distances on unsigned permutations, summarized in figure 3.8.

Proving the relations depicted in figure 3.8 is quite straightforward, either by simulating one operation using another or by noting that one is a restriction of another, and we leave this as an exercise to the reader. Other obvious relations, such as $2 \cdot d_1 \leq 2 \cdot d_2$ whenever $d_1 \leq d_2$, are not drawn.

4 Distances between Signed Permutations

Signed permutations allow taking the relative orientation of homologous markers into account and constitute a more biologically relevant model for genomes; for instance, reversals always change the strand of the reversed segment, so that the orientation of a gene inside the segment is changed as well. Therefore, whenever the orientation of genes is known and mutations that affect orientations are considered, it is better to use signed permutations as a model. Remarkably, some problems that are intractable for unsigned permutations become tractable for signed ones. Such a notable example is the reversal distance (see section 4.2).

4.1 Conserved Interval Distance

Bergeron and Stoye [45] introduced a variant of the breakpoint distance based on what they call *conserved segments*, or *conserved intervals* [49]. It will be the starting point of our study of distances between signed permutations, which will make it easier to subsequently introduce other distances that rely on the same concepts.

- Introduced by Bergeron and Stoye [45].
- Complexity: polynomial.
- Best algorithm: the conserved interval distance can be computed in $O(n)$ time (see Bergeron and Stoye [45]). There is no associated sorting problem.
- Diameter: not studied.

Oppositely to most distances studied in genome rearrangements, the conserved interval distance is not based on a transformation, and hence there is no associated sorting problem. Recall that an interval $I = \{|\pi_a|, \dots, |\pi_b|\}$ of a permutation π is a common interval if it is an interval of the identity permutation ι. Let m_I and M_I denote, respectively, the smallest and largest elements of an interval I, that is, $m_I = \min_{i \in [a,b]} |\pi_i|$ and $M_I = \max_{i \in [a,b]} |\pi_i|$.

Definition 4.1 A common interval $I = \{|\pi_a|, \ldots, |\pi_b|\}$ is **conserved** if either $\pi_a = m_I$ and $\pi_b = M_I$, or $\pi_a = -M_I$ and $\pi_b = -m_I$.

The fact that a common interval is conserved means that it is separated from the rest of the permutation, in the sense that the rearrangements that transform this permutation into the identity never overlap such an interval.

Therefore, the number of conserved intervals between two permutations π and σ (as defined by the number of conserved intervals in $\sigma^{-1} \circ \pi$) is a measure of similarity (of how much is conserved) between the two permutations. We denote as $fs(\pi)$ the number of conserved intervals of a permutation π. However, this quantity does not define a distance, and that is why the distance definition is slightly different.

Definition 4.2 For two permutations π and σ, the **conserved interval distance** $\mathrm{cid}(\pi, \sigma)$ is defined by $\mathrm{cid}(\pi, \sigma) = fs(\pi) + fs(\sigma) - 2fs(\sigma^{-1} \circ \pi)$.

It is a metric, and can be computed in linear time with a simple progression through the permutation (see Bergeron and Stoye [45]).

4.2 Signed Reversal Distance

Sorting by reversals is undoubtedly the most famous problem in the realm of genome rearrangements. This is certainly due to the existence of a polynomial-time algorithm for signed permutations (by contrast with the unsigned variant, which is **NP**-hard) that provides fast and accurate solutions for practical applications.

- Introduced by Bafna and Pevzner [29].
- Complexity: polynomial.
- Best algorithm: the signed reversal distance can be computed in $O(n)$ time (see Bader et al. [26]), whereas sorting by signed reversals can be done in $O(n^{3/2})$ time (using an algorithm by Tannier et al. [352] and a data structure by Han [194]).
- Diameter: $n + 1$, attained by any permutation whose breakpoint graph has only one cycle and whose elements are all positive (deduced from the distance formula; see section 4.2.2).

4.2.1 Reversals

Definition 4.3 For any permutation π in $S_n^{\pm}$, the **signed reversal** $\rho(i, j)$ with $1 \leq i \leq j \leq n$ applied to π reverses the closed interval determined by i and j and reverses the sign of all its elements, transforming π into $\pi \circ \rho(i, j)$. Therefore, $\rho(i, j)$ is the following permutation:

$$\begin{pmatrix} 1 \cdots i-1 & \underline{i \quad i+1 \quad \cdots \quad j-1 \quad j} & j+1 \cdots n \\ 1 \cdots i-1 & -j \quad -(j-1) \quad \cdots \quad -(i+1) \quad -i & j+1 \cdots n \end{pmatrix}.$$

Since the first mention of the problem of sorting signed permutations by reversals by Bafna and Pevzner [29], a polynomial-time algorithm has been published by Hannenhalli and Pevzner [199], and subsequent improvements have been achieved by Berman and Hannenhalli [56], Kaplan et al. [228], Bader et al. [26], Tannier and Sagot [351], Tannier et al. [352], and Han [194].

Bergeron [44] and Bergeron et al. [47, 49] made a great effort to make this problem simpler and teachable. A survey on the distance computation problem has been written by Bergeron et al. [50]. We give here an idea of how the distance formula and the sequence of reversals are computed.

4.2.2 The Distance Formula

4.2.2.1 The Breakpoint Graph in the Signed Case The *breakpoint graph* $BG(\pi)$ of a signed permutation is a slightly different version of the breakpoint graph of an unsigned permutation.

Definition 4.4 The **breakpoint graph** of a signed permutation π is the graph $BG(\pi) = (V, E)$, whose vertex set V contains, for $1 \leq g \leq n$, two vertices g_t and g_h, called the **tail** and the **head** of the gene g, plus two vertices denoted 0_h and $n+1_t$. The edge set E of $BG(\pi)$ is the union of two perfect matchings of V, respectively denoted by R, the **reality edges**, and D, the **desire edges**:

- D contains the edges $\{g_h, (g+1)_t\}$, for $0 \leq g \leq n$.
- R contains an edge from π_{i_h} if π_i is nonnegative, and from π_{i_t} otherwise, to π_{i+1_t} if π_{i+1} is nonnegative, and to π_{i+1_h} otherwise, for $0 \leq i \leq n$.

An example of a breakpoint graph is figure 4.1. The breakpoint graph of a signed permutation is exactly the breakpoint graph of the unsigned permutation obtained

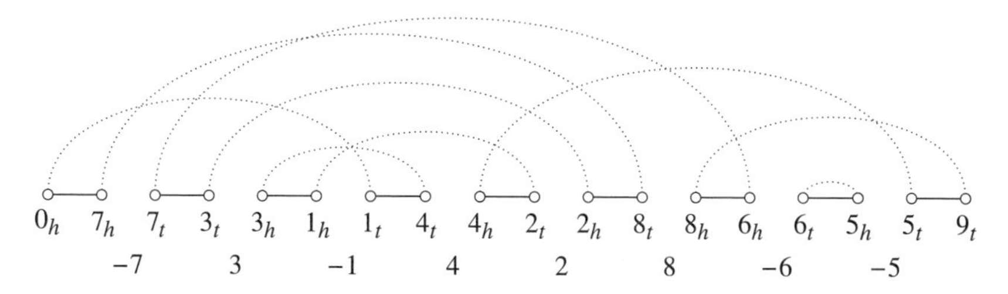

Figure 4.1
The breakpoint graph of the permutation (−7 3 −1 4 2 8 −6 −5)

by replacing every signed element with two unsigned elements, except that we omit edges of the form $\{g_t, g_h\}$.

It is easy to check that every vertex of $BG(\pi)$ has degree 2 (it has one incident edge in R and one in D), so the breakpoint graph is a set of disjoint alternating cycles, which makes the disjoint cycle decomposition trivial, as is the case for the cycle graph. The number of cycles in $BG(\pi)$ is denoted by $c(BG(\pi))$.

4.2.2.2 Unoriented Components The decomposition of a permutation into components plays a crucial role for the distance formula.

Definition 4.5 A **component** of a permutation is a conserved interval which is not the union of other conserved intervals. A point (π_i, π_{i+1}) **belongs** to a component C if both π_i and π_{i+1} are in C and C is the smallest component that contains them. A point is **oriented** if its elements have opposite signs. A component is said to be **oriented** if an oriented point belongs to it.

The structure of the family of all components is described by the following property (recall definition 2.19).

Property 4.1 [50] For two different components of a signed permutation, one of the following holds:

- They are disjoint.
- One is contained in the other and they have different extremities.
- They overlap on one element.

Definition 4.6 A sequence of successively overlapping components is called a **chain of components**. A chain C is **maximal** if no chain contains the components of C in addition to other components.

Definition 4.7 The **PQ-tree of components** of a permutation π is defined as follows:

- Each unoriented component is represented by a **P-node**.
- Each maximal chain of unoriented components (possibly a singleton) is represented by a **Q-node** whose (ordered) children are the P-nodes that represent the components of this chain.
- A Q-node is the child of the smallest component that contains the chain it represents.

An example of PQ-tree of components of a permutation is shown in Figure 4.2.

Definition 4.8 A leaf in the PQ-tree of unoriented components is **simple** if the path up to, but excluding the next vertex of degree at least 3 in the tree, contains no other P-node than itself.

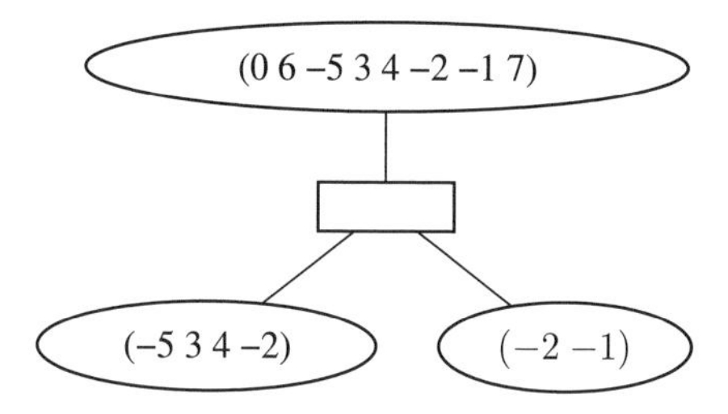

Figure 4.2
The PQ-tree of components of permutation (0 6 −5 3 4 −2 −1 7)

Let $t(\pi)$ be the number of leaves of the tree of unoriented components, plus one if there is an odd number of leaves and none is simple. Let $srd(\pi)$ denote the reversal distance of a signed permutation π. We can now state the distance formula.

Theorem 4.1 [199, 50] For any permutation π in $S_n^{\pm}$, we have

$$srd(\pi) = n + 1 - c(BG(\pi)) + t(\pi).$$

It means that $t(\pi)$ operations are required to eliminate all unoriented components, and $n + 1 - c(BG(\pi))$ are needed to sort oriented components. The reversal distance can be computed in linear time, thanks to this formula and to the computation of all the components in a traversal of the permutation.

4.2.3 The Scenario of Reversals

Once the linear-time computation of the distance formula has been stated, there is a trivial $O(n^4)$ algorithm that computes a sequence of length $srd(\pi)$: try every possible signed reversal ρ at each step, until one such that $srd(\pi \circ \rho) = srd(\pi) - 1$ is found. Such a reversal is called **safe**. Here is a description of the improvements of this time complexity. The procedure of finding sorting sequences of reversals is done in two steps in all solutions that are provided in the literature. It follows the division of the distance formula into two parameters: the first step consists in eliminating all unoriented components, and the second step consists in sorting the remaining oriented components.

The first step was given its best solution by Kaplan et al. [228], whose algorithm runs in linear time when coupled with the linear distance computation [26], and it is based on Hannenhalli and Pevzner's [199] early results on cutting and merging components.

The second step is the bottleneck of the whole procedure. At this point, if we assume that there is no unoriented component, the distance is $srd(\pi) = n + 1 - c(BG(\pi))$, so a safe reversal is one that increases $c(BG(\pi))$ and does not create unoriented components (since that would increase $t(\pi)$).

Improvements on the running time of the second step are achieved by using the concept of **oriented** reversals, which are reversals that increase $c(BG(\pi))$ (note that they are not necessarily safe). Finding an oriented reversal is an easy task: the interval spanned by any oriented desire edge of the breakpoint graph defines one. The hard part is to make sure that it does not increase the number of unoriented components.

The quadratic algorithms designed by Berman and Hannenhalli [56] and by Kaplan et al. [228] are based on the linear-time recognition of safe reversals. No better algorithm is known so far to recognize safe reversals, but Tannier and Sagot [351] proved that the recognition of a safe reversal at each step is not necessary: identifying oriented reversals is sufficient.

A sequence of oriented reversals $\rho_1, \ldots, \rho_k$ is said to be **maximal** if there is no oriented reversal in $\pi \circ \rho_1 \circ \cdots \circ \rho_k$. In particular, a sorting sequence is maximal, but the converse is not true. The algorithm is based on the following theorem by Tannier et al. [352].

Theorem 4.2 [352] If S is a maximal but not a sorting sequence of oriented reversals for a permutation, then there exists a nonempty sequence S' of oriented reversals such that S may be split into two parts, $S = S_1, S_2$, and S_1, S', S_2 is a sequence of oriented reversals.

This allows the construction of sequences of oriented reversals instead of safe reversals, and increases the size of those sequences by inserting reversals in the sequence instead of appending them.

This algorithm, with a classical data structure to represent permutations (as an array, for example) still has an $O(n^2)$ time complexity, because at each step it has to test the presence of an oriented reversal, and apply it to the permutation. The slight modification of a data structure invented by Kaplan and Verbin [227] allows picking and applying an oriented reversal in $O(\sqrt{n \log n})$ time, and using this, Tannier and Sagot's algorithm achieves $O(n^{3/2}\sqrt{\log n})$ time complexity.

Han [194] has announced another data structure that allows picking and applying an oriented reversal in $O(\sqrt{n})$ time, and a similar slight modification can probably decrease the complexity of the overall method to $O(n^{3/2})$.

4.2.4 The Space of All Optimal Solutions

Almost all studies on sorting sequences of reversals have been devoted to giving only one sequence, though it has been remarked that there are often plenty of them (it may be over several millions even for $n \leq 10$). A few studies have tried to fill this deficiency.

An algorithm to enumerate all safe reversals at a given step has been designed and implemented by Siepel [340]. A structure of the space of optimal solutions has been discovered by Bergeron et al. [48], and the algorithmics related to this structure have been studied by Braga et al. [87].

4.2.5 Experimental Results

Bader et al. [26] provide some experimental tests of their algorithm to compute the reversal distance. Bergeron and Strasbourg [46] have performed some experiments on an $O(n^3)$ algorithm to sort by reversals. Attempts to parallelize these algorithms have been made by She and Chen [338] and Kaplan and Verbin [227].

4.3 Variants of Sorting by Reversals

4.3.1 Perfect Signed Reversal Distance

Even though an optimal sequence of signed reversals can be found in polynomial time, the number of such sequences can be exponential, and hence additional criteria are needed to help in choosing one solution over another. The "perfect" variant of sorting by reversals is an attempt to add biological constraints to discriminate among the solutions.

- Introduced by Figeac and Varré [175].
- Complexity: **NP**-hard (Figeac and Varré [175]).
- Diameter: not studied.

Definition 4.9 A reversal $\rho_{i,j}$ **breaks** an interval I of a permutation π if I and the interval $[|\pi_i|, \ldots, |\pi_j|]$ overlap. If S is a subset of the set of common intervals of π, an S-**perfect** scenario of reversals is a sorting sequence of π that does not break any interval of S. The **S-perfect reversal distance** $perfd_S(\pi)$ is the smallest number of reversals in an S-perfect sequence.

A family S of common intervals is called **nested** if no pair of intervals from S overlaps.

Theorem 4.3 [175] Computing the S-perfect reversal distance is **NP**-hard, even if S is nested.

Note that the complexity of the particular case where S is the whole set of common intervals is not known. This result is proved by a reduction from unsigned sorting by reversals.

Perfect sorting by reversals may easily be achieved by optimally sorting all common intervals in a bottom-up search of the tree of strong common intervals, sorting the strong segments one by one. Whenever there is no ambiguity concerning the direction in which to sort a segment at each step, the procedure is polynomial. Bérard et al. [41, 42] and Diekmann et al. [142] have studied some classes of permutations for which this is the case: a family S of common intervals is called **separable** if every strong interval of S is the union of two overlapping intervals from S.

Theorem 4.4 [41] Computing the S-perfect reversal distance is polynomial if S is separable.

4.3.2 Prefix Reversals (Burnt Pancakes)

An analogue of the pancake-flipping problem (see section 3.4) has been studied in the signed case, and people usually refer to this variant as that of sorting *burnt pancakes*: all pancakes are burnt on one side, and are assigned a negative number when the burnt side is faceup, and positive otherwise. The goal is therefore to flip pancakes so that they not only end up sorted in increasing size order, but also with all burnt sides facedown.

- Introduced by Gates and Papadimitriou [183].
- Complexity: unknown.
- Diameter: unknown. Lies between $\frac{3}{2}n$ and $2n - 2$ (see Cohen and Blum [118]).

Definition 4.10 A **prefix reversal** is a signed reversal (definition 4.3) $\rho(i, j)$ with $i = 1$.

Results are not numerous for this variant. Only a few bounds for the diameter have been published [118]. Cohen and Blum [118] conjecture that the "negative identity permutation" $(-1\ -2\ \cdots\ -n)$ is the hardest permutation to sort. Heydari and Sudborough [212] proved that this precise permutation requires $3(n+1)/2$ reversals to be sorted, so this is the conjectured diameter. The proof of this conjecture would imply the same function as an upper bound for the diameter in the unsigned case as well.

4.3.3 Reversals That Are Symmetric Around a Point

This constraint is particularly relevant for rearrangements in the prokaryote world. Indeed, most of the observed reversals are symmetric around an origin or terminus of replication.

Definition 4.11 Given a circular permutation π°, a reversal is **symmetric around the point** $\pi^\circ_i \pi^\circ_{i+1}$ if it acts on the interval $(\pi^\circ_{i-k} \cdots \pi^\circ_i\ \pi^\circ_{i+1} \cdots \pi^\circ_{i+k+1})$, for some natural k (indices are calculated modulo $n+1$).

The only combinatorial paper taking this constraint into account seems to be the one by Ohlebusch et al. [283]. They compute the reversal distance in linear time when the only allowed reversals are symmetric around an origin or teminus of replication (the model is quite restrictive), and also compute the median for three permutations in linear time, a remarkable result since most median problems (which will be discussed in detail in chapter 13) are **NP**-hard.

Ajana et al. [4] used the problem of sorting by reversals and the exploration of all solutions to test whether reversals really occurred around an origin or terminus of replication.

4.3.4 Weighted Reversals

Just as in the unsigned case, some methods have been developed for weighted signed reversals, in order to model the natural selection of small inversions of genomes. It has been studied with a weight function $f(l) = l^{\alpha}$, where l is the size of a reversal and α is a given nonnegative parameter.

- Introduced by Swidan et al. [346].
- Complexity: unknown. There exists an approximation algorithm with a constant ratio for $\alpha \geq 2$ by Swidan et al. [346].
- Diameter: unknown; Swidan et al. [346] computed asymptotic values.

We summarize a few results of Swidan et al. [346] in table 4.1. There are lower and upper asymptotic bounds for the diameter, as well as some approximation algorithms with asymptotic ratios.

4.3.5 Fixed-Length Reversals

This problem may be seen as a particular case of the weighted reversals problem, with a special weight function that has only one nonzero value.

- Introduced by Qi et al. [304].
- Feasibility solved for even-length reversals in signed circular permutations (see Qi et al. [304]).

As in definition 3.20, we define, for a fixed integer k, a k-**reversal** as a reversal $\rho(i, j)$ with $j - i = k$. Since k-reversals do not necessarily generate the hyperoctahedral group, the first issue concerning sorting by k-reversals is to determine whenever

Table 4.1
Bounds on the diameter and approximation ratios for sorting signed permutations by weighted reversals

Value of α	Lower bound	Upper bound	Approximation ratio
$0 \leq \alpha < 1$	$\Omega(n)$	$O(n \log n)$	
$\alpha = 1$	$\Omega(n \log n)$	$O(n \log^2 n)$	$O(\log n)$
$1 < \alpha < 2$	$\Omega(n^{\alpha})$	$\Theta(n^{\alpha})$	$O(\log n)$
$\alpha \geq 2$	$\Omega(n^2)$	$\Theta(n^2)$	$O(1)$

Retrieved from Swidan et al. [346].

it is feasible. Qi et al. [304] count the number of classes that are generated by k-reversals, for all even k.

4.4 Combined Operations

The goal of rearrangement problems is to be able to combine several operations, giving weight to each of them according to its probability of occurrence. More work has been conducted on combined operations in the signed case than in the unsigned case, because these problems are attempts to come back to the biological application, and signed permutations seem to be a more realistic model.

4.4.1 Reversals and Transpositions

Sankoff [315] included reversals and transpositions in his methods, but gave no theoretical results. Blanchette et al. [66] then wrote a software called DERANGE (see chapter 15) that made it possible to give parameters for the weights of reversals and transpositions. According to the experiments they made, the most probable weights were 1 for reversals and 2 for transpositions. Note that a transposition may be simulated by three reversals, so an algorithm with weight 1 for reversals against weight 3 for transpositions would not give better solutions than sorting by reversals only.

Several variants of the problem have been studied according to the relative weights α that are given to transpositions (if weight 1 is given to a reversal). We summarize a few results in table 4.2.

For $\alpha = 1$, the permutation $(-1\ -2\ \cdots\ -n)$ can be sorted in $\lfloor \frac{n}{2} \rfloor + 2$ operations, as proved by Walter et al. [363] and Meidanis et al. [266]. Those authors conjecture that it is the value of the diameter.

4.4.2 Reversals, Transpositions, Transreversals, Revrevs

A special variant considers the operation of **transreversal**, that is, a transposition followed by a reversal on the transposed segment (both operations are counted as only

Table 4.2
A summary of results concerning a few variants of sorting by signed reversals (with weight 1) and transpositions (with various weights)

Weight of a transposition	Introduced by	Complexity	Best approximation ratio
$\alpha = 1$	[363]	unknown	2 [191, 252, 363]
$1 \leq \alpha \leq 2$	[27]	unknown	1.5 [27]
$\alpha = 2$	[162]	unknown	$(1 + \varepsilon)$ [162]

one operation), as well as **revrev**, that is, two reversals of contiguous segments. Lin et al. [256] gave a 1.75-approximation algorithm if transreversals and revrev operations are allowed.

Hartman and Sharan [203] have adapted the 1.5-approximation for sorting by transpositions to the signed case, and this improves the approximation ratio to 1.5 for sorting by reversals, transpositions, and transreversals (all with the same weight), as well as the problem where revrevs are allowed.

4.5 Double Cut-and-Joins

The double cut-and-join operation was first proposed by Yancopoulos et al. [375] as a unifying operation for computing genomic distances, and has gained an increasing popularity ever since.

- Introduced by Yancopoulos et al. [375].
- Complexity: polynomial.
- Best algorithm: both the sorting problem and the distance computation problems can be solved in $O(n)$ time (see Bergeron et al. [52]).

The best recent progress in genome rearrangements has probably been the remark that reversals, transpositions, and block interchanges, with respective weights 1, 2, and 2, can all be modeled by a single operation, called the **double cut-and-join** operation by Yancopoulos et al. [375] and the 2-**break rearrangement** operation by Alekseyev and Pevzner [8]. It is better defined on multichromosomal models (see part III), because the result of such a rearrangement may contain several chromosomes, and forbidding this possibility simply restricts the possible operations to reversals.

A succession of two double cut-and-joins, the first one of them creating a new chromosome, simulates one block interchange, so for signed permutations, the problem is equivalent to sorting by reversals and block interchanges, where block interchanges weigh twice as much as reversals in the objective function.

This model has the advantages of being both biologically relevant (these cut-and-join operations are likely to be close to what sometimes happens to genomes) and computationally easy, because in this case, the **DCJ distance**, between a signed permutation π and Id, denoted by $dcj(\pi)$, is proved by Yancopoulos et al. [375] to be

$$dcj(\pi) = p(\pi) - c(BG(\pi)),$$

where p is the number of points of the permutation, and c is the number of cycles of its breakpoint graph. The distance can therefore be computed in $O(n)$ time. A sorting sequence can also be computed in $O(n)$ time [52].

Mira and Meidanis [269] have considered the same problem from the point of view of the "algebraic formalism" studied by Meidanis and Dias [264] (see chapter 11), and have obtained the same results. Lin et al. [256] worked on minimizing the number of block interchanges in a scenario sorting a signed permutation by reversals and block interchanges. The algorithm to achieve this goal simply uses reversals to sort oriented components, and block interchanges to sort unoriented components of a permutation. It has time complexity $O(n^2)$.

5 Rearrangements of Partial Orders

5.1 Genomes as Partially Ordered Sets

Oppositely to permutations, the model we shall consider here does not rely on a total order on a chromosome (inherent in its representation as a permutation). Indeed, despite the increase in the number of sequencing projects, most genomes have not yet been completely sequenced. For these partially sequenced or assembled genomes, only partial gene maps are available (recombination analysis, physical mapping, etc.), which may have a low resolution, missing genes or markers, or conflicting ordering information among each other. Zheng et al. [380] introduced a new representation of a genome in terms of a partially ordered set (often abbreviated as "poset"). In this model, any linear extension of a poset represents a possible total order of the genome. In the context of posets, a genome rearrangement problem is to find a linear extension in each poset such that a criterion (number of reversals, number of breakpoints, etc.) is optimized. The current chapter is devoted to handling rearrangements of gene partial orders.

5.2 Partially Ordered Sets

5.2.1 Basic Definitions

Most notations and definitions are borrowed from the book by Davey and Priestley [132].

Definition 5.1 A **partial order** on a set P is a binary relation $\leq$ such that, for all x, y, and z in P:

1. $x \leq x$,
2. $x \leq y$ and $y \leq x$ imply $x = y$, and
3. $x \leq y$ and $y \leq z$ imply $x \leq z$.

These three relations are referred to as **reflexivity**, **antisymmetry**, and **transitivity**, respectively.

Definition 5.2 A set P equipped with a partial order relation $\leq$ is said to be a **partially ordered set** (also called a **poset**).

When it is necessary to specify the partial order relation, we write $(P, \leq)$. Notice that $x \leq y$ and $y \geq x$ are used interchangeably, and $x \nleq y$ means "$x \leq y$ *is false*." The symbol $\|$ is used to denote **noncomparability**: $x \| y$ if and only if $x \nleq y$ and $y \nleq x$. In some contexts, the partial order defined above is called a **nonstrict partial order**. A partial order relation $\leq$ on P gives rise to a relation $<$ of strict inequality: $x < y$ in P if and only if $x \leq y$ and $x \neq y$. Strict partial orders differ from partial orders only in whether each element is required to be unrelated, or required to be related, to itself.

Definition 5.3 Let P be a poset and $x, y \in P$. Then x is **covered** by y (or y **covers** x), denoted $x \lessdot y$, if $x < y$ and $x \leq z \leq y$ implies $x = z$.

Definition 5.4 A **totally ordered set** (or **linearly ordered set**) is a poset $(P, \leq)$ which has the property that every two elements of P are comparable (i.e., for all $x, y \in P$, either $x \leq y$ or $y \leq x$.

A totally ordered set is sometimes called a **chain**, especially when it is considered as a subset of some other poset. The poset P is an **antichain** if, for all $x, y \in P$, $x \leq y$ only if $x = y$ (i.e., every two elements of P are comparable). The **height** of P is the maximum size of a chain in it, and the **width** of P is defined to be the largest antichain in it. Dilworth's theorem characterizes the width of any poset in terms of a partition of the order into a minimum number of chains.

Theorem 5.1 [144] In any poset P, the size of a maximum antichain in P is equal to the number of chains needed to cover its elements.

A poset Q is called an **extension** of a poset P if the elements of P and Q are the same, and the set of relations of P is a subset of the set of relations of Q (i.e., for all $x, y \in P$, if $x < y$ in P, then $x < y$ in Q, but not necessarily conversely. Q is called a **linear extension** of P if Q is an extension of P and also a linear order. In computer science, algorithms for finding linear extensions of partial orders are called **topological sorting**.

Weak orders are strict partial orders in which incomparability is an equivalence relation (an illustration is given in figure 5.1). The equivalence classes of this "incomparability relation" partition the elements of P, and are totally ordered by $<$. Conversely, any total order on a partition of P gives rise to a weak ordering in which $x < y$ if and only if there exist sets P' and P'' in the partition with $x \in P'$, $y \in P''$,

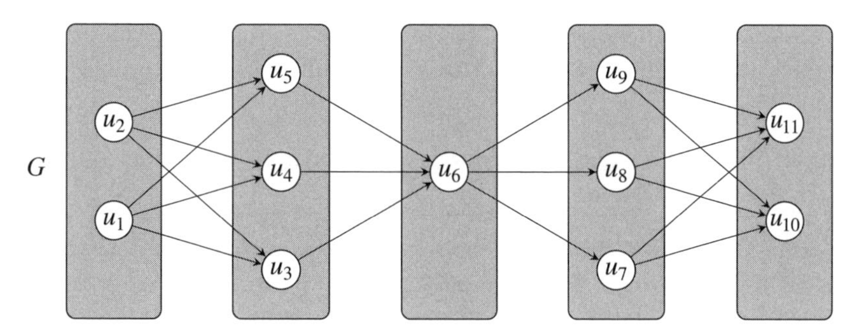

Figure 5.1
A weak order. Incomparability viewed as equivalence classes is indicated by shaded rectangles

and $P' < P''$ in the total order. Clearly, linear orders are weak orders in which the incomparability classes are singletons. Two weak orders $<_1$ and $<_2$ on P are **compatible** if $x <_1 y$ implies $x <_2 y$ whenever elements x and y of P are comparable in both $<_1$ and $<_2$.

Theorem 5.2 [78] Any two compatible weak orders have a common linear extension.

Let P be a poset and $Q \subseteq P$. Q is a **down-set** if, whenever $x \in Q$, $y \in P$ and $y \leq x$, we have $y \in Q$. And Q is an **up-set** if, whenever $x \in Q$, $y \in P$ and $y \geq x$, we have $y \in Q$. The **lower set** of Q is defined to be the set $\{y \in P : x \geq y \text{ for some } x \in Q\}$ and is denoted by $\downarrow Q$. An element $x \in Q$ is a **maximal element** of Q if $x \leq y$ and $y \in Q$ imply $x = y$. The set of all maximal elements of Q is denoted by max Q. A **minimal element** of Q and min Q are defined by reversing the order.

5.2.2 Representing Posets

There is a convenient way to represent a poset P by its **Hasse diagram**, defined as follows:

1. to each element $x \in P$, associate a point p_x of the Euclidean plane;
2. for each covering pair $x \lessdot y$, p_x and p_y are joined by a line segment, and the point representing x has a smaller vertical coordinate than the point representing y (p_x is located "below" p_y).

For algorithmic efficiency considerations, it is better not to give the full poset P, but instead the smallest directed acyclic graph (DAG) whose transitive closure is P (consider all arcs to be directed up in the Hasse diagram).

5.2.3 Topological Sorting

The **topological sorting** problem is the problem of computing a linear extension of a poset (usually given in the form of a DAG). Topological sorting on a DAG comes down to finding a linear ordering of its vertices in which each vertex appears before all nodes to which it has outbound edges. Three important facts about topological sorting are the following:

- Only DAGs can have linear extensions, since any directed cycle is an inherent contradiction to a linear order of tasks.
- Every DAG can be topologically sorted.
- DAGs typically allow many topological sorts (up to $n!$ if the DAG consists of n isolated vertices; i.e., every two elements of the corresponding poset are incomparable).

The usual algorithms for topological sorting have running time that is linear in the number of nodes plus the number of edges (i.e., $\Theta(n+m)$; see Cormen et al. [121]).

As observed before, the number of linear extensions can grow exponentially in the size of the DAG, and it turns out that even the problem of enumerating linear extensions is hard. This combinatorial explosion is a limitation if one is interested in finding all linear extensions, one linear extension satisfying a given property, or one optimizing a criterion. Algorithms for listing all linear extensions in a DAG are usually based on backtracking. They build all possible orderings from left to right, where each of the in-degree zero vertices are candidates for the next vertex.

5.3 Constructing a Poset

Map integration is the process of combining two or more maps constructed from different kinds of data or using different methodologies into a consensus map. Assuming there are no conflicting order relations, this is done by taking the union of the different gene maps. See figure 5.2 for an example.

In practical applications, different maps of the same genome occasionally conflict, either because $a < b$ in one data set and $b < a$ in the other, or because a gene is assigned to different chromosomes in the two data sets (Yap et al. [376]; Zheng and Sankoff [379]).

One simple way to avoid any cycles in the construction of the DAG is to delete all order relations that conflict with at least one other order relation (see Zheng and Sankoff [379]). This algorithm may, however, be too strict for practical considerations.

Another way of coping with conflicts is by determining which set of arcs or vertices would, if removed, eliminate the cycle. Finding a smallest such set, however, may not always have a biological explanation, but it would be a most parsimonious solution. More formally, a set of arcs (vertices) that may be removed from a directed graph to obtain an acyclic directed graph is called a **directed feedback arc set** (**feedback vertex**

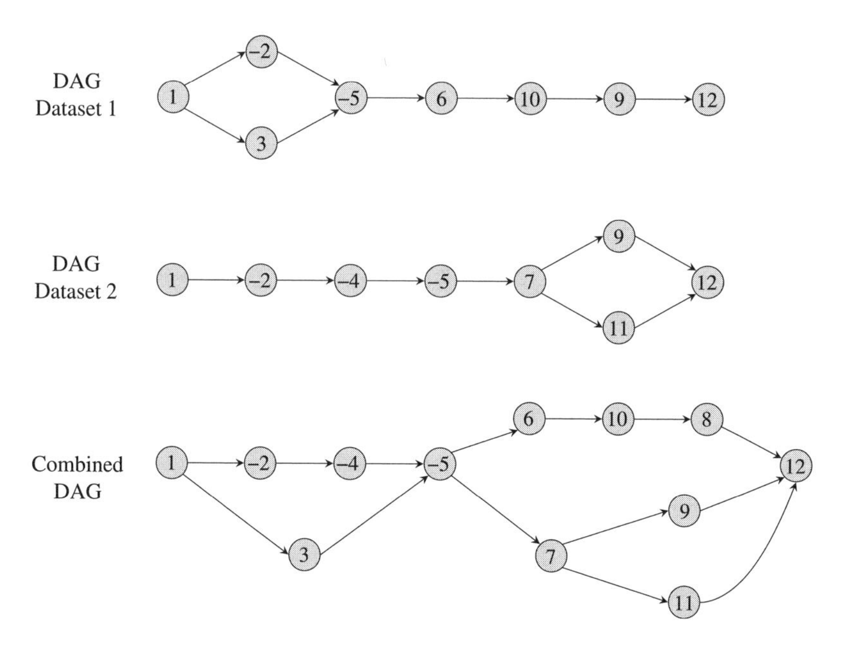

Figure 5.2
Two partially ordered genomes over the same set of genes (both given as DAGs) together with the combined DAG. A possible linear extension of the combined DAG is (1, −2, 3, −4, −5, 6, 7, 9, 10, 11, 8, 12)

set); see figure 5.3 for a simple illustration. The problem of finding a minimum cardinality feedback arc (vertex) set is called the MINIMUM FEEDBACK ARC (MINIMUM DIRECTED FEEDBACK VERTEX) problem. Both the minimum feedback arc and the minimum directed feedback vertex problems are **NP**-complete (see Garey and Johnson [181] and Kann [225] for a proof that they both actually are **APX**-hard) and are known to be approximable only within ratio $O(\log n \log \log n)$ (see Seymour [334]), where n is the number of vertices of the graph. Besides this bad complexity news, integrated graphs that represent genetic maps tend to be sparse (each vertex is connected to a small number of vertices, see Yap et al. [376]), so that parameterized techniques (or even brute-force approaches) may yield fast yet accurate solutions.

5.4 Reversal Distance

Let P and Q be two posets defined over the same set of signed genes. The **reversal distance** between P and Q is defined to be the minimum reversal distance between any linear extension of P and any linear extension of Q. That is,

$$rd(P, Q) = \min\{rd(\pi_P, \pi_Q) : \pi_P \in L(P) \text{ and } \pi_Q \in L(Q)\}.$$

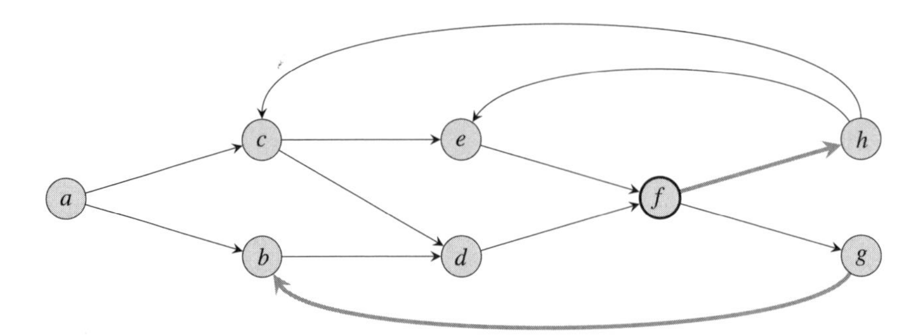

Figure 5.3
Obtaining a DAG from a (cyclic) directed graph G by either deleting two arcs (e.g., the two arcs (f,h) and (g,b)) or deleting vertex f

- Introduced by Zheng et al. [380].
- Complexity: **NP**-complete (see Fu and Jiang [180]).

Zheng et al. [380] proposed a depth-first branch-and-bound algorithm for computing the reversal distance between two posets.

5.5 Breakpoint Distance

Let P and Q be two posets defined over the same set of signed genes. The **breakpoint distance** between P and Q is defined to be the minimum breakpoint distance between any linear extension of P and any linear extension of Q. That is,

$$bd(P,Q) = \min\{bd(\pi_P, \pi_Q) : \pi_P \in L(P) \text{ and } \pi_Q \in L(Q)\}.$$

- Introduced by Blin et al. [72].
- Complexity: **NP**-complete (see Blin et al. [72]; Fu and Jiang [180]).

5.5.1 Exact Algorithms

Only a few results have been reported so far (see Blin et al. [72]), and these algorithms are concerned with computing the maximum number of adjacencies between a poset P and a total order.

Blin et al. [72] proposed a dynamic programming algorithm for arbitrary poset P. The main idea is to compute, for every antichain $Q \subseteq P$ and $x \in \max Q$, the maximum number $T(Q,x)$ of adjacencies obtained in a linear extension of $\downarrow Q$ ending in x. This number can be computed according to the following recursive formula:

$$\text{For all } Q \subseteq P,\ x \in \max Q, \quad T(Q,x) = \max_{y \in Q'} T(Q',y) + \begin{cases} 1 & \text{if } |x-y| = 1, \\ 0 & \text{otherwise} \end{cases}$$

where Q' is the set of maximal elements of the smallest down-set containing Q stripped of x, that is, $\max(\downarrow Q \backslash \{x\})$. This algorithm runs in $O(n\, a_P)$ time, where $n = |P|$ and a_P is the number of antichains in P. But a_P can be as large as 2^n, and the above dynamic programming procedure yields a combinatorial algorithm, impracticable in most cases. However, Blin et al. [72] observed that P is usually obtained by combining a relatively small number of genetic maps (weak orders). In particular, if P is obtained by combining m weak orders, each of height at most h and width at most w, then $a_P = O((h\,2^w)^m)$, and hence the total running time of the algorithm is $O((h\,2^w)^m\, n)$.

A linear time algorithm exists if P is a weak order (a single genetic map). This follows from the fact that the maximal antichains of P are totally ordered by the partial order (i.e., one can compute for all antichain $Q \in P$ the maximum number of adjacencies obtained in a linear extension of $\downarrow Q$ by dynamic programming.

5.5.2 Heuristics for Computing the Breakpoint Distance

Let P and Q be two posets on the same set of signed genes. Computing the breakpoint distance between P and Q is equivalent to the problem of finding two linear extensions that contain the maximum number of common adjacencies. Two heuristics have been proposed to compute the breakpoint distance between two partial orders defined on the same set of signed genes.

A first heuristic was proposed by Blin et al. [72] for the special case that one of the two posets is a total order. Starting from a DAG representing the transitive closure of the partial order, the algorithm consists in iteratively finding a longest valid direct or indirect path in the poset, incorporating it into the DAG, and computing the transitive closure.

Another heuristic, proposed by Fu and Jiang [180], does not require one of the two posets to be a total order. At the heart of this algorithm is the MINIMUM DOUBLE FEEDBACK VERTEX SET problem: given two directed graphs D_1 and D_2 on the same vertex set V, find a minimum cardinality subset $V' \subseteq V$ whose deletion leaves both D_1 and D_2 acyclic. The approach is a two-step procedure. First, for both P and Q, a directed graph, referred to as the adjacency order graph, is constructed; its vertex set is the set of all possible common adjacencies in P and Q, and its arcs denote compatibility of possible common adjacencies. Next, an approximation algorithm for a minimum cardinality feedback vertex set is used, taking as input the adjacency order graphs of P and Q. Clearly, the better the approximation algorithm for finding a double feedback vertex set, the better the heuristics. Extending a previous result of Demetrescu and Finocchi [134], Fu and Jiang [180] proved that the MINIMUM DOUBLE FEEDBACK VERTEX SET problem is approximable within ratio 2ℓ, where ℓ is the length of the longest cycle in any of the two input graphs.

6 Graph-Theoretic and Linear Algebra Formulations

This chapter illustrates how genome rearrangement problems spread to a wide mathematical field, including long-standing problems in algebra. In this chapter, we will examine a few structures that are in some sense at a higher level than structures such as the breakpoint graph or the cycle graph.

6.1 Simple Permutations and the Interleaving Graph

Observing that "long cycles" in the breakpoint graph of a permutation make it hard to analyze the various sorting problems they were concerned with, Hannenhalli and Pevzner [199] introduced equivalent transformations of permutations into simple permutations. This transformation, which can be achieved in linear time (see Gog and Bader [186]), preserves the signed reversal distance, and allowed them to restrict their attention to permutations with a breakpoint graph free of long cycles. Together with an additional structure known as the interleaving graph, their approach led to the first exact polynomial time algorithm for sorting by signed reversals.

Definition 6.1 A signed permutation π is called a k-**permutation** (with $k \geq 1$ in $\mathbb{N}$) if its breakpoint graph contains only cycles with k desire edges and k reality edges.

Definition 6.2 Desire edges $\{\pi_i, \pi_j\}$ $(i < j)$ and $\{\pi_k, \pi_l\}$ $(k < l)$ in $BG(\pi)$ **interleave** if intervals $[i, j]$ and $[k, l]$ intersect but neither of them contains the other.

Definition 6.3 Two cycles C_1, C_2 in $BG(\pi)$ **interleave** if there exists a pair of desire edges e_1 in C_1 and e_2 in C_2 that interleave.

Definition 6.4 The **interleaving graph** of a permutation π is the graph $IL(\pi)$ whose vertices are the cycles of $BG(\pi)$ and in which two vertices are joined by an edge if the corresponding cycles interleave.

Definition 6.5 A desire edge of $BG(\pi)$ is **oriented** if it joins two gene tails or two gene heads. A cycle of $BG(\pi)$ is **oriented** if it contains an oriented desire edge. A vertex of $IL(\pi)$ is **oriented** if the cycle it represents is oriented.

Interleaving graphs are in consequence graphs with orientations defined on every vertex.

6.2 The Overlap Graph

Kaplan et al. [228] introduced another structure based on the breakpoint graph that allowed them to improve the results obtained by Hannenhalli and Pevzner [199] using the interleaving graph. This new graph, which they called the *overlap graph*, allowed them to obtain an $O(n^2)$ algorithm for sorting by signed reversals, a significant improvement over the $O(n^4)$ algorithm of Hannenhalli and Pevzner [199].

Definition 6.6 The **overlap graph** of a permutation π is the graph $OV(\pi)$ whose vertices are the $n+1$ desire edges of $BG(\pi)$, and in which there is an edge between vertices v_i and v_j if the corresponding desire edges interleave.

Figure 6.1 shows an example of an overlap graph. The overlap graph of a permutation has oriented and unoriented vertices. Isolated vertices are unoriented, and correspond to adjacencies of the permutation.

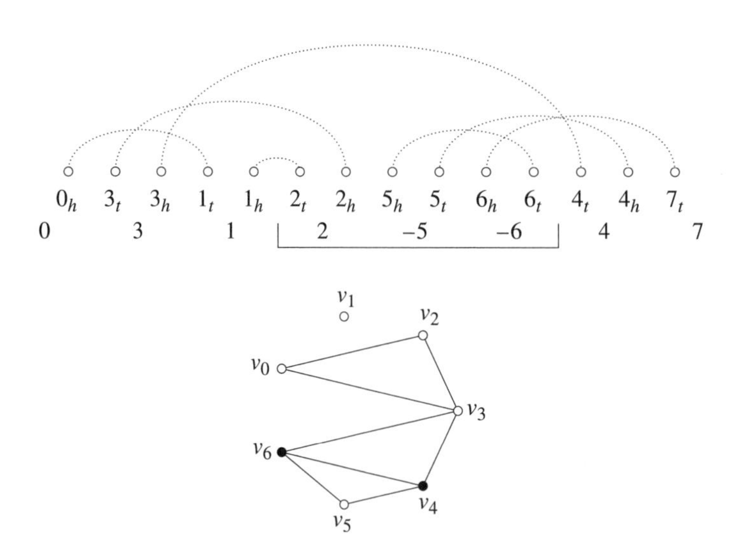

Figure 6.1
The reversal $\rho(v_1) = \rho_{3,5}$ in permutation (0 3 1 6 5 −2 4 7), and the corresponding local complementation of vertex v_1 in the overlap graph. Oriented vertices of the overlap graph are drawn in black, and unoriented vertices are white

6.3 The Local Complementation of a Graph

Let G be a graph whose vertices are labeled as oriented or unoriented (this applies, among others, to overlap and interleaving graphs).

Definition 6.7 The **local complementation** of the subgraph induced by V is the operation that consists in adding an edge between $x, y \in V$ if there is no edge $x, y \in G$, deleting x, y if there is an edge x, y in G, and switching the orientation of all vertices in V.

If v is an oriented vertex of G, we denote by G/v the result of the local complementation of the "closed neighbourhood" $N_G(v) \cup \{v\}$ of v, which we call the local complementation of v. Note that in G/v, v is unoriented and isolated. An example of local complementation is given in figure 6.1.

The relation between sorting by reversals and local complementation is given by the following lemma (see Hannenhalli and Pevzner [199]; Kaplan et al. [288]).

Lemma 6.1 [199, 288] For a permutation π, if v is an oriented vertex of the overlap graph, $OV(\pi \circ \rho(v)) = OV(\pi)/v$, and if v is an oriented vertex of the interleaving graph, $IL(\pi \circ \rho(v)) = IL(\pi)/v$.

The **local complementation game** is the following: given a graph G with oriented and nonoriented vertices, find a sequence of local complementation of oriented vertices such that at the end, the graph has only unoriented isolated vertices. A graph is said to be **tight** if such a sequence exists.

As we have seen, this graph theoretic formulation captures the oriented component of sorting by reversals. The theorem of Hannenhalli and Pevzner [199] yields the following:

Theorem 6.1 [199] A graph G is tight if and only if it has no component with only unoriented vertices.

6.4 The Matrix Tightness Problem

Given an $n \times n$ symmetric matrix M over the binary field $GF[2]$ (all such matrices are the adjacency matrices of some undirected graph on n vertices, with oriented and unoriented vertices), an element i $(1 \leq i \leq n)$ is called **oriented** if $M_{i,i} = 1$. Let $M_{*,i}$ be the column vector corresponding to column i in M, and $M_{i,*}$ be the line vector corresponding to row i in M. The **elimination** of an oriented element i is defined as the transformation of M into $M \backslash i = M - M_{*,v} M_{v,*}$.

Eliminating an element of a matrix corresponds to performing Gaussian elimination on the element on the diagonal. A matrix is called **tight** if it can be transformed into the 0 matrix by elimination operations.

Lemma 6.2 [199] If $M(G)$ is the adjacency matrix of a graph G with oriented and unoriented vertices, and v is an oriented vertex, then $M(G \backslash v) = M(G) \backslash v$.

Thus this algebraic formulation again captures the oriented component of sorting by reversals. The theorem of Hannenhalli and Pevzner [199] can be formulated in the following way. If $G(M)$ is the graph whose adjacency matrix is M, then we call the **components** of M the connected components of G. A component of M is said to be **oriented** if it has an oriented element.

Theorem 6.2 [199] A symmetric matrix M over $GF[2]$ is tight if and only if it has no unoriented component.

Hartman and Verbin [204] have generalized this result to the matrices over any field of characteristic 2. Here, an element is said to be **oriented** if it is different from the null element.

Theorem 6.3 [204] A symmetric matrix M over a field of characteristic 2 is tight if and only if it has no unoriented component.

6.5 Extension to Sorting by Transpositions

Hartman and Verbin [204] have again generalized this formulation to capture some equivalent to sorting by transpositions instead of sorting by reversals.

Given a ring R and an $n \times n$ matrix over R, an element i ($1 \leq i \leq n$) is called **oriented** if M_{ii} is invertible in R.

The **tightness problem** is to determine if a matrix is tight. It is equivalent to deciding, given a matrix M, if it can be factorized in the form $M = PLUP^{-1}$, where P is a permutation matrix, L is a lower triangular matrix, and U is an upper triangular matrix. Thus the matrix tightness problem can be formulated in classical algebraic terms: a matrix is tight if and only if it is equivalent to a positive definite matrix.

Lempel [244] and Seroussi and Lempel [332] have obtained some sufficient conditions for this tightness problem, and their result is generalized by our previous remarks (i.e., there is a necessary and sufficient condition of the matrix tightness for symmetric matrices over a field of characteristic 2 that we can test in polynomial time). Generalizing and partly solving earlier algebraic results is one of the merits of combinatorics of genome rearrangements.

The general complexity of the tightness problem is not known, and the nice remark of Hartman and Verbin [204] is that sorting 3-permutations by transpositions can be formulated as a matrix tightness problem. Indeed, in the overlap graph of a 3-permutation, each cycle of $BG(\pi)$ is represented by three vertices. The interleaving graph does not capture enough information to model all the interleaving possibilities of 3-cycles. But the overlap graph is redundant, as the adjacencies of one vertex are determined by the adjacencies of the other two vertices that represent the same cycle.

For two cycles C_1 and C_2 of $BG(\pi)$, let u_1 and v_1 be two of the three vertices representing C_1, and u_2 and v_2 be two of the three vertices representing C_2 (arbitrary chosen). Let $edge(xy)$ denote the Boolean function whose value is 1 if there is an edge between vertices x and y, and 0 otherwise. Now let $M(C_1, C_2)$ be the 2×2 matrix over $GF[2]$:

$$\begin{pmatrix} edge(u_1v_2) & edge(u_1v_1) \\ edge(u_2v_2) & edge(u_2v_1) \end{pmatrix}.$$

Furthermore, let $M(\pi)$ be the matrix whose rows and columns are indexed by the cycles of $BG(\pi)$, built as follows:

- If $i \neq j$, then $M(\pi)_{i,j}$ is the matrix $M(i, j)$;
- $M(\pi)_{i,i}$ is the 2×2 identity matrix over $GF[2]$ if i is an interleaving cycle, and the null matrix over $GF[2]$ otherwise.

Then $M(\pi)$ is a Hermitian matrix over a 16-element ring whose elements are 2×2 matrices over $GF[2]$. The conjecture of Christie [115] cited on page 36 has an equivalent under matrix tightness vocabulary:

Property 6.1 A permutation has transposition distance $\frac{tb(\pi)}{3}$ if and only if $M(\pi)$ is tight.

6.6 The Intermediate Case of Directed Local Complementation

However, this alternative formulation has not proved efficient enough to achieve a breakthrough in the problem of sorting by transpositions. One of the most promising ways to do so may be to find some intermediate formulation between the tightness problem when M is symmetric over $GF[2]$ (which corresponds to the polynomial-time-solvable problem of sorting signed permutations by reversals) and when M is Hermitian over the 16-element ring described above (which corresponds to the problem of sorting unsigned permutations by transpositions, of unknown complexity).

Hartman and Verbin [204] propose the following nice open problem, which may be called **directed graph tightness**, and which corresponds to the matrix tightness problem when M is a matrix over $GF[2]$, not necessarily symmetric.

This corresponds to the tightness problem on directed graphs, where the local complementation of an oriented vertex v results in the following changes:

- Invert the orientation of all the vertices that join v with two edges (both orientations);

- Delete all edges incident to v;
- If there is an edge from i to v and from v to j, then add an edge from i to j if it does not exist, and remove it if it exists.

A directed graph is called **tight** if there is a sequence of such operations that transforms it into a stable set with only unoriented vertices.

It is an open problem to find a good algorithm to decide, given a directed graph with oriented and unoriented vertices, if this graph is tight.

II MODELS HANDLING DUPLICATIONS: STRINGS

7 Generalities

The combinatorial study of genome rearrangements started with permutations, but permutations lack the possibility of taking *duplications* into account. Duplications are a major evolutionary event, believed to be one of the most important mechanisms for novel generations in evolution, and almost all data sets on eukaryotes contain duplicated genes (see, e.g., Ohno [284]). An appropriate tool for studying genomes with duplicated genes was therefore needed, and *strings* are a very natural generalization of permutations that fit this purpose well. They allow the consideration of four possible rearrangement events in addition to the ones we have already seen: insertions, deletions, duplications, and replacements. We will see here in part II that **NP**-completeness and even inapproximability results are very numerous, and they are the main combinatorial part of this domain. The subject was briefly surveyed in 2005 by El-Mabrouk [156].

7.1 Biological Motivations

Duplications can occur at several levels, ranging from the duplication of a single gene or small segment of DNA to the duplication of a whole chromosome; even whole genome duplications are known to occur. These evolutionary events result in genomes in which some markers are undifferentiable, and we call them *duplicated genes* or simply *duplicates*.

Given a set of genomes, all copies of a given gene among those genomes are said to be **homologous**, which means that they originate from a common ancestral gene, and form a **gene family**. The presence of two copies of a gene in a set of genomes may be explained by **speciation** events (i.e., the appearance of two distinct species, each genome carrying the gene); it can also be explained by **duplication** events, which result in two copies of a gene in the same genome. The relationships between the copies of a gene in a gene family can thus be of several types. Two copies of a gene are said to be **orthologous** if they derive from a speciation event, and **paralogous** if they derive from a duplication event. Given two genomes and a gene family, a distinction is made between **out-paralogs**, which are paralogous gene copies that derive from a duplication that occurred *before* the last common ancestor of the two genomes, and **in-paralogs**, which derive from a duplication that occurred *after* the last common ancestor. Note that the word *gene* is a bit ambiguous, as it may refer either to a family (there are several copies of the same gene in the genomes) or to copies (two genes may derive from a duplication).

Those different situations motivate and justify the use of the models we will consider here in part II. Indeed, when comparing two sequences under the assumption that all copies of a given element in a single string are in-paralogs, the goal will be to identify the position of the unique ancestor. If there can be out-paralogs, then the goal will be to detect orthologs by matching some copies. The distances between two strings will vary according to which model is chosen. Every combinatorial problem we have seen so far can be reformulated in terms of strings, but the algorithmic treatment is usually completely different. For instance, the breakpoint graph, which is a ubiquitous object when dealing with permutations, is not used on strings, in spite of some attempts to define them in the case of whole genome duplications by Alekseyev and Pevzner [9] (see chapter 13).

7.2 Strings and Rearrangements on Strings

Definition 7.1 An **alphabet** $\mathcal{A}$ is a finite set of **gene families**, each represented by a character. A **string** on $\mathcal{A}$ is a sequence of members of gene families of $\mathcal{A}$, also called **genes** or **characters** of the string. A **signed string** is a sequence of signed members $+a$ or $-a$ of $\mathcal{A}$ ($+$ signs are often omitted). The ith element of a string S is denoted by S_i.

Any alphabet can be used for genome rearrangement problems, but just as we chose a canonical set for permutations, here we will use $\{1, 2, \ldots, k\}$ as the "standard" alphabet. $\mathcal{A}^*$ is the infinite set of all strings on $\mathcal{A}$, and $\mathcal{A}^n$ is the finite set of all strings of length n on $\mathcal{A}$.

Given a string S of length n on $\mathcal{A}$, we denote $occ(a, S)$ as the number of members of the gene family $a \in \mathcal{A}$ occurring in S, and $occ(S)$ as the maximum value $occ(a, S)$ over all $a \in \mathcal{A}$. Moreover, $f(S)$ denotes the number of gene families that have at least two members in S. A **duplicate** of some character $+a$ or $-a$ of S is another character $+a$ or $-a$ on the same string.

If $\mathcal{A}$ has two elements, strings on $\mathcal{A}$ are said to be **binary**; if $\mathcal{A}$ has three elements, strings on $\mathcal{A}$ are said to be **ternary**; and if $\mathcal{A}$ has k elements, strings are said to be k-**ary**.

Definition 7.2 A **substring** of a string S of length n is a string $S_iS_{i+1}\cdots S_j$, where $1 \leq i \leq j \leq n$, and $S_i, \ldots, S_j$ appear consecutively in S. A **prefix** is a substring containing the first element. A **suffix** is a substring containing the last element. A **subsequence** of a string S is a string obtained from S by deleting some elements.

This leads to the following parameters for two strings S and T of respective lengths n, m.

Definition 7.3 The **longest common prefix** of S and T is the string $L = (S_1\ S_2 \cdots S_p)$ such that for all $1 \leq i \leq p$, $S_i = T_i$ and p is maximal. Its length is denoted by $lcp(S, T)$.

Definition 7.4 The **longest common suffix** of S and T is the string $L = (S_{n-p}\ S_{n-p+1} \cdots S_n)$ such that for all $n - p \leq i \leq n$, $S_i = T_i$ and p is maximal. Its length is denoted by $lcs(S, T)$.

Definition 7.5 A **longest common substring** of S and T is a string that is a substring of S and T of maximum length. Its length is denoted by $LCS(S, T)$.

Definition 7.6 A **longest common subsequence** of S and T is a maximum length string that is a subsequence of both S and T.

These definitions can be illustrated by the following example: let $S = rearranging$, $T = rearrangement$, and $U = sorting$.

- The longest common prefix of *rearranging* and *rearrangement* has length 8, which matches their longest common substring, and
- The longest common suffix of *rearranging* and *sorting* has length 3, which matches their longest common substring.
- The longest common subsequence of *rearranging* and *rearrangement* has length 9, and the longest common subsequence of *rearranging* and *sorting* has length 4.

We will sometimes need to consider longest common substrings "up to a reversal" (i.e., the longest common substrings of S and the result of a complete reversal of T).

As in the case of permutations, we define *circular strings*. Let $\equiv^\circ$ be the equivalence relation between strings of size n on $\mathcal{A}$, defined as follows: $S \equiv^\circ T$ if there exists an integer h such that for all i in $\{1, \ldots, n\}$, $S_i = T_{(i+h \bmod n)+1}$. In other words, S and T are equivalent if S can be obtained by rotating the elements of T. An equivalence class under this relation is called a **circular string**. These strings are sometimes (but rarely) used in the following chapters to model circular chromosomes.

In contrast to permutations, strings are not required to have the same gene content. Thus the following rearrangements may be considered in addition to all the rearrangements defined for permutations.

Definition 7.7 A **block** is a string on the alphabet $\mathcal{A}$, usually much shorter than a genomic string S. A **block edit operation** (a **block edit** for short) on a string $S =$

$S_1S_2\cdots S_n$ is one of the following operations, where $1 \le i \le i+j \le n$, $1 \le k \le n$, and $1 \le h$.

- The **insertion** of a block $B = B_1B_2\cdots B_h$ at position i in S yields the string $S_1S_2\cdots S_{i-1}B_1B_2\cdots B_hS_i\cdots S_n$.
- The **deletion** of the block $S_iS_{i+1}\cdots S_{i+j}$ from S yields the string $S_1S_2\cdots S_{i-1}S_{i+j+1}\cdots S_n$.
- The **duplication** of the block $S_iS_{i+1}\cdots S_{i+j}$ of S at position k in S yields the string $S_1S_2\cdots S_{k-1}S_iS_{i+1}\cdots S_{i+j}S_k\cdots S_n$.
- The **replacement** of a block $S_iS_{i+1}\cdots S_{i+j}$ of S with a block $B = B_1B_2\cdots B_h$ yields the string $S_1S_2\cdots S_{i-1}B_1B_2\cdots B_hS_{i+j+1}\cdots S_n$.

7.3 Balanced Strings

These additional rearrangements are necessary if the strings being compared do not have the same gene content, but there is an interesting case in which strings actually do. In this case, insertions, deletions, duplications, and deletions may not be invoked, even if the strings may contain duplicated genes.

Definition 7.8 Two strings on the same alphabet that contain the same number of members of each gene family are said to be **balanced** or **related**.

Two balanced strings are obviously of same length. This property ensures that it is possible to transform one string into another without block edit operations as rearrangements. The notation $\mathcal{L}(a_1, \ldots, a_k)$ will be used to denote the set of all balanced strings containing a_i members of the gene family i, for $1 \le i \le k$.

A very nice, canonical way to express genome rearrangement problems on permutations was the general problem of sorting permutations under constraints. Unfortunately, the left-invariance property of all distances we studied in the context of permutations does not generalize to strings, and therefore does not allow us to propose such a canonical expression in that setting. Therefore, we will not refer to the rearrangement of strings using an operation ρ as "sorting (strings) by ρ," but rather as "rearranging (strings) by ρ."

Nevertheless, some authors have considered various sorting problems related to balanced strings, where the goal is to transform a given string S on $\mathcal{A}$ into an "identity string" (depending on S) or, more formally, into a string whose characters are ordered lexicographically and the number of occurrences of each character in S is preserved. That problem will be referred to as the *sorting* problem, and should not be confused with the *rearrangement* problem described above, since contrary to the case of permutations, they are not equivalent.

7.4 How to Deal with Multiple Copies?

It is not equally difficult to take into account multiple copies when considering two balanced or two general (that is, not necessarily balanced) strings. By convention, balanced strings are supposed to contain only out-paralogs, which means that each of the h members of some gene family present in each string S and T originates from one of the h members of the same gene family present on their last common ancestor. The difficulty is then to identify (that is, to match) the pairs of members, one on each string, that originate from the same member of the last common ancestor (i.e., they are orthologous). In contrast, general strings allow us to assume the existence of both out-paralogs and in-paralogs on each string, so that deletion and insertion events have to be considered in addition to the rearrangement events when comparing general strings. The assignment of ortholog pairs of genes, given two strings, yields the essential definition of a matching:

Definition 7.9 A **(gene) matching** $\mathcal{M}$ between two strings S in $\mathcal{A}^n$ and T in $\mathcal{A}^m$ is a set of pairs (S_i, T_j) such that the pairs $(i, j) \in \{1, 2, \dots, n\} \times \{1, 2, \dots, m\}$ are disjoint, and S_i and T_j belong to the same gene family. In this case, characters S_i and T_j are **matched** to each other.

When comparing two strings S and T, a matching between S and T aims at representing the common content of the strings, as supported by their last common ancestor and regardless of (but without losing touch with) the order of the characters. Any pair of matched characters is then assumed to correspond to orthologous genes, and the unmatched characters are assumed to be in-paralogs. Here rearrangement studies meet the important problem of ortholog identifications. The members of the same gene family that are present in the same string and that are matched are out-paralogs. In order to distinguish out-paralogs from each other, a *relabeling* may be performed, which gives new and distinct names (i.e., new and distinct characters from $\mathcal{A}$) to out-paralogs and renames the orthologs of each out-paralog accordingly. The last step of such a treatment of strings is obtaining a pruning.

Definition 7.10 Let $\mathcal{M}$ be a matching between two strings S and T over some alphabet $\mathcal{A}$. An **$\mathcal{M}$-pruning** of S and T is the pair (S, T) of strings obtained from S and T by removing all characters that are not matched and relabeling the remaining characters according to $\mathcal{M}$. The strings S and T are called the **$\mathcal{M}$-pruned strings** of S and T, respectively.

By convention, we use the terms *pruning* and *pruned string* when the matching does not need to be identified by a notation. The good news at this stage is that if we assume that the relabeling is done in such a way that the characters in S and T are

$\{1, 2, \ldots, |\mathcal{M}|\}$, then both pruned strings are permutations of $|\mathcal{M}|$ elements and may be compared using the usual distances between permutations.

Now, turning back to the initial question: "How to deal with multiple copies?," two answers are available: either define a collection of possible rearrangements and compute the minimum number of operations needed to transform one string into the other, or reduce strings to permutations using matchings and prunings, then find the distance between the permutations. These two approaches, called here the *block edit model* and the *match-and-prune model*, are differently expressed on balanced and general strings. On balanced strings, none of the models includes insertions, duplications, replacements, or deletions, so that each gene on one string has an orthologous gene on the other one. This orthologous gene is not indicated a priori, but is deduced from the result of the comparison between the strings. On general strings, the block edit model includes insertions, duplications, replacements, and deletions, but is by no means more precise concerning the matching between orthologous genes. On the contrary, in this case the match-and-prune model proposes several types of matchings, each of which corresponds to particular assumptions on the last common ancestor's composition. All matchings of a given type are then used to obtain prunings, and prunings are compared as permutations in order to find the best of them.

8 Distances between Arbitrary Strings

An evolutionary scenario must take two categories of events into account: events that modify the *order* of genes in a genome without modifying its composition in terms of gene content and multiplicity, and events that modify the *content* of the genome without modifying the gene order. A natural question arises here about the existence of events that would modify both aspects. It has a positive answer (such events exist, such as block replacement; see section 8.2), but very few works focus on this type of events; see section 8.2.1).

As indicated in chapter 7, two main ideas are used to deal with arbitrary (that is, not necessarily balanced) strings. The first idea aims at transforming strings into permutations, using the *match-and-prune model*, so as to minimize the distance or maximize the similarity between the resulting permutations. Section 8.1 presents this model.

The second idea is based on counting the number of operations, belonging to a given class, that are needed to transform one string into another. Inserting, duplicating, replacing, or deleting a character or, more generally, a substring are operations that allow us to change the content of a string as needed. This is subsequently called the *block edit model*, surveyed in section 8.2. These approaches are presented here in the reverse chronological order of their appearance in the scientific literature.

A noticeable particularity of this chapter is the notion of *similarity*, which appears here not only as an intuitive opposite to *distance*, but also as a quantifiable feature that we evaluate using a criterion that is no longer a *minimum* but a *maximum*. Sometimes, and we point out the case as soon as it appears, authors attempt to transform a quite natural definition of similarity into a less natural but certainly more mathematically rigorous notion of distance. However, it should be noted that the notion of distance used by several authors is sometimes not really a *metric*, but has at least two of the three properties of a metric. In all cases, when the distinction between distances and similarities is not necessary, we simply call them *measures*.

In this chapter, the two strings S and T that we compare are built over the usual alphabet $\mathcal{A} = \{1, 2, \ldots, k\}$ and have respective lengths n and m.

8.1 The Match-and-Prune Model

The match-and-prune model addresses the following question, which arises naturally when one is trying to discover the relationships between two genomes S and T: How can we take into account, when comparing genomes with duplicates, the fact that the structure of the last common ancestor of S and T plays an important role in the evolutionary distance between the two genomes? Since this structure is unknown, unless we have very good reasons to conclude that we can afford to keep it unknown, we need to be able to model it.

Three (sub)models are used to achieve this goal, and they differ essentially by the underlying assumptions that are made about the content of the last common ancestor. However, they all rely on the same method to compute measures (distances and similarities) between strings, which takes into account only their common genes, as identified by the composition of the last common ancestor. Those models are the following:

1. the **exemplar model**, in which the last common ancestor is assumed to contain exactly one member of each gene family $a \in \mathcal{A}$ that has members in both S and T;
2. the **intermediate model**, in which the last common ancestor is assumed to contain $h_a \geq 1$ members of each gene family $a \in \mathcal{A}$ that is common to S and T;
3. the **full model**, in which the last common ancestor is assumed to contain as many members as possible (that is, $min\{occ(a, S), occ(a, T)\}$) from any gene family $a \in \mathcal{A}$ that is common to S and T.

In all cases, exactly h_a (where $h_a = 1$, $h_a \geq 1$, or $h_a = min\{occ(a, S), occ(a, T)\}$) members of the gene family a in each string are identified as corresponding to ancestral members. Those members in S and T form pairs of orthologs according to the ancestral member to which they correspond, and thus altogether form a matching.

The history of these models starts with a paper by Sankoff [316], who built the foundations of the exemplar model and, at the same time, of the most general match-and-prune model. Besides its biological motivations, Sankoff's idea has two important features that make it attractive: reducing to 1 the cardinality of each gene family in each string implies that (1) the resulting strings are permutations, and computing distances between permutations is both already studied and achievable in polynomial time for many biologically relevant distances; and (2) the one-to-one correspondence between genes that belong to the same family in both strings is obvious, and can therefore avoid further complications. The next model to be defined was the full model, whose first ideas were also suggested by Sankoff [316]; it was more precisely defined by Tang and Moret [347] for balanced strings. The most recent model is the intermediate model, proposed by Angibaud et al. [16].

As already mentioned, each model is defined by a particular type of matching. To formally present them, we use, in the remainder of this section, the following conventions and notations for the objects we manipulate.

- The strings S and T are signed strings.
- The $\mathcal{M}$-pruning of two strings S, T, where $\mathcal{M}$ is a given matching, is denoted by (S, T).
- Rather than repeatedly using the word *respectively* to describe alternatives—as in "the exemplar (intermediate, full, respectively) model satisfies property A (B, C, respectively)"—we use the "/" separator that is assumed to preserve the rank in the list when two lists are linked together—as in "the exemplar/intermediate/full model satisfies property A/B/C."

Definition 8.1 Let S, T be two strings over the alphabet $\mathcal{A}$. An **exemplar/intermediate/full matching** $\mathcal{M}$ between S and T is a matching between S and T such that exactly one/at least one/as many as possible pairs of each gene family occurring in both S and T exist in $\mathcal{M}$.

Each type of matching yields a different type of pruning (see figure 8.1).

Definition 8.2 Let S, T be two strings over the alphabet $\mathcal{A}$. An **exemplar/intermediate/full pruning** between S and T is an $\mathcal{M}$-pruning, where $\mathcal{M}$ is an exemplar/intermediate/full matching.

We are now ready to explain how distances and similarities are computed between general strings, using the three models. Recall that the optimum value of a measure is the *minimum value* when the measure is a distance and the *maximum value* when the measure is a similarity.

Definition 8.3 Given two strings S and T and a measure ms, the **exemplar/intermediate/full measure** between S and T is the optimum value of ms over all possible exemplar/intermediate/full prunings (S, T) of S and T.

For any given measure, computing the exemplar/intermediate/full measure constitutes three (usually different) optimization problems. Indeed, although the three models range, with respect to the cardinality of the matching between the two strings, from a simpler one to a more complex one, no linear classification from simple to difficult may be performed on the resulting problems. However, there is at least one precise case in which all three models are equivalent.

Remark 8.1 When one of the strings S, T contains no more than one member of each gene family—that is, $occ(S) = 1$ or $occ(T) = 1$—the exemplar, intermediate, and full models are equivalent.

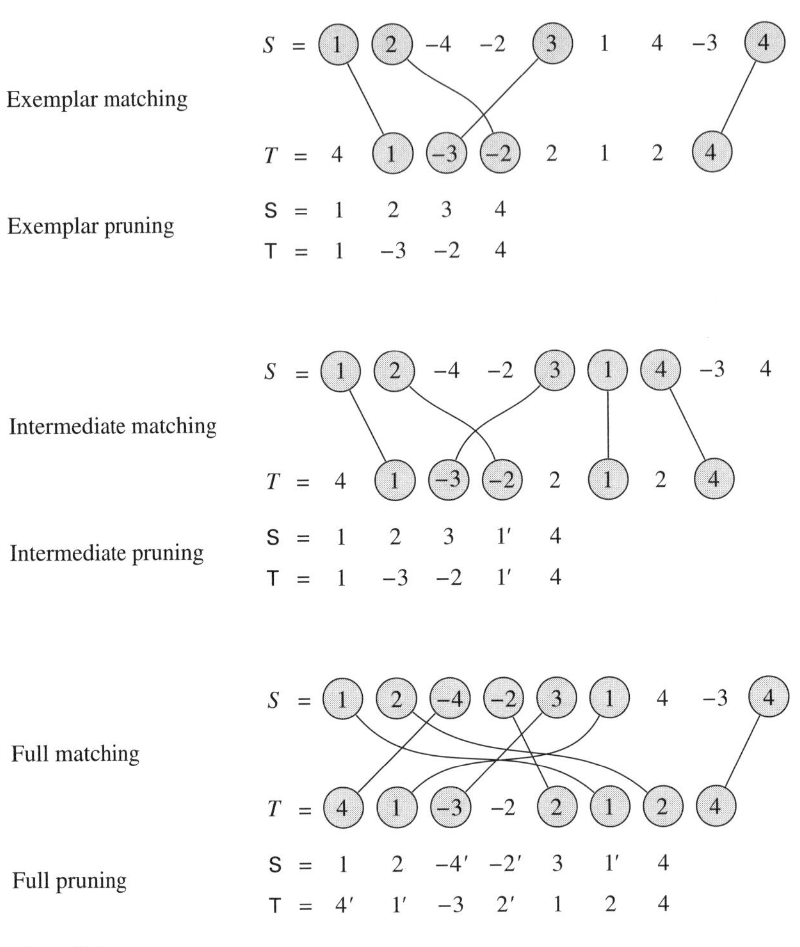

Figure 8.1
Matchings and prunings for the strings $S = 1\,2\,{-4}\,{-2}\,3\,1\,4\,{-3}\,4$ and $T = 4\,1\,{-3}\,{-2}\,2\,1\,2\,4$

As a consequence, all the **NP**-completeness or inapproximability results obtained for one of the models in this case extend to the two other models.

In the rest of this section, we present the different results available on the match-and-prune model. This model allows us to avoid the difficulty of dealing with multiple copies by reducing string comparison to permutation comparison, but this reduction rarely yields polynomial-time solvable problems.

8.1.1 Breakpoint Distance

Together with the signed reversal distance presented in the next subsection, the breakpoint distance is one of the first applications of Sankoff's exemplar model [316]. The two distances are defined on different bases: the reversal distance counts the minimum number of operations to transform one string into the other, whereas the breakpoint distance counts the structural differences between the two strings. However, they are closely related: their values differ only slightly, and they both satisfy a property of monotonicity that allows a similar algorithmic treatment.

- Introduced by Sankoff [316] (exemplar), by Angibaud et al. [16] (intermediate), and by Blin et al. [69] (full).
- Complexity: **NP**-complete (see Bryant [89]) and **APX**-hard (see Angibaud et al. [19]) (exemplar, intermediate, and full). Not approximable at all (see Chen et al. [108]) (exemplar and intermediate).
- Exact algorithms: Sankoff [316], Nguyen et al. [280] (exemplar and intermediate), Blin et al. [69] (full), Angibaud et al. [17] (exemplar, intermediate, full).

The definition of the breakpoint distance for signed permutations is given in section 2.6.1 and involves linear extensions of permutations.

Remark 8.2 Several authors use a slightly different notion of breakpoint distance, which does not consider the linear extensions of the permutations, but rather the permutations in their precise form, to compute the number of breakpoints. Whereas the definition we use implies that the breakpoint distance between two permutations is 0 if and only if the permutations are identical, this variant admits a 0 value if and only if permutations are identical *up to a complete reversal.*

Under all thee models (exemplar, intermediate, full), any pruning yields two permutations, thereby ensuring that the exemplar/intermediate/matching breakpoint distance is well defined. Although the three resulting problems are different in general, two of them are identical in this particular case:

Theorem 8.1 [15] Let S, T be two strings over the alphabet $\mathcal{A}$. The exemplar breakpoint distance between S, T is equal to the intermediate breakpoint distance between S, T.

As an example, consider the two strings S and T in figure 8.1. For the exemplar pruning $(\mathsf{S},\mathsf{T}) = (1\ 2\ 3\ 4, 1\ {-3}\ {-2}\ 4)$ the breakpoint distance between S and T is 2; for the intermediate pruning $(\mathsf{S},\mathsf{T}) = (1\ 2\ 3\ 1'\ 4, 1\ {-3}\ {-2}\ 1'\ 4)$ the breakpoint distance is also 2; and for the full pruning $(\mathsf{S},\mathsf{T}) = (1\ 2\ {-4'}\ {-2'}\ 3\ 1'\ 4, 4'\ 1'\ {-3}\ 2'\ 1\ 2\ 4)$ the breakpoint distance is 5. Since no exemplar/intermediate/full pruning improves these distances, the exemplar/intermediate/full breakpoint distance between S and T is 2/2/5.

We now turn our attention to pegged strings, which were introduced by Sankoff [316] and are particularly interesting both when looking for an exact algorithm and for showing the **NP**-completeness of computing the exemplar breakpoint distance.

Definition 8.4 A character c of a string S is a **singleton** of S if no other member of the gene family of c occurs on S.

Definition 8.5 A string S over the alphabet $\mathcal{A}$ is **pegged** if each interval between two characters in the same gene family contains at least one singleton.

Pegged strings have the interesting property that singletons act as markers helping to uniquely identify each occurrence of a non-singleton by its position with respect to these markers.

8.1.1.1 Lower and Upper Bounds Theorem 3.17 in section 3.3.1 and lemma 3.9 in section 3.3.2 provide lower and upper bounds for the breakpoint distance between two permutations that is based on the reversal distance of two permutations. Such bounds also exist for general strings, and are given in section 8.1.2.1.

In the current subsection, we present lower and upper bounds for the full breakpoint distance, based on a generalization of the reversal distance that we call the *exclusive block edit distance with reversals*. These results are due to Blin et al. [69].

Definition 8.6 Let $S = S_1 S_2 \cdots S_n$ and T be two strings over the alphabet $\mathcal{A}$. The **exclusive block insertion** of a block $B = B_1 B_2 \cdots B_h$ over $\mathcal{A}$ at position i in S yields the string

$$S' = S_1 S_2 \cdots S_{i-1} B_1 B_2 \cdots B_h S_i S_{i+1} \cdots S_n$$

if $occ(a, S') \leq occ(a, T)$ for each $a \in \mathcal{A}$ occurring in B. Otherwise, the operation is forbidden.

Definition 8.7 Let $S = S_1 S_2 \cdots S_n$ and T be two strings over the alphabet $\mathcal{A}$. The **exclusive block deletion** of the block $B = S_i S_{i+1} \cdots S_{i+j}$ from S yields the string

$$S' = S_1 S_2 \cdots S_{i-1} S_{i+j+1} \cdots S_n$$

if $occ(a, S') \geq occ(a, T)$ for all $a \in \mathcal{A}$ occurring in B. Otherwise, the operation is forbidden.

In other words, exclusive block insertions and deletions must not increase the disparity between the number of occurrences on S and on T of each gene family. When one focuses on an arbitrary gene family a, one notices that members of a may be involved either exclusively in deletion events or exclusively in insertion events, but not in both types of operations. This is why we term these operations *exclusive*.

Definition 8.8 Let S and T be two strings over the alphabet $\mathcal{A}$. The **exclusive block edit distance with reversals** between S and T is the minimum number of reversals, exclusive block insertions, and exclusive block deletions needed to transform S into T.

With the notation $fbd(S,T)$ and $xbed(S,T)$ to denote, respectively, the full breakpoint distance and the exclusive block edit distance with reversals between S, T Blin et al. (69) prove that

$$\frac{xbed(S,T)}{2} \leq fbd(S,T) \leq 2 \times xbed(S,T).$$

8.1.1.2 Computational Complexity The breakpoint distance is the first measure presented under the match-and-prune model, and it is representative of the type of results one obtains under this model. **NP**-completeness and **APX**-hardness results are shared by the major part of the measures, even on very restricted sets of instances, such as those with $occ(S) = 1$, $occ(T) = 2$.

Bryant [89] investigated the **NP**-completeness of computing the exemplar breakpoint distance, and proved the following result, which concerns all models because of remark 8.1.

Theorem 8.2 [89] Under the exemplar/intermediate/full model, computing the breakpoint distance between two strings S and T is an **NP**-complete problem, even when $occ(S) = 1$, $occ(T) = 2$ and both genomes are pegged.

When strings are unsigned, the **NP**-completeness still holds under the exemplar/intermediate model, as shown by Nguyen [279]. Blin et al. [69] prove the **NP**-completeness in the case where $f(S) = f(T) = 1$ under the full model. The strongest inapproximability result known to the date refers to remark 8.1.

Theorem 8.3 [19] Computing the exemplar/intermediate/full breakpoint distance between two strings S and T is **APX**-hard, even when $occ(S) = 1$ and $occ(T) = 2$.

Further results are available on the exemplar/intermediate version of the problem. The first results implying the inapproximability under this model assume, without loss of generality, that S is longer than T, so that $n = max(|S|, |T|)$.

Theorem 8.4 [108] The exemplar/intermediate breakpoint distance cannot be approximated within a factor $c \log n$, for some constant $c > 0$.

Even when strings S and T are very particular, it is a hard task to approximate their exemplar/intermediate breakpoint distance.

Theorem 8.5 [279] Computing the exemplar/intermediate breakpoint distance between two strings S and T is **APX**-hard, even when strings are unsigned and pegged.

Chen et al. [108] obtained a particularly important inapproximability result: even when $occ(S) = 3$ and $occ(T) = 3$, the exemplar/intermediate breakpoint distance is not approximable at all. This is a consequence of a related **NP**-complete problem presented in section 8.1.1.5.

8.1.1.3 Approximation Algorithms Not surprisingly, finding algorithms with bounded error is difficult. However, good and quite simple heuristics exist (see section 8.1.8). The algorithm in this subsection, due to Chen et al. [108], is an approximation algorithm for a particular case of the exemplar/intermediate version of the problem.

Definition 8.9 A string S is h-**span**, where h is an integer, if for each gene family $a \in \mathcal{A}$, the leftmost and the rightmost occurrences of a or $-a$ in S are at distance at most h along S.

Recalling that $\mathcal{A} = \{1, 2, \ldots, k\}$, Chen et al. [108] propose a $2(1 + \log k)$-approximation algorithm to compute the exemplar breakpoint distance for strings S and T where S is $O(\log k)$-span. When $occ(S) = 1$, S is h-span for every h, so the same algorithm solves the problem when $occ(S) = 1$ and T is arbitrary.

8.1.1.4 Exact Algorithms As we have already mentioned, the three available models lead to two different optimization problems, but these problems have common features that may sometimes be used to develop similar algorithms. Two main approaches exist for computing the breakpoint distance exactly, and they constitute frameworks for further algorithms to compute different distances.

Branch-and-Bound Sankoff [316] proposes an exact algorithm to compute the exemplar breakpoint distance that is based on a branch-and-bound technique. The algorithm attempts to increasingly build all different exemplar prunings of S and T. The increase of a partial pruning is performed by adding one pair of characters in the same gene family at each step, and is canceled as soon as the breakpoint distance of the current pruning exceeds the best current distance obtained for a whole exemplar pruning. Such a canceling decision is correct if the following property of monotonicity holds.

Definition 8.10 Let *dist* be an arbitrary distance between two signed permutations. Then *dist* is **monotonic** if, for each pair of strings S and T, deleting all occurrences of any gene family from S and T does not increase the exemplar *dist* distance between S and T.

The breakpoint distance is monotonic, so Sankoff's algorithm correctly stops the analysis of a useless pruning. However, the number of exemplar prunings may be very high, as confirmed by the following result.

Theorem 8.6 [316] The number of different exemplar prunings is upper-bounded by

$$\prod_{a \in \mathcal{A}} occ(a, S)\ occ(s, T),$$

where equality holds if and only if both strings are pegged.

Nguyen et al. [280] propose to reduce the running time of Sankoff's algorithm by identifying independent disjoint subsets of gene families, performing Sankoff's algorithm on each subset and then merging the resulting partial exemplar prunings. Since not all pairs of strings S, T admit such a partition into independent disjoint subsets, a heuristic is devised by forcing a partition.

The algorithm used by Blin et al. [69] to compute the full breakpoint distance also follows a branch-and-bound strategy, whose validity is once again ensured by the monotonicity properties of the breakpoint distance. The algorithm starts with an empty pruning of S, T and investigates all the possibilities to extend it, as long as the current pruning does not exceed the best current distance obtained for a whole full pruning. A suffix tree is used to store the substrings of T and to search for those substrings that are common to S and T, which provide the candidates for extension.

Linear Pseudo-Boolean (LPB) Programming Angibaud et al. [17] propose to formulate the problem as a linear pseudo-Boolean problem (or LPB problem), that is, a linear program whose variables take 0 or 1 values, and which is solved using an LPB solver. Moreover, they show that only minor modifications have to be performed on the program in order to change the model (exemplar, intermediate, or full) and/or the computed measure (from breakpoint distance to adjacency similarity; see section 8.1.3).

8.1.1.5 Variants We are now interested in the problem of deciding whether the breakpoint distance between two strings S and T is 0 or not. Although seemingly not so complicated and even a little bit arbitrary, this problem may help deduce inapproximability results.

- Introduced by Chen et al. [108] (exemplar) and by Angibaud et al. [19] (intermediate, full).
- Complexity: **NP**-complete (see Chen et al. [108]) (exemplar and intermediate). Polynomial (see Angibaud et al. [19]) (full).
- Exact algorithms: $O(nm \log\log(nm))$ time (see Angibaud et al. [19]) (full).

The **NP**-completeness of deciding whether the exemplar/intermediate breakpoint distance between two strings is 0 or not has the following important consequence: unless $\mathbf{P} = \mathbf{NP}$, no polynomial-time algorithm may be found to f-approximate the exemplar/intermediate breakpoint distance, whatever the function f. **NP**-completeness also holds for the two particular cases $occ(S) = occ(T) = 3$ (see Chen et al. [108]) and $occ(S) = 2$, $occ(T)$ arbitrary (see Angibaud et al. [19]), so that the same affirmation holds for these particular cases.

For the exemplar/intermediate model, Angibaud et al. [19] proposed an exact algorithm running in $O(p(r)\,1.6182^{2r})$ time, when $occ(S) = occ(T) = 2$, where r is upper-bounded by the number of gene families that occur exactly twice in S and in T, and $p(r)$ is a polynomial function.

8.1.2 Signed Reversal Distance

The signed reversal distance is the second distance considered by Sankoff [316] to illustrate his theory of exemplar distances. Under the full model, the signed reversal distance is very well studied on balanced strings (see chapter 9) and not studied at all on general strings. No specific result exists for the intermediate model.

- Introduced by Sankoff [316] (exemplar) and by Chen et al. [106] (full).
- Complexity: **NP**-complete, by Bryant [89] (exemplar); **APX**-hard by Angibaud et al. [19] (exemplar, intermediate, and full). Not approximable at all, by Chen et al. [108] (exemplar).
- Exact algorithms: Sankoff [316] (exemplar).

Due to its similarity to the breakpoint distance, the results on the signed reversal distance are mainly easy variants of the results on the breakpoint distance. Many of these results are proved similarly for the two distances, but some of them could easily be deduced from the lower and upper bounds (we present now).

8.1.2.1 Lower and Upper Bounds Under each of the exemplar, intermediate, and full models, the signed reversal and breakpoint distances are equal up to a multiplicative factor 2, as shown below. Without loss of generality, the simple proof below is performed only for the exemplar case. The intermediate and full cases are identical.

Given two strings S and T, let $ebd(S,T)$ and $erd(S,T)$ be their exemplar breakpoint and signed reversal distance, respectively. Furthermore, let $bd(\mathsf{S},\mathsf{T})$ and $rd(\mathsf{S},\mathsf{T})$ be the breakpoint and signed reversal distance obtained on the exemplar pruning (S,T) of S, T. Consider the sequence of $rd(\mathsf{S},\mathsf{T})$ reversals that transform S into T and note that the first reversal always creates two new breakpoints with respect to S; all the other reversals create at least one, and at most two, breakpoints each. Thus we have the following easy results:

$$rd(\mathsf{S},\mathsf{T}) \geq bd(\mathsf{S},\mathsf{T})/2$$

$$rd(\mathsf{S},\mathsf{T}) < bd(\mathsf{S},\mathsf{T}).$$

It is easy to show that the first formula holds for exemplar distances (Nguyen [279] gives it without any proof):

$$erd(S, T) \geq ebd(S, T)/2. \tag{8.1}$$

To see this, let $(\mathsf{S}', \mathsf{T}')$ and $(\mathsf{S}'', \mathsf{T}'')$ be the exemplar prunings yielding the minimum values of the reversal and breakpoint distances between S and T, respectively. Then

$$ebd(S, T) = bd(\mathsf{S}'', \mathsf{T}'') \leq bd(\mathsf{S}', \mathsf{T}') \leq 2\, rd(\mathsf{S}', \mathsf{T}') = 2\, erd(S, T),$$

and we are done. We further have

$$ebd(S, T) > erd(S, T). \tag{8.2}$$

Using a similar easy reasoning,

$$ebd(S, T) = bd(\mathsf{S}'', \mathsf{T}'') > rd(\mathsf{S}'', \mathsf{T}'') \geq rd(\mathsf{S}', \mathsf{T}') = erd(S, T).$$

We then deduce that the two exemplar distances are identical up to a factor 2.

8.1.2.2 Computational Complexity When compared with the result on the exemplar breakpoint distance, Bryant's **NP**-completeness result on the exemplar signed reversal distance is slightly weaker.

Theorem 8.7 [89] Computing the exemplar signed reversal distance between two strings S and T is an **NP**-complete problem, even when $occ(S) = occ(T) = 2$ and both genomes are pegged.

In particular, this result does not allow deducing the **NP**-completeness under the intermediate or full model. However, the inapproximability results proved by Angibaud et al. [19] (theorem 8.3), Chen et al. [108] (theorem 8.4), and Nguyen [279] (theorem 8.5) remain true, because of equations (8.1), (8.2), and their homologs for the intermediate and full distances.

8.1.2.3 Approximation Algorithms Since the exemplar breakpoint and signed reversal distances are equal up to a factor 2, Chen's approximation algorithm for the breakpoint distance [108] (see section 8.1.1.3) also gives an $O(\log k)$ approximation for the exemplar signed reversal distance, when S is $O(\log k)$-span. Again, when $occ(S) = 1$, S is h-span for every h; thus the same algorithm solves the problem when $occ(S) = 1$ and T is arbitrary.

8.1.2.4 Exact Algorithms Sankoff's algorithm [316], presented in section 8.1.1.4, to compute the exemplar breakpoint distance is also applicable to the signed reversal distance, which is also monotonic and easy to compute. However, the improvement proposed by Nguyen et al. [280], as well as the LPB approach of Angibaud et al. [17], do not apply for the signed reversal distance. In particular, no algorithm is known for the intermediate and full models.

8.1.2.5 Variants The problem of deciding whether the exemplar/intermediate/full signed reversal distance between two strings S and T is 0 or not inherits from results on the corresponding breakpoint problem, since $ebd(S,T) = 0$ if and only if $erd(S,T) = 0$ (and similarly for the two other models) and because of equations (8.1), (8.2), and their homologs for the intermediate and full models.

- Introduced by Chen et al. [108] (exemplar).
- Complexity: **NP**-complete, by Chen et al. [108] (exemplar and intermediate). Polynomial by Angibaud et al. [19] (full).

Thus, unless $\mathbf{P} = \mathbf{NP}$, no polynomial algorithm may be found to f-approximate the exemplar/intermediate reversal distance, whatever the function f. The same holds for the two particular cases $occ(S) = occ(T) = 3$ and $occ(S) = 2$, $occ(T)$ arbitrary, by inheritance from the corresponding breakpoint problem.

The problem of computing the signed reversal distance has some other variants. Swenson et al. [345] attempt to approximate the true evolutionary edit distance between two strings S and T by computing a full pruning that minimizes the number of insertions, deletions, and reversals. However, they propose only a heuristic for this related problem.

8.1.3 Adjacency Similarity

We deal in this subsection with a similarity measure that is the complement of the breakpoint distance. The basis of this measure is the preserved adjacency between two consecutive characters in S and T.

- Introduced by Angibaud et al. [17] (exemplar, full) and Angibaud et al. [15] (intermediate).
- Complexity: **NP**-complete, **W[1]**-hard, and admits no $n^{1-\varepsilon}$-approximation (see Chen et al. [109]) (exemplar, intermediate, full).
- Exact algorithms: Angibaud et al. [17] (exemplar, full) and Angibaud et al. [15] (intermediate).
- Best approximation ratio: 4 for balanced genomes (see Angibaud et al. [19]) (full).

The definition of the adjacency similarity is natural, and uses linear extensions (see page 20) to ensure a complementarity with the breakpoint distance.

Definition 8.11 The **adjacency similarity** between two signed permutations π and σ is the number of adjacencies between the linear extensions of π and σ.

It is easy to transform this measure into an exemplar/intermediate/full measure. The resulting problems of computing the exemplar/intermediate/full adjacency similarity all differ from each other.

8.1.3.1 Computational Complexity Chen et al. [109] present a polynomial Turing reduction from the INDEPENDENT SET problem on a graph G to the problem of computing the exemplar adjacency similarity between two unsigned strings S, T with $occ(S) = 1$, $occ(T) = 2$. This reduction has a particularity that allows one to further deduce several complexity results from already known results on INDEPENDENT SET. Namely, an independent set of size k exists in G if and only if the exemplar adjacency similarity between S and T is k. Since the measure to optimize in both problems is k, this Turing reduction may be used to deduce inapproximability results as well.

Indeed, since INDEPENDENT SET is **NP**-complete, nonapproximable within a factor of $n^{1-\varepsilon}$ [205], and **W**[**1**]-complete [148] with respect to the parameter *size of the independent set*, the same results hold for computing the exemplar adjacency similarity between S and T, where the **W**[**1**]-hardness holds with respect to the parameter *value of the exemplar adjacency similarity*. Remark 8.1 thus allows us to formulate the following result.

Theorem 8.8 [109] Computing the exemplar/intermediate/full adjacency similarity between two strings S and T is an **NP**-complete, nonapproximable within a factor of $n^{1-\varepsilon}$, and a **W**[**1**]-hard problem, even when $occ(S) = 1$, $occ(T) = 2$, and both strings are unsigned.

Recall that in the parameterized complexity theory, the **W**[**1**]-hardness of a problem with respect to some parameter p ensures that the problem cannot be solved in $O(f(p)\,g(n))$ running time, where n is the size of the problem, $g(n)$ is a polynomial in n, and $f(p)$ is an arbitrary function (see appendix B).

8.1.3.2 Approximation Algorithms Angibaud et al. [19] propose the following three approximation algorithms for a particular case of the full model, in which we consider two balanced strings S and T with $occ(S) = occ(T) = k$. Although these three algorithms have a common feature—they are obtained by reduction to a problem for which approximation algorithms are known—they are different in that the additional problems they use are distinct.

When $k = 2$, the problem is reduced to the MAX-2-CSP problem, which requires finding an assignment for a set of literals, so as to maximize the number of satisfied formulas, from a given set of Boolean formulas with at most two literals each. The

1.1442-approximation algorithm for MAX-2-CSP by Charikar et al. [101] yields a 1.1442-approximation algorithm for our problem.

When $k = 3$, the problem is reduced to the INDEPENDENT SET problem in a graph with degree $\Delta \leq 12$, which can be solved by a $(\frac{\Delta+3}{5} + \varepsilon)$-approximation algorithm due to Berman and Fürer [55]. This gives a $(3 + \varepsilon)$-approximation algorithm for our problem (where ε is as small as needed).

When k is arbitrary, the problem is reduced to the MAXIMUM WEIGHTED 2-INTERVAL PATTERN problem, defined by Crochemore et al. [127], which seeks a maximum weight subset of comparable 2-intervals from a given set of 2-intervals. A 2-**interval** is a pair of intervals, and several sets of relations are defined to compare pairs of 2-intervals (see Crochemore et al. [127]). The 4-approximation algorithm for this problem yields a 4-approximation algorithm for computing the full adjacency similarity of two balanced strings.

8.1.3.3 Exact Algorithms Starting with the idea of computing the breakpoint distance under a given model (exemplar, intermediate, or full), using a linear pseudo-Boolean (LPB) program, Angibaud et al. [17] use the following (already noted) complementarity between the breakpoint distance and the adjacency similarity.

Lemma 8.1 [17] Let $\mathcal{M}$ be a matching between two strings S and T, and let (S, T) be the corresponding $\mathcal{M}$-pruning. Then

$$bd(\mathsf{S}, \mathsf{T}) + as(\mathsf{S}, \mathsf{T}) = |\mathcal{M}| + 1, \tag{8.3}$$

where bd and as denote, respectively, the breakpoint distance and the adjacency similarity between two permutations.

To illustrate this lemma, consider again the strings S and T in figure 8.1. The exemplar pruning in the figure has two breakpoints and three adjacencies. The intermediate pruning has two breakpoints and four adjacencies; and the full pruning has five breakpoints and three adjacencies.

All exemplar matchings $\mathcal{M}$ of two given strings S and T have the same cardinality, so $bd(\mathsf{S}, \mathsf{T}) + as(\mathsf{S}, \mathsf{T})$ is a constant over all exemplar prunings (S, T) of S and T. Full matchings share this property of exemplar matchings (i.e., they all have the same cardinality); therefore $bd(\mathsf{S}, \mathsf{T}) + as(\mathsf{S}, \mathsf{T})$ is a constant over all full prunings as well. Consequently, under the exemplar and full models, minimizing $bd(\mathsf{S}, \mathsf{T})$ and maximizing $as(\mathsf{S}, \mathsf{T})$ are equivalent problems. This is not the case for the intermediate model, whose allowed matchings have different cardinalities.

Moreover, Angibaud et al. [17] [15] notice that the adjacency similarity problem is better suited to the LPB approach than the breakpoint distance problem for two reasons: the constraints in the resulting program are simpler and less numerous, and the running time of the program is strongly reduced.

Therefore, they give LPB programs for computing the exemplar, intermediate, and full adjacency similarity of two strings S and T, as well as for computing the corresponding prunings. The exemplar and full prunings are then used to compute the exemplar and full breakpoint distances according to equation (8.3), and to deduce the value of the intermediate breakpoint distance according to theorem 8.1. In this way, all breakpoint distances are obtained via the values or the matchings corresponding to the adjacency similarities. Good heuristics to solve the problem exist as well, as will be discussed in section 8.1.8.

8.1.4 Common Intervals Similarity

Common intervals are a natural generalization of adjacencies, as they identify subsets of characters that appear contiguously, but possibly in a different order, in both strings.

- Introduced by Chauve et al. [104] (exemplar, full) and by Angibaud et al. [16] (intermediate).
- Complexity: **NP**-complete (see Chauve et al. [104]) and **APX**-hard (see Angibaud et al. [19]) (exemplar, intermediate, and full).
- Exact algorithms: see Angibaud et al. [16] (exemplar, intermediate, full).

Recall (see section 2.6.2) that an **interval** with extremities i, j, $i \leq j$, of a permutation π, is the set $\{|\pi_i|, |\pi_{i+1}|, \ldots, |\pi_{j-1}|, |\pi_j|\}$ of unsigned characters located between positions i and j (included) in π. A **common interval** of two permutations is an interval of both.

Definition 8.12 The **common intervals similarity** between two signed permutations π and σ is the number of common intervals of π and σ.

Computing the number of common intervals of two permutations on n elements is done in $O(n+K)$ (see Uno and Yagiura [361]), where K is the number of common intervals. Heber and Stoye [210] achieve the same running time for $h \geq 3$ permutations (in this case, n is the total size of the h permutations).

The exemplar, intermediate and full common intervals similarities are defined on strings as the maximum common intervals similarity over all possible prunings satisfying the model.

8.1.4.1 Computational Complexity

The classical optimization problem called VERTEX COVER is a very useful tool in proving hardness results. The decision problem associated with it is **NP**-complete and allows deducing via a Turing reduction and using remark 8.1 that:

Theorem 8.9 [104] Computing the exemplar/intermediate/full common intervals similarity between two strings S and T is an **NP**-complete problem, even if $occ(S) = 1$ and $occ(T) = 2$.

The variant of VERTEX COVER where the input graph is 3-regular (i.e., each vertex has degree 3) is **APX**-hard, and yields, via an L-reduction, the following result.

Theorem 8.10 [19] Computing the exemplar/intermediate/full common interval similarity between two strings S and T is an **APX**-hard problem, even if $occ(S) = 1$ and $occ(T) = 2$.

Once again, as soon as duplicates are allowed in S or T, the problem of computing the exemplar/intermediate/matching distance becomes very hard.

8.1.4.2 Exact Algorithms Angibaud et al. [16] propose another pseudo-Boolean framework whose aim is to compute the common intervals similarity under each of the three models. Again, transforming the LPB program for one model into the LPB program for another model is an easy task and involves only one set of similar constraints (as shown by the authors). Section 8.1.8 presents three good heuristics for this problem.

8.1.4.3 Variants Bourque et al. [83] attempt to solve the exemplar common interval similarity between S and T using the following approach. First, identify the intervals of *all* possible exemplar prunings of S and (separately) of T. Then, select the intervals that occur both on a pruning of S and on a pruning of T. Finally, select a maximum number of compatible intervals (i.e., common intervals in the *same* pruning of S and T). The drawback of this approach is that, with the algorithm proposed by Bourque et al. [83], some families may be completely removed from the strings during the second step, so that in the third step the pruning is not necessarily exemplar. Note also that the third step uses a heuristic, so that the entire algorithm is a heuristic as well.

The exemplar, intermediate, and full models reduce the problem of comparing strings to the problem of comparing permutations, so that the only definition one needs is that of a common interval of two permutations. However, common intervals are also defined for strings, as proposed by Didier [141] and by Schmidt and Stoye [330].

Definition 8.13 An **interval** I of a string S is a set $\{|S_i|, |S_{i+1}|, \ldots, |S_{j-1}|, |S_j|\}$ of unsigned characters for some i, j such that $1 \leq i \leq j \leq n$. In this case, $[i, j]$ is a **location** of I in S. A **maximal location** $[i, j]$ of I in S is a location with the property that S_{i-1} and S_{j+1} (whenever they exist) do not belong to I.

Intervals of permutations can be defined by either their location or their content. In the case of strings, however, the distinction between location and content matters.

Definitions based on both concepts have been investigated: Didier [141] gives the following definition of a common interval.

Definition 8.14 Let I be an interval of S and T. Let $[i, j]$ and $[h, l]$ be two maximal locations of I on S and T, respectively. Then the pair $([i, j], [h, l])$ is a **common interval** of S and T.

This definition identifies a common interval not only by its set of elements but also by its locations on S and T. The contrary holds with Schmidt and Stoye's [330] definition.

Definition 8.15 A **common $\mathcal{CS}$-factor** of two strings S and T is an interval of both S and T.

Didier [141] gives an $O(n^2 \log n)$ running algorithm to find the common intervals of two strings S and T of maximum length n. Schmidt and Stoye [330] compute the common $\mathcal{CS}$-factors and their locations (thus also solving Didier's problem) with an $O(n^2)$ running-time algorithm for two strings and an $O(hn^2)$ algorithm for h-strings (the extension of the definition to $h \geq 3$ strings is obvious).

8.1.5 Conserved Intervals Similarity

Conserved intervals in permutations are a particular case of common intervals, introduced with the aim of identifying parts of the string S that should not be broken when transforming S into T by a sequence of biological operations.

- Introduced by Bourque et al. [83] (exemplar) and Angibaud et al. [19] (intermediate and full).
- Complexity: **APX**-hard (see Angibaud et al. [19]) (exemplar, intermediate and full).
- Exact algorithms: can be devised from results by Angibaud et al. [16].

Conserved intervals were introduced by Bergeron and Stoye [45], following an initial, slightly stronger definition given by Bergeron et al. [49]. In section 4.1, conserved intervals are defined in the special case where one of the permutations is the identity permutation. Here, we give the definition without this assumption in order to be able to use it for strings, for which no left-invariance concept allows the assumption that one permutation is the identity.

Definition 8.16 A **conserved interval** of two signed permutations π and σ is a common interval with extremities $i < j$ in π and $h < l$ in σ such that either $\pi_i = \sigma_h$ and $\pi_j = \sigma_l$ holds, or $\pi_i = -\sigma_l$ and $\pi_j = -\sigma_h$ holds.

A conserved interval is thus a common interval of π and σ whose extremities accurately separate it from the rest of the permutation.

Definition 8.17 The **conserved intervals similarity** between two signed permutations π and σ is the number of conserved intervals of π and σ.

To illustrate this definition, consider the strings S and T in figure 8.1 and the alternative exemplar pruning $(\mathsf{S}, \mathsf{T}) = (1\ -4\ -2\ 3, 1\ -3\ 2\ 4)$. The conserved intervals similarity of S and T has value 3, which is given by the intervals with extremities (recall these are *indices*, not elements of the permutations) $i = 2$, $j = 3$; $i = 2$, $j = 4$; and $i = 3$, $j = 4$ of S.

Computing the number of conserved intervals between two permutations on n elements can be done in $O(n)$ time using, for instance, the algorithm of Bergeron and Stoye [45]. This complexity results from the fact that one does not need to *enumerate* all intervals (their number may be $O(n^2)$).

8.1.5.1 Computational Complexity

Results regarding the computation of the exemplar/intermediate/full conserved intervals similarity confirm the expectations supported by the previous similar problems. Angibaud et al. [19] and remark 8.1 allow the formulation of the following theorem.

Theorem 8.11 [19] Computing the exemplar/intermediate/full conserved interval similarity between two strings S and T is **APX**-hard, even when $occ(S) = 1$ and $occ(T) = 2$.

Note that in Angibaud et al. [19], singletons are considered as conserved intervals, which is not the case in definition 8.16. However, the inapproximability results (as well as the algorithms we refer to in the next subsection) remain valid.

8.1.5.2 Exact Algorithms

We are not aware of any exact algorithm to compute the conserved intervals similarity between two strings, but the LPB programs given by Angibaud et al. [16] for the common intervals similarity can easily be modified to include the particularity of conserved intervals.

8.1.5.3 Variants

Bourque et al. [83] investigate the problem of computing the exemplar conserved similarity, and propose a heuristic for a variant of the problem that requires the obtained pruning of S and T to contain 0 or 1 pair of genes in each gene family. See section 8.1.4.3 for several comments on the algorithm (which is similar for common and conserved intervals).

8.1.6 Conserved Intervals Distance

The distance we study now has a particularity: it is defined between two sets $\mathcal{S}$ and $\mathcal{T}$ of strings over the alphabet $\mathcal{A}$, rather than between two strings. This generalization is not possible for rearrangement distances in general, but it is possible in this case because no scenario is required.

- Introduced by Chen et al. [107] (exemplar) and by Blin and Rizzi [68] (full).
- Complexity: **NP**-complete, even when $\mathcal{S} = \{S\}$ and $\mathcal{T} = \{T\}$ (see Blin and Rizzi [68]) (exemplar, intermediate, full). Not approximable even when $\mathcal{S} = \{S\}$ and $\mathcal{T} = \{T\}$ (see Chen et al. [107]) (exemplar).

Defining a conserved interval of an arbitrary set $\mathcal{P}$ of permutations is an easy task, by similarity with the case $|\mathcal{P}| = 2$. This definition comes from Bergeron and Stoye [45].

Definition 8.18 A **conserved interval** of a set $\mathcal{P}$ of permutations over the alphabet $\mathcal{A}$ is any set $C \subseteq \mathcal{A}$ with the property that C is a conserved interval of each pair of permutations in $\mathcal{P}$.

Consequently, the extremities of the interval are either a, b (in this order) or $-b$, $-a$ (in this order) in each permutation, for some a and b such that $|a|, |b| \in \mathcal{A}$. Moreover, the set of elements lying between those extremities is the same in all permutations. It is worth noting here that the above definition allows $\mathcal{P}$ to contain only one permutation. In this case, every interval of size at least 2 is a conserved interval.

Bergeron and Stoye [45] give an algorithm with $O(|\mathcal{P}|n)$ time complexity to compute the number of conserved intervals of a set $\mathcal{P}$ of permutations of n elements. They also transform the measure of similarity between two or more permutations into a distance (which really satisfies the three properties of a metric). Let $N_{\mathcal{P}}$ be the number of conserved intervals of a set $\mathcal{P}$ of permutations. The distance is defined similarly to definition 4.2, and generalized to sets of permutations.

Definition 8.19 The **conserved intervals distance** between two sets of permutations on n elements $\mathcal{P}$ and $\mathcal{Q}$ is defined by

$$cid(\mathcal{P}, \mathcal{Q}) = N_{\mathcal{P}} + N_{\mathcal{Q}} - 2N_{\mathcal{P} \cup \mathcal{Q}}. \tag{8.4}$$

Now, the definition of a conserved intervals distance between two sets of strings is conceivable only if we define a pruning of two sets of strings. We first need the definition of a matching on a set of strings.

Definition 8.20 Let $\mathcal{U} = \{U^1, U^2, \dots, U^t\}$ be a set of strings over an alphabet $\mathcal{A}$. A t**-dimensional matching** between the strings in $\mathcal{U}$ is a set of disjoint t-tuples $(j^1, j^2, \dots, j^t)$ such that $1 \leq j^i \leq |U^i|$ for all $i \in \{1, 2, \dots, t\}$ and $U^1_{j^1}, U^2_{j^2}, \dots, U^t_{j^t}$ belong to the same gene family.

Furthermore, a matching may fall into one of the three categories with which the reader should be familiar by now.

Definition 8.21 A t-dimensional matching $\mathcal{M}$ of $\mathcal{U}$ is **exemplar/intermediate/full** if it contains exactly one/at least one/as many as possible t-tuples of each gene family that occurs in all strings.

The definition of a pruning is now an easy exercise. For the exemplar model, it is given by Chen et al. [107], whereas for the other models such a definition is missing, either because the corresponding distance was considered only for two singletons (as for the full model, see Blin and Rizzi [68]) or because the distance was not yet considered at all (as for the intermediate model).

Definition 8.22 Let $\mathcal{S}$ and $\mathcal{T}$ be two sets of strings over an alphabet $\mathcal{A}$ and let $\mathcal{M}$ be an exemplar/intermediate/full matching between the strings in $\mathcal{S} \cup \mathcal{T}$. The $\mathcal{M}$-**pruning** of $\mathcal{S}$ and $\mathcal{T}$ is the pair $(\mathcal{S}^*, \mathcal{T}^*)$ of sets of strings obtained respectively from $\mathcal{S}$ and $\mathcal{T}$ by removing all characters that do not occur in $\mathcal{M}$ and by relabeling the remaining characters with distinct labels, according to $\mathcal{M}$.

Moreover, the $\mathcal{M}$-pruning is **exemplar/intermediate/full** according to the type of $\mathcal{M}$. The conserved intervals distance is now defined in a quite similar fashion to the distances between two strings (see Chen et al. [107]):

Definition 8.23 Let $\mathcal{S}$ and $\mathcal{T}$ be two sets of strings over the alphabet $\mathcal{A}$. The **exemplar/intermediate/full conserved intervals distance** between $\mathcal{S}$ and $\mathcal{T}$ is the minimum conserved intervals distance obtained over all exemplar/intermediate/full prunings $(\mathcal{S}^*, \mathcal{T}^*)$ of $\mathcal{S}$ and $\mathcal{T}$.

8.1.6.1 Computational Complexity

Blin and Rizzi [68] consider the particular case where both $\mathcal{S}$ and $\mathcal{T}$ are singletons, and prove its **NP**-completeness:

Theorem 8.12 [68] Computing the exemplar/intermediate/full conserved intervals distance between two sets $\mathcal{S}$ and $\mathcal{T}$ of strings is **NP**-complete even when $\mathcal{S} = \{S\}$, $\mathcal{T} = \{T\}$, and $occ(S) = 1$.

According to Blin et al. [73], this result can be extended to the case $occ(S) = 1$ and $occ(T) = 2$. Blin and Rizzi [68] also show that, under the full model, computing the conserved intervals distance remains **NP**-complete even when $\mathcal{S} = \{S\}$, $\mathcal{T} = \{T\}$, and $f(S) = f(T) = 1$. As far as inapproximability results are concerned, Chen et al. [107] first prove:

Theorem 8.13 [107] It is **NP**-complete to approximate the exemplar conserved intervals distance within a factor of $c \log n$ for some constant $c > 0$, even when both sets of strings have cardinality 2.

Then, by inquiring about the complexity of deciding whether this distance equals 0 or not for two given strings, they deduce that the exemplar conserved intervals distance is not approximable at all, even when both sets of strings have cardinality 1. (See section 8.1.6.3.)

8.1.6.2 Exact Algorithms One could be tempted to say that, similarly to the conserved intervals similarity, the particular case where the two sets of strings are singletons could be handled by the exemplar/intermediate/full variants of the LPB programs given by Angibaud et al. [16] to compute the common intervals similarity. They should just be modified so as to select conserved intervals instead of common intervals and to minimize the function issued from equation (8.4) instead of maximizing the number of intervals. The problem with this approach is that the function to minimize in that case is no longer linear, since $N_{\{S\}} = \binom{n}{2}$ when $|S| = n$.

Instead, the exemplar and full variants of this particular case of the problem can be computed using the following observation. Recall that $\mathcal{S} = \{S\}$ and $\mathcal{T} = \{T\}$: any pruning of $\mathcal{S}$ and $\mathcal{T}$ is then a pruning of S and T, and is therefore a pair (S, T) of permutations. If $\mathcal{M}$ is the matching that yields this pruning, equation (8.4) becomes

$$cid(\{\mathsf{S}\}, \{\mathsf{T}\}) = 2\binom{|\mathcal{M}|}{2} - 2\ cis(\mathsf{S}, \mathsf{T}),$$

since $N_{\{\mathsf{S}\}} = N_{\{\mathsf{T}\}} = \binom{|\mathcal{M}|}{2}$ and $N_{\{\mathsf{S}\} \cup \{\mathsf{T}\}}$ is the number of conserved intervals (or the conserved intervals similarity) between S and T, which we denote as $cis(\mathsf{S}, \mathsf{T})$. Consequently, we have

$$cid(\mathsf{S}, \mathsf{T}) + 2\ cis(\mathsf{S}, \mathsf{T}) = 2\binom{|\mathcal{M}|}{2},$$

and the right-hand side of this equation is a constant over all exemplar prunings of S and T. Therefore, minimizing the conserved intervals distance and maximizing the conserved intervals similarity are equivalent problems under this model, and the same holds for the full model. Obviously, the claim is not valid for the intermediate model. Section 8.1.8 presents a heuristic for this problem, studied by Blin and Rizzi [68].

8.1.6.3 Variants Several examples above show that the difficulty of approximating a distance is easily argued if one first proves the **NP**-completeness of testing whether this distance equals 0 or not. When applied to the conserved intervals distance, this technique again works under the exemplar model.

- Introduced by Chen et al. [107] (exemplar).
- Complexity: **NP**-complete, even when $\mathcal{S} = \{S\}$ and $\mathcal{T} = \{T\}$ (see Chen et al. [107]) (exemplar).

Based on the **NP**-completeness result of Chen et al. [107], one easily infers that unless **P** = **NP**, no polynomial-time algorithm may be found to approximate the exemplar conserved intervals distance with a bounded ratio. Since Chen et al.'s result [107] holds for the particular case where $\mathcal{S} = \{S\}$ and $\mathcal{T} = \{T\}$ with $occ(S) = occ(T) = 3$, the inapproximability result holds in this case as well.

8.1.7 MAD and SAD Numbers

The measures to estimate the (dis)similarity between two permutations that we have mentioned so far fall into two categories: they estimate either the *distance* or the *similarity* between the two permutations. The two measures in this subsection, both defined by Sankoff and Haque [321], belong to neither category: unlike distances, their value is never 0; and unlike similarities, their value grows as the dissimilarity of the permutations grows.

The following properties hold for both the so-called MAD and SAD numbers, when the usual models (i.e., exemplar, intermediate, and full) are considered on a pair of strings.

- Introduced by Chauve et al. [104] (exemplar and full).
- Complexity: **NP**-complete and **APX**-hard (see Chauve et al. [104]) (exemplar, intermediate, and full).

Let $\pi = (\pi_1\ \pi_2\ \cdots\ \pi_n)$ and $\sigma = (\sigma_1\ \sigma_2\ \cdots\ \sigma_n)$ be two permutations, and let σ^π be the permutation obtained from σ by renaming the elements of π so as to obtain the identity permutation ι, and then renaming the elements of σ accordingly.

The MAD and SAD numbers measure how far genes have to move from their initial position in one genome in order to yield the other genome, and this measure focuses either on each gene (MAD) or on all genes altogether (SAD).

Definition 8.24 The **maximum adjacency disruption number** (or **MAD number**) of two permutations π and σ on n elements is

$$MAD(\pi, \sigma) = \max_{1 \le i \le n-1} \max\{|\sigma_i^\pi - \sigma_{i+1}^\pi|, |\pi_i^\sigma - \pi_{i+1}^\sigma|\}.$$

More intuitively, the MAD number of π and σ is the largest gap between two consecutive elements in σ^π or in π^σ.

Definition 8.25 The **summed adjacency disruption number** (or **SAD number**) of two permutations π and σ on n elements is

$$SAD(\pi, \sigma) = \sum_{i=1}^{n-1} (|\sigma_i^\pi - \sigma_{i+1}^\pi| + |\pi_i^\sigma - \pi_{i+1}^\sigma|).$$

More intuitively, the SAD number is the sum of all gaps between two consecutive elements in σ^{π} and in π^{σ}. When π and σ are identical, the MAD number equals 1 and the SAD number equals $2(n-1)$.

Definition 8.26 Let S and T be two strings over the alphabet $\mathcal{A}$. The **exemplar/intermediate/full MAD number** of S and T is the minimum MAD number obtained over all exemplar/intermediate/full prunings (S, T) of S and T.

Definition 8.27 Let S and T be two strings over the alphabet $\mathcal{A}$. The **exemplar/intermediate/full SAD number** of S and T is the minimum SAD number obtained over all exemplar/intermediate/full pruning (S, T) of S and T.

Chauve et al. [104] prove the following inapproximability result, which implies the **APX**-hardness of computing the MAD number under all three models (because of remark 8.1).

Theorem 8.14 [104] Unless $\mathbf{P} = \mathbf{NP}$, no $(2 - \varepsilon)$-approximation algorithm $(\varepsilon > 0)$ exists for computing the exemplar/intermediate/full MAD number of two strings, even when $occ(S) = 1$ and $occ(T) = 9$.

In turn, the **APX**-hardness of computing the SAD number is a consequence of theorem 8.15 (recall that $n = |S|$ and assume without loss of generality that $n = min(|S|, |T|)$).

Theorem 8.15 [104] Unless $\mathbf{P} = \mathbf{NP}$, there is a constant $c > 0$ such that no $(c \log n)$-approximation algorithm exists for computing the exemplar/intermediate/full SAD number of two strings S and T, even when $occ(S) = 1$.

8.1.8 Heuristics

We focus in this subsection on heuristics to deal with the aforementioned problems. The motivation for our choice to present heuristics rather than exact algorithms (or rather than both heuristics and exact algorithms) relies on their universality: *these* heuristics are identical for all distances and need only minor changes to handle one model rather than another.

8.1.8.1 Description The three heuristics presented in this section use the notion of a *longest common substring, up to a complete reversal* defined in section 7.2, and are all based on the following easy idea. Assuming temporarily that one aims at finding a full matching between S and T, which intuitively preserves the most conserved regions between both strings, an easy way to find such a matching is given by the ILCS heuristic (see algorithm 8.1), where we assume that each longest common substring found on S and T is identified by *one* precise occurrence on each of S and T. Figure 8.2 shows an example.

Algorithm 8.1
ILCS Heuristic (full matching)

Input: two genomes S and T

Output: a matching between S and T

1. Compute a longest common substring L of S and T, up to a complete reversal, exclusively made of unmatched characters from S and T.
2. Match the characters of S and T belonging to the occurrences of L according to their positions in L.
3. Iterate the process until all possible characters have been matched.
4. Remove all unmatched characters.
5. Compute the required distance on the resulting pruning (S, T).

As far as we know, this idea was first proposed by Tichy [357] and has often been used since (see subsection 8.1.8.2). Angibaud et al. [18, 15] proceeded to a large number of time-consuming distance computations, and noticed that even small changes in the ILCS algorithm might improve the execution time, the quality of the result and its applicability to various models. The IILCS heuristic (see algorithm 8.2), proposed by Angibaud et al. [18], is such a variant of ILCS where the removal of characters that cannot be matched is done *before* starting a new iteration.

This new heuristic allows one to obtain in step 2 one or several pairs of matched characters in each gene family, according to the model, and to discard in step 3 all characters that become useless. Besides the flexibility introduced by this variant of ILCS in regard to the model, an improvement of the results may also be expected, as IILCS better takes into account the final goal of matching characters, which is to identify as many conserved regions as possible in the resulting pruning, and not in S and T. Indeed, the resulting pruning has consecutive characters that were not consecutive in the initial strings, and thus has conserved regions that possibly were not conserved in the initial strings. The early removal of characters by IILCS allows non-adjacent characters in S or T to become adjacent at the end of some iteration, if the characters between them are not matched. New longest common substrings may then be formed in this way, thus improving the identification of common regions in the final pruning (see figure 8.2).

The argument these heuristics rely on is that long common substrings are strongly conserved regions that strongly affect the values of all measures, either distances or similarities. Such an argument is supported by the good performances of these heuristics (see below), but cannot be invoked when the longest common substrings are short (i.e., not exceeding some given length h). Consequently, it could be reasonable to stop the execution of the IILCS heuristic when the threshold h is reached for the length of the longest common substring, and then to apply some exact (and thus exponential) algorithm to optimally match the remaining characters according to the problem $\mathcal{P}$ to solve. Problem $\mathcal{P}$ is defined by the measure to compute and the model to use. This idea yields the hybrid method $\mathrm{HYB}_{\mathcal{P}}(h)$ in algorithm 8.3, proposed by Angibaud et al. [18].

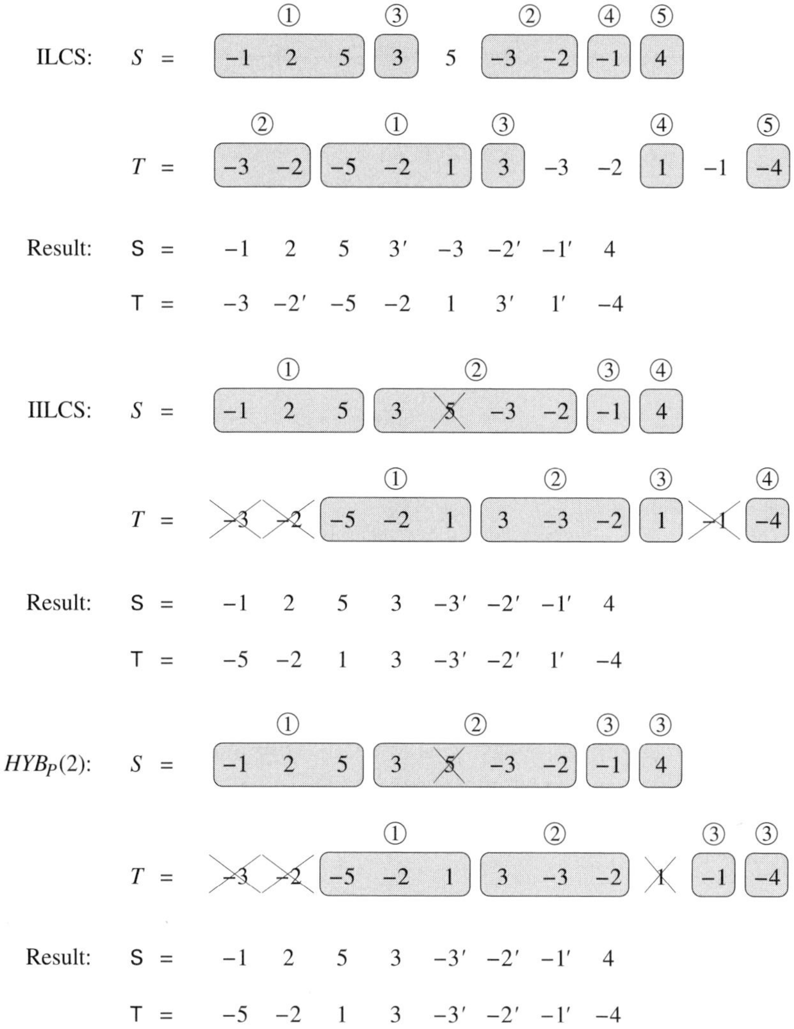

ILCS: S = [−1 2 5]① [3]③ 5 [−3 −2]② [−1]④ [4]⑤

T = [−3 −2]② [−5 −2 1]① [3]③ −3 −2 [1]④ −1 [−4]⑤

Result: S = −1 2 5 3′ −3 −2′ −1′ 4

T = −3 −2′ −5 −2 1 3′ 1′ −4

IILCS: S = [−1 2 5]① [3 ~~5~~ −3 −2]② [−1]③ [4]④

T = ~~−3~~ ~~−2~~ [−5 −2 1]① [3 −3 −2]② [1]③ ~~−1~~ [−4]④

Result: S = −1 2 5 3 −3′ −2′ −1′ 4

T = −5 −2 1 3 −3′ −2′ 1′ −4

$HYB_{\mathcal{P}}(2)$: S = [−1 2 5]① [3 ~~5~~ −3 −2]② [−1]③ [4]③

T = ~~−3~~ ~~−2~~ [−5 −2 1]① [3 −3 −2]② ~~1~~ [−1]③ [−4]③

Result: S = −1 2 5 3 −3′ −2′ −1′ 4

T = −5 −2 1 3 −3′ −2′ −1′ −4

Figure 8.2
Execution and results of the three heuristics, seeking a full matching, on the strings $S = -1\,2\,5\,3\,5\,-3\,-2\,-1\,4$ and $T = -3\,-2\,-5\,-2\,1\,3\,-3\,-2\,1\,-1\,-4$. As an example, the problem $\mathcal{P}$ in the HYB heuristic seeks to compute the conserved intervals similarity. The circled numbers indicate in which order the longest common substrings were identified, except for the ③ in the HYB heuristic, which in this case means that the matchings were decided simultaneously by the exact algorithm evoked in the last step of the HYB heuristic

Algorithm 8.2
IILCS Heuristic (exemplar/intermediate/full matching)

Input: two genomes S and T

Output: a matching between S and T

1. Compute a longest common substring L of S and T, up to a complete reversal, exclusively made of unmatched characters from S and T.
2. Match (all or part of) the characters of S and T belonging to the occurrences of L according to their positions in L, so as to fit the exemplar/intermediate/full model constraints.
3. Remove all characters of S and T that are not and cannot be matched (keep the notation S and T for the resulting strings).
4. Iterate the process until all possible characters have been matched.
5. Compute the required distance on the resulting pruning, now denoted (S, T).

Algorithm 8.3
$\mathrm{HYB}_{\mathcal{P}}(h)$ Heuristic (exemplar/intermediate/full matching)

Input: two genomes S and T

Output: a matching between S and T

1. Run the IILCS heuristic on S and T until no longest common substring of size at least h exists.
2. Relabel the matched characters with new, distinct characters.
3. Call an exact algorithm $\mathcal{A}_{\mathcal{P}}$ to solve problem $\mathcal{P}$ on the resulting strings S' and T'.

Since IILCS and $\mathrm{HYB}_{\mathcal{P}}(h)$ perform identically until S' and T' are obtained, and since the exact algorithm will always provide the best result for S' and T', one can ensure that the hybrid method will always give a better result than IILCS (see figure 8.2 for an example). However, the running time of $\mathrm{HYB}_{\mathcal{P}}(h)$ is exponential in the worst case (as opposed to the polynomial running time of ILCS and IILCS), and this is its main drawback.

8.1.8.2 Applications All these heuristics are based on the easy, very intuitive idea that longest common substrings capture a lot of the similarities between strings, and on the easy, very intuitive greedy approach. It therefore comes as no surprise that they appeared independently and with small variants in different contexts.

The ILCS idea of Tichy [357] is presented as a possible way, eventually rejected as inefficient, to compute a specific covering set. Shapira and Storer [337] (improved version of [335]) apply it, by forbidding the complete reversal, to evaluate the string edit distance with transpositions (see section 8.2.4). Chrobak et al. [117] use the same variant of the heuristic to address a related problem, the MINIMUM COMMON STRING PARTITION PROBLEM (see section 9.1.1). Swenson et al. [345] use the full matching obtained by the ILCS heuristic to approximate the true evolutionary distance between two genomes. Blin and Rizzi [68] use ILCS to approximate the conserved intervals distance between two singleton sets, and Blin et al. [70] build phylogenetic trees based on several distances issued from classical measures (breakpoint distance, common intervals similarity, conserved intervals similarity); these distances are evaluated using the ILCS heuristic.

The IILCS and HYB$_{\mathcal{P}}(h)$ heuristics were introduced more recently, and were systematically evaluated together with ILCS by Angibaud et al. [18, 15] on several problems and data sets for which exact results are known. These evaluations show that the heuristics perform very well on experimental data.

8.2 The Block Edit Model

Except in section 8.2.5, the strings in this section are unsigned. Recall that the operations of **block duplication**, **deletion**, **insertion**, and **replacement** may be considered to edit strings (page 94), and they are called *block edit operations*.

A special case of a block is a one-character-long block, which we call a character. Consequently, a block edit operation involving a one-character-long block is called a **character edit operation** (or a **character edit** for short). Two standard distances defined with character edits only are the Hamming and the Levenshtein distances.

Definition 8.28 The **Hamming distance** between two strings S and T of equal length is the minimum number of character replacements needed to transform S into T. The **Levenshtein** (or **string edit**) **distance** between two strings S and T is the minimum number of character insertions, deletions, and replacements needed to transform S into T.

As an example, for the strings $S = 1\ 2\ 4\ 2\ 3\ 1\ 4$ and $T = 1\ 4\ 2\ 2\ 4\ 3\ 1$, the Hamming distance is equal to 5 (character replacements are needed at positions 2, 3, 5, 6, 7) and the Levenshtein distance is equal to 4 (delete each 4 and then insert them again at the appropriate places).

In the block edit model, the distance between two strings S and T is defined as the minimum number of operations, among a given set of allowed operations including at least one block edit, needed to transform S into T. Such a distance is called a **block edit distance**, and the corresponding set of allowed operations is a **block edit collection**. Of course, as long as no constraint is imposed on the blocks involved in the operations, not every possible block edit collection defines an interesting distance. In the following, block edit collections will be given in their minimal form, which implies that when a block edit is allowed, the corresponding character edit is allowed as well.

8.2.1 Block Covering Distance

The problem of (possibly partially) covering two strings S and T with pairs of blocks (one in each string) so as to minimize the total pairwise distances between blocks was introduced in its most general form by Lopresti and Tomkins [257], under the name *block edit distance*. However, though the term *edit* suggests that one string is transformed into the other using a set of block operations, no such transformation is performed here.

Although slightly different from our purposes, this distance is definitely interesting for three reasons. First, for the generality of its statement, which is based on an arbitrary distance between blocks. This is quite rare in the more recent studies, which often focus on a unique, precise distance. Second, because pairing blocks with different contents suggests the use of a block replacement operation, which is the only operation that changes, at the same time, both the content and the character order at a given place in the string. And third, because the **NP**-completeness results on this distance support (without rigorously proving, however) the idea that computing a distance based on block edits becomes **NP**-complete as soon as transpositions (that is, block moving) are allowed.

- Introduced by Lopresti and Tomkins [257].
- Complexity: **NP**-complete (see Lopresti and Tomkins [257]).

Let S and T be two unsigned strings on the alphabet $\mathcal{A}$ and let *dist* be a function with real values defined on any pair of blocks of S and T. In this context, blocks are identified by their positions and their lengths.

Definition 8.29 A **t-size block family** A of S is a collection of t blocks of S. The set of all t-size block families of S is denoted by $\mathcal{F}_t(S)$.

If $A \in \mathcal{F}_t(S)$ and $B \in \mathcal{F}_t(S)$ are given for an arbitrary $t \geq 1$, a one-to-one correspondence between the blocks in A and the blocks in B is represented as a permutation $\sigma \in S_t$. Consequently, block $A^{(i)}$ in A and block $B^{(\sigma_i)}$ in B form a pair.

The constraints on the block families considered in order to compare S and T are identified as a quadruple $C^S D^S$-$C^T D^T$, where $C^q \in \{C, \bar{C}\}$ and $D^q \in \{D, \bar{D}\}$ for each $q \in \{S, T\}$ describe the required properties of the block families on S and T. Notation C means that the block family is required to cover the string, and notation $\bar{C}$ signifies that there is no such constraint. Notation D means that the block family is required to contain disjoint blocks of the corresponding string, and notation $\bar{D}$ signifies that no such constraint is imposed.

Definition 8.30 The **$C^S D^S$-$C^T D^T$-block covering distance** of S and T is given by

$$bcd(S,T) = \min_{t \geq 1} \min_{\substack{A \in \mathcal{F}_t(S) \\ B \in \mathcal{F}_t(T),}} \min_{\sigma \in S_t} \sum_{i=1}^{t} dist(A^{(i)}, B^{(\sigma_i)}),$$

where each A and each B is required to satisfy the constraints given by $C^S D^S$ and $C^T D^T$, respectively.

Table 8.1
Complexity of computing the block covering distance

	CD	$C\bar{D}$	$\bar{C}D$	$\overline{CD}$
CD	**NP**-complete	**NP**-complete	**NP**-complete	$O(n^2m)$
$C\bar{D}$		**NP**-complete	**NP**-complete	$O(n^2m^2)$
$\bar{C}D$			**NP**-complete	$O(n^2m)$
$\overline{CD}$				$O(n^2m^2)$

From [257].

Equivalently, the block covering distance between S and T is the minimum value obtained by choosing a constrained set of block pairs from S and T, respectively, and computing the total distance over these pairs. The constraints on S and T are independent of each other, thus allowing an important flexibility in the statement of particular cases of the problem.

Note that, depending on the various parameters in the definition, certain versions of the distance may not be interesting (or well defined) unless supplementary constraints are formulated. Keeping in mind that *dist* may have negative values and that blocks in a family may not be disjoint, Lopresti and Tomkins [257] add the constraint that any two pairs of blocks that are put in correspondence by σ have to be distinct in terms of sequences, in order to avoid measure diverging.

8.2.1.1 Computational Complexity Lopresti and Tomkins [257] investigate the various cases with respect to the constraints on S and T, and prove the **NP**-completeness of computing the block covering distance when both strings are constrained with respect to C and/or D. See table 8.1 and recall that n, m are the lengths of S and T, respectively.

These proofs of **NP**-completeness use a specific function *dist*, so that one cannot deduce from these results the **NP**-completeness of important particular cases such as the case where *dist* is the Levenshtein distance. A separate proof is then necessary.

Theorem 8.16 [257] Computing the CD-CD block covering distance is **NP**-complete when *dist* is the Levenshtein distance.

Remark 8.3 It is worth noting here that several authors (e.g., Shapira and Storer [336] and Muthukrishnan and Sahinalp [278]) simply refer to the **NP**-completeness results of Lopresti and Tomkins [257] with the aim of justifying the **NP**-completeness of related block edit problems (see the rest of this chapter). As long as no complete proof supports these affirmations, we choose to consider that this information is not irrefutable. But we admit that the **NP**-completeness is conceivable.

8.2.1.2 Easy Cases When the distance *dist* between blocks is arbitrary, the problem is polynomial (Lopresti and Tomkins [257]) if at least one of the strings is unconstrained. The running times of the algorithms proposed by Lopresti and Tomkins [257] are presented in table 8.1.

Interesting particular cases are obtained by forcing the blocks to have a specific form, which is easily obtained from the standard statement of the problem by appropriately choosing the function *dist*. Tichy [357] studies the problem of finding a minimum set of identical blocks of S and T that covers exactly once each character $T[i]$ that is common to S and T. Once the maximal blocks of T that exclusively contain characters not in S are copied at the end of S in an appropriate (but easy to find) order, the problem becomes a variant of the block covering distance problem: just ask that the paired blocks be identical by defining *dist* equal to 1 for identical blocks and $+\infty$ otherwise, and impose constraints $\overline{CD}$-CD. Tichy [357] obtains a linear algorithm for this variant of the problem.

8.2.1.3 Variants The unsigned MINIMUM COMMON STRING PARTITION presented in section 9.1.1 is the variant of the CD-CD-block covering distance where *dist* has value 1 for identical blocks and both S and T are required to be balanced (otherwise, their block covering distance is $+\infty$). Using common intervals for strings (see section 8.1.4.3) instead of identical strings, Blin et al. [71] obtain a variant of the problem for which they show the **NP**-completeness and give a heuristic.

8.2.2 Symmetric Block Edit Distance

Many block edit distances may be obtained by (sometimes slightly) changing the block operations allowed on the two strings S and T (recall that these operations form the block edit collection associated with a block edit distance). These distances are collected here into three classes, which are presented in table 8.2.

The symmetric block edit distance is based on some powerful block operations (see table 8.2), such as transpositions (see definition 3.1) and reversals (see definition 3.14), that are defined on strings similarly to permutations, but also uses, in order to ensure the symmetry of the resulting distance, a constrained definition of block deletion, the *block uncopy*.

- Introduced by Cormode et al. [123].
- Complexity: unknown (see remark 8.3).
- Best approximation ratio: $O(\log n \log^* n)$, where $\log^* n$ is the number of times the log function has to be applied to n to produce a constant.

The operation of block uncopy used by this distance is the opposite of a duplication.

Table 8.2
Block edit distances and their block edit collections

Distance	Character insertion	Character deletion	Character replace-ment	Block duplication	Block uncopy	Block deletion	Trans-position	Reversal	Copy reversal	Section
Symmetric block edit distance [123]	•	•	•	•	•		•	•		8.2.2
Variant 1 [35]	•	•	•	•	•		•			8.2.2
Variant 2 [278]			•					•		8.2.2
Large block edit distance [336]	•	•		•		•	•		•	8.2.3
Variant 1 [336]	•	•		•		•	•			8.2.3
Variant 2 [161]	•	•	•	•		•	•			8.2.3
Variant 3 [336]	•					•				8.2.3
Variant 4 [336]						•				8.2.3
String edit distance with transpositions [122]	•	•	•				•			8.2.4
Variant [337]	•	•					•			8.2.4

Definition 8.31 Let S be a string and $S_i S_{i+1} \cdots S_{i+h-1}$ and $S_j S_{j+1} \cdots S_{j+h-1}$ be two disjoint occurrences of the same substring in S. The operation of **block uncopy** $S_i S_{i+1} \cdots S_{i+h-1}$ from S consists in deleting $S_i S_{i+1} \cdots S_{i+h-1}$ from S.

Each operation in the block edit collection is therefore reversible, since its opposite is also in the block edit collection, and this justifies the term *symmetric* we used to identify this distance.

For example, transforming $S = 2\ 1\ 2\ 3\ 1\ 4$ into $T = 2\ 4\ 1$ can be done by deleting element 3, uncopying the first block 2 1, and reversing the block 1 4. To transform T into S, the opposite operations are available: reverse the block 4 1, copy the block 2 1 at position 1, and insert 3 at position 4.

Remark 8.4 Cormode et al. [123], as well as other authors cited in this section, refer to a transposition as a **block move**.

8.2.2.1 Approximation Algorithms The technique used by Cormode et al. [123] to obtain the $O(\log n \log^* n)$-approximation algorithm is based on embedding strings into vector spaces. Given a distance *dist* between strings and $\mathcal{S}$ a space provided with a distance L, a function H defined on strings and with values in $\mathcal{S}$ is found such that $dist(S, T)$ is equal, up to an estimable factor, to $L(H(S), H(T))$.

Such a transformation is obtained by (1) defining a hierarchical structure of a string S using a parsing technique called Symmetric LCP, and (2) defining the binary vector $H(S)$ associated to the string S according to this structure. The metric L is then, when we focus on computing the symmetric block edit distance $sbed(S, T)$, simply the Hamming distance hd. Cormode et al. [123] show that

$$\frac{1}{3}\, sbed(S, T) \leq hd(H(S), H(T)) \leq O(sbed(S, T) \log n \log^* n).$$

The algorithm has $O(n \log n)$ running time.

8.2.2.2 Variants Batu and Sahinalp [35] consider the variant of the symmetric block edit distance where reversals are forbidden, and give a $O(\log n (\log^* n)^2)$-approximation algorithm.

Muthukrishnan and Sahinalp [278] identify a polynomial-time solvable variant of the block edit problem, in which not only the allowed operations are reduced to character replacements and reversals, but the operations are further required not to overlap. It is a consequence of the operations allowed in this case that strings must have equal length. Their algorithm computes this distance exactly in $O(n \log^2 n)$ time.

8.2.3 Large Block Edit Distance

Intuitively, the operation of deleting an arbitrary block from a string is much more natural, and seemingly more reasonable when strings represent genomes, than the operation of block uncopy used in the block edit distance. However, this change in the block edit collection (see table 8.2) has an important impact on the resulting distance. It becomes possible to transform S into T by first removing S and then successively inserting the characters of T into an initially empty string, which is certainly not a common way to transform one genome into another during the evolution. Also, it is possible to find a minimum sequence of operations to transform S into T that is not reversible, in the sense that we cannot transform T into S using the opposite operations. For instance, when $S = 1\ 1\ 2\ 2$ and $T = 1\ 1$, we transform S into T with one operation by deleting the block 2 2, but we transform T into S with two operations by inserting 2 twice. The large block edit distance is no longer a metric, since it is not symmetric.

Note also that character replacements are forbidden and reversals are replaced by copy reversals, and these are two other, less crucial, changes with respect to the symmetric block edit distance. A **copy reversal** collapses the copy of a block immediately followed by a reversal into a single operation.

- Introduced by Shapira and Storer [336].
- Complexity: see remark 8.3.
- Best approximation ratio: 3.5, by Shapira and Storer [336].

8.2.3.1 Approximation Algorithms The 3.5-approximation algorithm of Shapira and Storer [336] uses a greedy technique. The idea is to consider a string U, which is initially S, and to which characters are consecutively added at the end while T is examined. At each step, a longest prefix of T that is present in U up to a reversal is searched for. If such a string of length 2 or more is found, then a new character representing the corresponding copy or copy-reversal operation is added to U. Otherwise, the first character of T is added to U. This character represents either an operation of character copy, if the character was already present in U, or an operation of character insertion into T with respect to S in the contrary case. String T then loses the prefix used in the current step, and another step begins.

8.2.3.2 Variants Shapira and Storer [336] also consider the variant of the large block edit distance where copy reversals are forbidden, and show that a similar algorithm achieves an approximation ratio of 4.

Ergun et al. [161] consider a distance where the block edit collection contains the same operations as in Shapira anad Storer's variant above, as well as the character replacement (see table 8.2). They propose a linear-time 12-approximation algorithm for this problem.

It is easily seen that a character replacement may be simulated using a character insertion and a character deletion (which is a block deletion), so that a way to obtain T from S in Shapira and Storer's variant [336] is to simulate every character replacement in the variant of Ergun et al. [161] by a character insertion and a character deletion. Therefore,

$$lbed_S(S,T) \leq lbed_E(S,T) + n_E^{cr}(S,T) \leq 2\ lbed_E(S,T),$$

where $lbed_S$ and $lbed_E$ represent the distances in the Shapira and Storer model and the Ergun et al. model, respectively, and $n_E^{cr}(S,T)$ is the number of character replacements counted in $lbed_E(S,T)$. Consequently, due to the inclusion of the block edit collections, the following estimation holds:

$$lbed_S(S,T) \leq 2\ lbed_E(S,T) \leq 2\ lbed_S(S,T). \tag{8.5}$$

It is then easy to deduce that the 4-approximation algorithm for the Shapira and Storer model gives a 8-approximation ratio for the Ergun et al. model, thus improving the ratio of 12 found in [161].

Finally, for the very restricted variants of the large block edit distance where block deletions only, or both block deletions and character insertions, form the block edit collection, the problem is polynomial and solvable with a standard dynamic program (see Shapira and Storer [336]).

8.2.4 String Edit Distance with Transpositions

This distance may be seen as a variant of several distances seen before, but deserves to be presented separately because of its relationship to the minimum common string partition and the reversal distance (see chapter 9). The operations allowed here are the character edits and the transposition.

- Introduced by Cormode and Muthukrishnan [122].
- Complexity: Shapira and Storer [337] (improved version of [335]) claim the **NP**-completeness of the problem, based on their result (see section 8.2.4.2) on the problem that does not allow character replacements.
- Best approximation ratio: $O(\log n \log^* n)$, by Cormode and Muthukrishnan [122], where $\log^* n$ is the number of times the log function has to be applied to n to produce a constant.

8.2.4.1 Approximation Algorithms The string edit distance with transpositions is simply an extension of the Levenshtein distance, which does not allow rearranging characters, by adding a rearrangement operation, the transposition. Although computing the Levenshtein distance of two strings is a well-known polynomial problem, the best approximation algorithm when transpositions are added remains the one in [122], which uses a parsing technique and has $O(n \log n)$ running time.

8.2.4.2 Variants Shapira and Storer [337] consider the variant of the problem where only character insertions, character deletions, and transpositions are allowed. Interestingly, for this variant a proof of **NP**-completeness is provided.

Theorem 8.17 [337] Computing the minimum number of character insertions, character deletions, and transpositions to transform a string S into a string T is **NP**-complete.

The proof is by reduction from the BIN-PACKING problem. The best approximation ratio of $O(\log n \log^* n)$ is obtained by Cormode and Muthukrishnan's algorithm [122], which still works in this case because character replacements are simulated by character insertions and deletions, so that a formula similar to inequality (8.5) holds.

8.2.5 Signed Strings

In this section, strings are signed, so that a reversal reverses not only the order of the characters but also their signs.

Marron et al. [263] are concerned with the problem of finding the distance between a given signed string S and the target string $T = 1\ 2\ 3 \ldots n$ when the block edit collection consists of character insertions, character deletions, and reversals. A 10-approximation algorithm is proposed, which first finds a cover of T with maximal common blocks of S and T (up to a reversal) in order to assign unique names to duplicates, and then uses El-Mabrouk's algorithm [154] for rearranging permutations with different gene content but no gene duplicates.

Using the same operations and the same initial idea, Swenson et al. [345] consider the problem of computing the distance between two arbitrary signed strings, and give an $O(n^2)$ running time heuristic for it.

9 Distances between Balanced Strings

Deletions, insertions, duplications, and replacement events are necessary to compare arbitrary strings, because the gene content of the strings is different. But the special case where the gene content is equal is interesting, because it allows us to handle duplicated genes and not to take into account these edit operations in the scenarios. In this case, we require that every character has the same number of occurrences in the considered strings, which means that we assume all gene copies on one genome are out-paralogs. In other words, in this case we deal with *balanced* strings (see definition 7.8). It turns out that many results specific to this case exist, and thus it appears natural that they should be summarized in a separate chapter.

In this chapter, we will assume that all strings are balanced, and belong to the class $\mathcal{L}(a_1, \ldots, a_k)$, unless explicitly stated otherwise. As a consequence, for any two input strings S and T, we have $|S| = |T| = n$, $occ(S) = occ(T)$, and $f(S) = f(T)$. (Recall that $occ(S)$ is the number of duplicates in S of the largest family, and $f(S)$ is the number of gene families with at least two duplicated genes in S). When the alphabet is binary (i.e., when strings belong to the class $\mathcal{L}(a_1, a_2)$), we will always assume that the alphabet $\mathcal{A}$ is $\{0, 1\}$. Moreover, we reuse the notation introduced in chapter 3 in the context of permutations. For instance, whereas $td(\pi, \sigma)$ denoted the transposition distance between permutations π and σ, $td(S, T)$ will denote the transposition distance between strings S and T. Notions that are not obviously translated will be redefined.

9.1 Minimum Common String Partition Problems

The so-called **minimum common string partition** problem (or MCSP, for short) may be seen as an attempt to compute a breakpoint distance between strings, as this was the easiest measure of dissimilarity between permutations. For the sake of clarity, and though the results are sometimes very close, we will develop the three existing variants (unsigned MCSP, signed MCSP, and reversed MCSP) separately.

9.1.1 Unsigned MCSP

• Introduced independently by Chen et al. [106] and Swenson et al. [345] (who refer to it as SEQUENCE COVER).

• Complexity:

— **APX**-hard even when $occ(S) = 2$ (see Goldstein et al. [188]).
— **NP**-hard even when $f(S) = 1$ (see Blin et al. [69]).

• Best approximation ratio: $O(\log n \log^* n)$, where $\log^* n$ is the number of times the log function must be applied to n to produce a constant (see Cormode and Muthukrishnan [122] and Kolman and Waleń [239]). When $occ(S) = c$, the ratio is

— 1.1037 when $c = 2$ (see Goldstein et al. [188]),
— 4 when $c = 3$ (see Goldstein et al. [188]), and
— $4c$ otherwise (see Kolman and Waleń [239]).

First, we begin with some definitions that will help formalize the MCSP problem.

Definition 9.1 A **partition** of a string S is a set of substrings $\{S_1, S_2, \ldots, S_p\}$ such that the concatenation $S_1\ S_2 \cdots S_p$ is S.

Definition 9.2 A **common partition** of two strings S and T is a pair of partitions $S(1), \ldots, S(p)$ of S and $T(1), \ldots, T(p)$ of T, for which there exists a permutation π such that $S(i) = T(\pi_i)$, for all $1 \leq i \leq p$. The common partition is said to be **minimum** if there is no other common partition partitioning S and T into less than p blocks.

As a simple illustration of the above definition, for $S = abaab$ and $T = ababa$, (ab, ab, a) is a common partition, and (aba, ab) is the minimum common partition.

Formulated as a decision problem, MCSP is stated as follows:

MINIMUM COMMON STRING PARTITION (MCSP)

INSTANCE: Two unsigned balanced strings S and T, a positive integer K.

QUESTION: Is there a common partition of S and T with at most K blocks?

Approximating MCSP When $occ(S) = 2$, there is a 3-approximation (see Chrobak et al. [117]). However, the best result so far, due to Goldstein et al. [188], is a 1.1037-approximation for 2-MCSP. Goldstein et al. [188] gave a 4-approximation for the case $occ(S) = 3$.

In the general case, when $occ(S) = c$, Kolman [238] designed a $2c^2$-approximation algorithm. However, the best result so far, due to Kolman and Waleń [239], is a $4c$-approximation.

In the general case, too, an approximation ratio of $O(\log n \log^* n)$ exists. This ratio is derived from a series of equivalences (up to a constant factor) between different problems. This series of equivalences is given in the following theorem.

Theorem 9.1 [337, 226]

• The minimum number of character insertions, character deletions, character replacements, and transpositions to transform S into T is equal, up to a constant factor, to the same problem where replacements are forbidden.

• The minimum number of character insertions, character deletions, and transpositions to transform S into T is equal, up to a constant factor, to the same problem where only transpositions are allowed.

• The minimum number of transpositions to transform S into T is equal, up to a constant factor, to the minimum number of strings of a common string partition of S and T.

Thanks to the above theorem, we conclude that if an approximation algorithm of ratio r exists for the first problem mentioned above, then an approximation algorithm exists for MCSP whose ratio is $c \cdot r$, where c is a constant. Cormode and Muthukrishnan [122] proved that there exists an approximation algorithm of ratio $O(\log n \log^* n)$ for the first problem mentioned above (see also subsection 8.2.4). Hence, there exists an approximation algorithm of ratio $O(\log n \log^* n)$ for MCSP.

Greedy Algorithms for MCSP Shapira and Storer [337] (in an improved version of [335]) proposed a greedy heuristic that allows us to find a minimum common partition between two balanced strings. Let $\mathcal{H}$ denote this heuristic. The main idea of $\mathcal{H}$ consists in iteratively finding—and matching—a longest common substring between the two strings, until every character is covered. In other words, $\mathcal{H}$ is nothing more than IILCS (see algorithm 8.2), where the "up to a complete reversal" part is ignored. Shapira and Storer [335] claimed that $\mathcal{H}$ achieves an approximation ratio of $O(\log n)$, which Chrobak et al. [117] later disproved: indeed, they showed that the approximation ratio of $\mathcal{H}$ lies between $\Omega(n^{0.43})$ and $O(n^{0.69})$, a contradiction with the result of Shapira and Storer [335]. (It turned out that the $O(\log n)$ ratio from Shapira and Storer was false; however, Shapira and Storer [337] later identified a subclass of (unsigned) instances for which $\mathcal{H}$ does not exceed the $O(\log n)$ ratio.) Kaplan and Shafrir [226] later displayed an infinite family of (unsigned) strings for which the approximation ratio of $\mathcal{H}$ is $\Omega(n^{0.46})$, thus improving the $O(\Omega^{0.43})$ ratio from Chrobak et al. [117]. Finally, note that He [206] slightly modified $\mathcal{H}$ and claimed better experimental performances; however, no rigorous analysis was made, and we mention this variant here only for the sake of completeness.

9.1.2 Signed MCSP

It is of course possible to define and study the variant of MCSP that applies to signed strings. In that case, the definition of a common partition should be modified: the permutation π in definition 9.2 may now be signed, and in that case $T_{-i} = -T_i$, where $-T_i$ means that the substring is reversed both in sign and in order. For instance, if $T_i = +a\ -b\ +d\ +d$, then $-T_i = -d\ -d\ +b\ -a$. Intuitively, SMCSP is the variant of MCSP where substrings are common up to a complete reversal (in sign and order).

SIGNED MINIMUM COMMON STRING PARTITION (SMCSP)

INSTANCE: Two signed balanced strings S and T, a positive integer K.

PROBLEM: Is there a common partition of S and T with at most K blocks?

• Introduced independently by Chen et al. [106] and Swenson et al. [345] (who refer to it as SEQUENCE COVER).

• Complexity:

— **APX**-hard even when $occ(S) = 2$ (see Goldstein et al. [188]).
— **NP**-hard even when $f(S) = 1$ (see Blin et al. [69]).

• Best approximation ratio: When $occ(S) = c$:

— 1.1037 when $c = 2$ (see Goldstein et al. [188]),
— 4 when $c = 3$ (see Goldstein et al. [188]), and
— $O(c)$ otherwise (see Kolman and Waleń [239]).

A ratio of $O(n^{0.69})$ also exists (see Chrobak et al. [117]).

Equivalence Between SMCSP and the Full Breakpoint Distance The size of the minimum common partition can be viewed as the minimum number of places where a rearrangement scenario has to break. In that sense, it is an analog of the breakpoint distance between signed permutations. More precisely, it is not difficult to see that both problems are equivalent. Indeed, let S and T be two signed balanced strings of length n. As with permutations (see page 20), we may define a **linear extension** of S and T by adding two artificial nonduplicated genes α and ω to their extremities. Therefore, the linear extension of S is the string $S' = \alpha\ S_1\ \cdots\ S_n\ \omega$ on $\mathcal{A} \cup \{\alpha, \omega\}$, and the linear extension of T is $T' = \alpha\ T_1\ \cdots\ T_n\ \omega$. Let $bd(S', T')$ denote the minimum number of breakpoints between the linear extensions S' and T' under the full model (i.e., after a relabeling according to a matching), and let $SMCSP(S', T')$ be the size of a minimum common partition between S' and T'. Then we have $\text{SMCSP}(S', T') = p$ if and only if $bd(S', T') = p - 1$.

Approximating SMCSP When $occ(S) = 2$, there is a 3-approximation (see Chrobak et al. [117]), but also a 1.5-approximation (see Chen et al. [106]). However, the best result so far, due to Goldstein et al. [188], is a 1.1037-approximation. Goldstein et al. [188] give a 4-approximation for 3-MCSP.

In the general case, when $occ(S) = c$, Kolman [238] designed a $2c^2$-approximation algorithm. However, the best result so far, due to Kolman and Waleń [239], is an $O(c)$-approximation. In the general case as well, however, there exists a ratio of $O(n^{0.69})$, due to the greedy algorithm presented and analyzed in the next subsection.

Greedy Algorithms for SMCSP In the *unsigned* case, Shapira and Storer [335] proposed a greedy heuristic that allows us to find a minimum common partition between two balanced strings, which was denoted $\mathcal{H}$ and discussed in section 9.1.1. Heuristic $\mathcal{H}$ can of course be extended to the *signed* case, in order to try to approximate SMCSP. It suffices, at each step of the algorithm, to look for a longest common substring between S and T, where two substrings S' of S and T' of T are considered to be common if $S' = T'$ or if $S' = -T'$, where $-T'$ means T' reversed both in sign and in order. This algorithm is exactly the heuristic IILCS (see algorithm 8.2).

The only results in the signed case are due to Chrobak et al. [117], who obtained the following bounds: $\mathcal{H}$ approximates SMCSP by a ratio lying between $\Omega(n^{0.43})$ and $O(n^{0.69})$. We note that the results of Shapira and Storer [335, 337] and Kaplan and Shafrir [226] that were given in section 9.1.1 were presented in [226] only in the *unsigned* case. Whether or not they apply to the signed case is not explicitly stated.

9.1.3 Reversed MCSP

- Introduced by Kolman [238].
- Complexity: Unknown.
- Best approximation ratio: when $occ(S) = c$, there is an $O(c)$-approximation (see Kolman and Waleń [239]).

A third, but less studied, problem is the reversed minimum common string partition problem (RMCSP). In this variant, strings are unsigned, but the definition of common partition is extended as follows: a **common possibly reversed partition** of two strings S and T is a pair of partitions $S_1, \ldots, S_p$ of S and $T_1, \ldots, T_p$ of T, for which there exists a permutation π such that for all $1 \leq i \leq p$, $S_i = T_{\pi_i}$ or $S_i = \overline{T}_{\pi_i}$, where $\overline{T_j}$ is the string T_j in the reverse order. For instance, $\overline{abbgega} = agegbba$, so these blocks may match in the RMCSP problem.

REVERSED MINIMUM COMMON STRING PARTITION (RMCSP)

INSTANCE: Two unsigned balanced strings S and T, a positive integer K.

PROBLEM: Is there a common possibly reversed partition of S and T with at most K blocks?

RMCSP has been introduced and defined mainly because its optimum is a constant multiplicative factor away from the optimum of the problem of rearranging balanced strings by unsigned reversals, which will be discussed in section 9.2 (see also figure 9.1). Thus, any approximation result for one problem gives an approximation result for the other. However, surprisingly, the complexity of RMCSP has not been studied. It can be seen as an equivalent to the breakpoint distance problem on unsigned strings (i.e., an analog to the number of strong breakpoints for permutations).

9.1.4 Full Breakpoint Distance

As mentioned in section 9.1.2, the problem of determining the minimum number of breakpoints under the full model is equivalent to the SMCSP problem, since both optima differ by exactly 1 (see also figure 9.1). Therefore, all complexity and approximation results that apply to SMCSP also apply here.

Since the problem of determining $bd(S,T)$ is **NP**-hard in the balanced case, Angibaud et al. [19] studied the complexity of the following decision problem: given two signed balanced strings S and T, decide whether $bd(S,T)=0$. They showed the following result.

Theorem 9.2 [19] Given two balanced strings S and T of length n, there exists an $O(n \log \log n)$ algorithm for determining whether $bd(S,T)=0$.

In the remainder of this chapter, we consider the following type of problem: given a set of operations on strings, each being given a certain cost, and two balanced strings S and T, find a scenario (i.e., a sequence of operations) that transforms S into T and minimizes the total cost. In order to avoid confusion, we will denote this problem as the **rearranging** problem.

Indeed, a possible variant is the **sorting** problem, where only one string S is given as input, and the goal is to find a scenario that lexicographically orders S. For instance, if $S = 011010111$, sorting S would output string $S' = 000111111$. We note that rearranging and sorting are equivalent problems for permutations; however, we wish to stress that, in the case of balanced strings, *rearranging* and *sorting* are distinct problems.

Moreover, a third variant exists: the **grouping** problem, introduced by Hurkens et al. [215]. It consists in transforming a string S into a string where all occurrences of the same character are consecutive; however, no lexicographic order between the characters is required. For instance, if $S = 011010111$, grouping S could lead either to string $S'_1 = 000111111$ or to string $S'_2 = 111111000$. Grouping can thus be seen as a weaker form of sorting.

9.2 Reversal Distance

9.2.1 Unsigned Reversals

- Introduced by Pevzner and Waterman [298].
- Complexity: *Rearranging* and *sorting* are **APX**-hard. This follows from hardness results on the same problem for permutations. There is a proof of **NP**-completeness for *rearranging* strings built on a binary alphabet, by Christie and Irving [116], using a reduction of 3-PARTITION, and an alternative proof by Radcliffe et al. [309] using a reduction of SORTING PERMUTATIONS BY (UNSIGNED) REVERSALS.

Sorting strings built on a binary or a ternary alphabet is polynomial-time solvable (see Christie and Irving [116] and Radcliffe et al. [309], respectively).

• Best approximation ratios: *Rearranging* by unsigned reversals can be $O(c)$-approximated when $occ(S) = c$ (see Kolman and Waleń [239]).

A polynomial-time approximation scheme (PTAS) exists for *rearranging dense* instances built over an alphabet of fixed size k (see Radcliffe et al. [309]).

• Diameter: The diameter for strings of length n over an alphabet of size k is $n - \lceil \frac{n}{k} \rceil$, as shown implicitly by Radcliffe et al. [309].

Bounds and Diameter Let $rdD_k(n)$ denote the diameter for the unsigned reversal distance between any two balanced strings of length n defined over an alphabet of size k. Christie and Irving [116] generalized the notion of breakpoint on unsigned permutations to binary strings, in order to derive a lower bound on the reversal distance between binary strings.

Breakpoints on binary strings have a particular definition, which is not the usual definition for general strings. Breakpoints are defined on linear extensions of strings (see page 136): strings are extended with two additional characters, denoted here α and ω. Let $f_{ab}(S)$ denote the number of times the substring ab occurs in S, where $a, b \in \{\alpha, 0, 1, \omega\}$. The number of **breakpoints** between S and T is

$$b(S, T) = \sum_{a,b \in \{\alpha,0,1,\omega\}} \max(0, f_{ab}(S) - f_{ab}(T)).$$

Note that $S = T$ implies $b(S, T) = 0$, but the converse is not necessarily true (e.g., $S = \alpha 101001 \omega$ and $T = \alpha 100101 \omega$).

Lemma 9.1 [116] Let S and T be two balanced binary strings of length n; we have

$$\left\lceil \frac{b(S, T)}{2} \right\rceil \leq rd(S, T) \leq \left\lfloor \frac{n}{2} \right\rfloor.$$

For any $p \geq 1$, let $E_p = 0^p 1^p$ and $C_p = (01)^p$. It is not difficult to see by lemma 9.1 that $rd(E_p, C_p) \geq p$ and $rd(0 \cdot E_p, 0 \cdot C_p) \geq p$; therefore, such pairs of strings reach the upper bound $p = \lfloor \frac{n}{2} \rfloor$. Since the lower and upper bounds of lemma 9.1 match, we get the following result.

Theorem 9.3 [116] For all $n \geq 1$, $rdD_2(n) = \lfloor \frac{n}{2} \rfloor$.

Moreover, Christie and Irving [116] proved that for any balanced strings of even length $2p \geq 6$, the only strings that have a rearranging distance equal to $rdD_2(2p)$ are E_p and C_p.

More generally, Radcliffe et al. [309] studied the maximum number of reversals between any two strings belonging to the class $\mathcal{L}(a_1, \ldots, a_k)$. They proved the following result.

Theorem 9.4 [309] The maximum number of reversals between any two balanced strings of length n belonging to the class $\mathcal{L}(a_1, \ldots, a_k)$ is $n - \max_i a_i$.

Although the authors did not explicitly mention it in their paper, the above result implies the following corollary.

Corollary 9.1 For all $n \geq k \geq 1$, $rdD_k(n) = n - \lceil \frac{n}{k} \rceil$.

Indeed, maximizing $n - \max_i a_i$ corresponds to minimizing $\max_i a_i$. However, since the strings are built over an alphabet of size k, they must use k different characters. It is thus easy to see that the minimum is achieved when each letter occurs (roughly) the same number of times, and in that case $\max_i a_i = \lceil \frac{n}{k} \rceil$.

Remark 9.1 The above corollary implies two previously known results.

- $rdD_2(n) = \lfloor \frac{n}{2} \rfloor$ (see theorem 9.3 above).
- $rdD_n(n) = n - 1$ (i.e., the diameter for *permutations* is equal to $n - 1$ (see section 4.2, where n refers to the initial permutation instead of its linear extension).

Rearranging Rearranging unsigned *permutations* by reversals is **APX**-hard, and hence rearranging unsigned balanced strings by reversals is **APX**-hard as well. The following result shows that the problem remains hard even when the strings are taken from an alphabet of restricted size.

Theorem 9.5 [116] Rearranging unsigned strings by reversals is **NP**-hard, even for binary alphabets.

Radcliffe et al. [309] studied the approximability of rearranging a subclass of instances, which they call **dense** instances. Dense instances are instances such that the reversal distance is at least $c \cdot n$, where $c > 0$ is a constant and n is the length of the strings. Recall that k is the size of the alphabet over which the two input strings are built.

Theorem 9.6 [309] For any fixed k, there exists a PTAS for rearranging *dense* instances.

Radcliffe et al. [309] conjecture that the same result holds for permutations. Approximation results were obtained through the study of RMCSP (see section 9.1.3), for which we know that the optimum is a constant multiplicative factor away from the optimum of the problem of rearranging balanced strings by unsigned reversals (see also figure 9.1). Hence, the approximation results presented in section 9.1.3 also hold here. That is, there exists an $O(c^2)$-approximation (see Kolman [238]), a result later improved by Kolman and Waleń [239], who gave an $O(c)$-approximation algorithm for the problem.

Sorting Since rearranging unsigned *permutations* by reversals is **APX**-hard (see section 3.3), and rearranging and sorting permutations are equivalent problems, we conclude that sorting unsigned balanced strings by reversals is **APX**-hard as well. However, this **APX**-hardness result holds for large alphabets, because it was shown on permutations. If we restrict the size of the alphabet, the problem becomes polynomial. Note that this does not hold for the *rearranging* problem (see the above paragraph and, in particular, theorem 9.5).

First, we need some notation and definitions. A **uniform block** in S is a maximal-length substring of S that contains a unique symbol. $B(S)$ denotes the number of uniform blocks in S, and $z(S)$ denotes the number of uniform blocks composed of 0s in S. For instance, if $S = 00110000333222$, then $S' = 333$ is a uniform block, $B(S) = 5$, and $z(S) = 2$.

For any string S, let $rsd(S)$ be the reversal *sorting* distance of S. Christie and Irving [116] related the value $rsd(S)$ to $z(S)$ for any binary string in the following way.

Lemma 9.2 [116] For any binary string S,

$$rsd(S) = \begin{cases} z(S) - 1 & \text{if } S_1 = 0, \\ z(S) & \text{otherwise.} \end{cases}$$

Since computing $z(S)$ can be done in linear time, the above result leads to the following theorem.

Theorem 9.7 [116] There is a linear-time algorithm for sorting unsigned binary strings by reversals on binary alphabets.

Radcliffe et al. [309] studied strings built over a *ternary* alphabet, in a fashion similar to the work of Christie and Irving [116] on binary alphabets. They also characterized the reversal distance for such strings, depending on the form of the input string. This led to the following theorem.

Theorem 9.8 [309] There is a linear-time algorithm for sorting unsigned strings by reversals on ternary alphabets.

Radcliffe et al. conjecture that for alphabets of fixed size k, there is a polynomial-time algorithm for sorting a string by reversals.

9.2.2 Signed Reversals

- Introduced independently by Radcliffe et al. [309] and Chen et al. [106].
- Complexity: *Rearranging* is **NP**-hard, even on binary alphabets (see Radcliffe et al. [309]). It is also **NP**-hard even when there is at most one negative and one positive occurrence of each symbol (see Radcliffe et al. [309]).

- Best approximation ratio: When $occ(S) = c$, an approximation ratio of $O(c)$ exists (see Kolman and Waleń [239]).

Rearranging: Complexity and Approximations As said above, the two main results are the following.

Theorem 9.9 [309]

- Rearranging signed strings by reversals is **NP**-hard, even on binary alphabets.
- Rearranging signed strings by reversals is **NP**-hard, even when there is at most one negative and one positive occurrence of each symbol.

Chen et al. [106] studied the relationship between rearranging strings by signed reversals and the SMCSP problem defined in section 9.1.2. They showed that the size of the optimal solution of SMCSP and the signed reversal rearranging distance differ only by a constant multiplicative factor (see also figure 9.2). We note that the result of Chen et al. [106] can be seen as a generalization of a well-known result on permutations, which relates the number of (signed) breakpoints and the number of (signed) reversals. It should not come as a surprise, since we know that SMCSP and counting the number of breakpoints are equivalent problems (see section 9.1.2).

The above result allows us to obtain the following approximation ratios for the signed reversal distance using the results on SMCSP: (1) there exists an $O(c)$-approximation when $occ(S) = c$ (see Kolman and Waleń [239]), and (2) a ratio of $O(n^{0.69})$ also exists (see Chrobak et al. [117]).

Heuristics Several heuristics have been proposed in the specific context of rearranging signed strings by reversals. The first is the heuristic based on the longest common substring, presented in section 9.1.2. The approximation obtained by this heuristic lies between $\Omega(n^{0.43})$ and $O(n^{0.69})$ for strings of length n (see Chrobak et al. [117]).

Some other groups of authors, such as Chen et al. [106] and Suksawatchon et al. [344], proposed their own heuristics for dealing with balanced strings. However, there is no rigorous combinatorial analysis of the performances of their algorithms, and we mention them here only for completeness.

9.2.3 Sorting by Reversals with Length-Weighted Costs

In this variant, each reversal is assigned a given cost by a function $f(l) = l^{\alpha}$, where l is the length of the reversal and $\alpha \geq 0$ is an input parameter. The goal is to *sort* strings with a minimum cost.

Note that the particular case $\alpha = 0$ corresponds to the classical problem of sorting by reversals, presented in section 9.2.1. In this section, we will assume that *all strings are binary*.

- Introduced by Bender et al. [38] (see also Bender et al. [39]).
- Complexity: Polynomial for unsigned linear binary strings, when $\alpha = 1$ and $\alpha \geq 2$. Polynomial for unsigned circular binary strings when $\alpha = 1$ (see table 9.2). Unknown in the other cases.
- Best approximation ratio: see table 9.2.
- Diameter: see table 9.1.

There are four variants, depending on whether the considered strings are (1) signed or unsigned, and (2) linear or circular.

Sorting Diameter for Binary Strings Results on the diameter for the *sorting* problem are similar in the four variants. They were obtained by Bender et al. [38] (in the *linear unsigned* case) and by Swidan et al. [346] (in the other cases), and are summarized in table 9.1.

Sorting Complexity and Approximations Here, the results differ in the four variants. They were obtained by Bender et al. [38] (in the *linear unsigned* case) and by Swidan et al. [346] (in the other cases), and are summarized in table 9.2.

Table 9.1
Diameter for sorting binary strings of length n

α	Sorting diameter
$0 \leq \alpha < 1$	$\Theta(n)$
$\alpha = 1$	$\Theta(n \log n)$
$1 < \alpha < 2$	$\Theta(n^{\alpha})$
$\alpha \geq 2$	$\Theta(n^2)$

From [39].

Table 9.2
Approximation ratios for sorting binary strings of length n

<table>
<tr><td rowspan="2">α</td><td colspan="2">Unsigned</td><td colspan="2">Signed</td></tr>
<tr><td>Linear</td><td>Circular</td><td>Linear</td><td>Circular</td></tr>
<tr><td>$0 \leq \alpha < 1$</td><td colspan="2">$O(1)$</td><td colspan="2">$O(1)$</td></tr>
<tr><td>$\alpha = 1$</td><td colspan="2">Polynomial</td><td colspan="2">3</td></tr>
<tr><td>$1 < \alpha < 2$</td><td>$O(1)$</td><td>Unknown</td><td>$O(1)$</td><td>Unknown</td></tr>
<tr><td>$\alpha \geq 2$</td><td>Polynomial</td><td>$O(1)$</td><td colspan="2">$O(1)$</td></tr>
</table>

From [39].

Table 9.3
Diameters for sorting unsigned binary strings of length n without reversals of length strictly greater than p

α	Lower bound	Upper bound
$0 \leq \alpha < 1$	$\Theta(n + n^2 p^{\alpha-2})$	
$\alpha = 1$	$\Theta\left(n \log n + \frac{n^2}{p}\right)$	$\Theta\left(n \log p + \frac{n^2}{p}\right)$
$1 < \alpha < 2$	$\Theta(n^2 p^{\alpha-2})$	
$\alpha \geq 2$	$\Theta(n^2)$	

Table 9.4
Approximation ratios for sorting unsigned binary strings of length n without reversals of length strictly greater than p, in the case $1 \leq \alpha < 2$

p	Approximation ratio
$p = \Omega(n)$	$O(1)$
Other values of p	$2 \log_2 n + 1$

Sorting by Restricted Length-Weighted Reversals More recently, Nguyen et al. [281] studied the same problem, but with the additional restriction that no reversal of length strictly greater than a given parameter p is allowed.

The results of Nguyen et al. [281] concerning the diameter are summarized in table 9.3; they are similar in the linear and circular cases. It can be seen that setting $p = n$ (i.e., no restriction exists on the length of a reversal), we obtain the same results as in table 9.2.

The results summarized in table 9.4 are concerned with approximation algorithms. They hold only for the case $1 \leq \alpha < 2$, for both linear and circular strings. When $\alpha < 1$, no result is known. When $\alpha \geq 2$, Nguyen et al. [281] indicate that the results are similar to the ones from Bender et al. [38] (see table 9.3), because short reversals are always preferred to long ones.

9.2.4 Prefix Reversals on Unsigned Strings (Pancake-Flipping)

- Introduced by Christie [115].
- Complexity: *Rearranging* by prefix reversals is **NP**-hard for binary strings (see Hurkens et al. [215]).

The *sorting* problem for binary and ternary strings is polynomial (see Christie [115] and Hurkens et al. [215], respectively).

The *grouping* problem for binary and ternary strings is polynomial (see Hurkens et al. [215]).

• Best approximation ratio: A PTAS exists for *rearranging dense* instances built over an alphabet of fixed size k (see Radcliffe et al. [309]).

Both sorting and grouping admit PTAS for alphabets of fixed size $k \geq 4$ (see Hurkens et al. [215]).

• Diameter: $n-1$ for binary strings (see Christie [115]). Lies between $n-1$ and $\lfloor \frac{4n}{3} \rfloor$ for ternary strings (see Hurkens et al. [215]). Unknown in the general case.

Let S and T be two balanced strings of length n, defined over an alphabet of size k. We denote the prefix reversal distance between S and T by $prd(S,T)$, and the diameter for the prefix reversal distance between any two such strings by $prD_k(n)$.

Recall that $B(S)$ denotes the number of uniform blocks in S, and $z(S)$ denotes the number of uniform blocks composed of 0s in S.

Bounds and Diameter Let S and T be two balanced strings of length n. Christie [115] proved the following two results.

Theorem 9.10 [115]

$$prd(S,T) \geq \begin{cases} |B(S) - B(T)| + 1 & \text{if } S_n \neq T_n, \\ |B(S) - B(T)| & \text{otherwise.} \end{cases}$$

Theorem 9.11 [115] $prd(S,T) \leq n-1$.

The two above results led Christie [115] to prove the following.

Theorem 9.12 [115] For all $n \geq 1$, $prdD_2(n) = n-1$.

We note that Christie [115] also gave a full characterization of those instances that are at distance exactly $prdD_2(n) = n-1$. For ternary alphabets, Hurkens et al. [215] gave the following bounds.

Theorem 9.13 [215] For all $n > 3$, $n-1 \leq prdD_3(n) \leq \lfloor \frac{4n}{3} \rfloor$.

Rearranging Not surprisingly, rearranging unsigned strings by prefix reversals is hard, as shown by Hurkens et al. [215].

Theorem 9.14 [215] Rearranging balanced unsigned strings by prefix reversals is **NP**-hard, even on binary alphabets.

As they did for unsigned reversals, Radcliffe et al. [309] studied the approximability of rearranging by prefix reversals for *dense* instances. They obtained the following result (recall that k is the size of the alphabet over which the two input strings are built).

Theorem 9.15 [309] For any fixed k, there exists a PTAS for rearranging dense instances.

Sorting Christie [115] studied the problem of *sorting* a binary string S by prefix reversals. The main results from Christie [115] are the following. Let $prsd(S)$ be the prefix reversal *sorting* distance for string S.

Theorem 9.16 [115]

$$prsd(S) = \begin{cases} 2(z(S) - 1) & \text{if } S_1 = 0, \\ 2z(S) - 1 & \text{otherwise.} \end{cases}$$

The direct consequence of the above theorem is the following.

Theorem 9.17 [115] There exists a linear-time algorithm for sorting any binary string by prefix reversals.

When $k = 3$ (i.e., for ternary strings), Hurkens et al. [215] characterized the prefix reversal distance, depending on the form of the input string. This led to the following theorem.

Theorem 9.18 [215] There exists a linear-time algorithm for sorting any ternary string by prefix reversals.

It remains open, however, whether there exists a polynomial-time algorithm for sorting strings defined over an alphabet of fixed size $k \geq 4$. Hurkens et al. [215], though, showed the following.

Theorem 9.19 [215] For any fixed $k \geq 4$, there is a PTAS for sorting k-ary strings by prefix reversals.

Grouping Let S be a string of length n, defined over an alphabet of size k. We denote the prefix reversal *grouping* distance for S by $prgd(S)$.

Theorem 9.20 [215] For any $k \geq 2$, and for any string S of length n built over an alphabet of size k, $n - k \leq prgd(S) \leq n - 2$.

As a consequence, $prgd(S) = n - 2$ for binary strings. In addition, Hurkens et al. [215] gave a simple greedy algorithm that yields a grouping scenario using exactly $n - 2$ prefix reversals.

Theorem 9.21 [215] There exists a polynomial-time algorithm for grouping any binary string by prefix reversals.

When $k = 3$, we know by theorem 9.20 that either $prgd(S) = n - 2$ or $prgd(S) = n - 3$. Hurkens et al. [215] gave a full characterization of strings achieving a grouping distance of $n - 3$ (resp. $n - 2$). This characterization yields a polynomial-time algorithm for deciding whether $prgd(S) = n - 2$ or $prgd(S) = n - 3$, and therefore the following theorem.

Theorem 9.22 [215] There exists a polynomial-time algorithm for grouping any ternary string by prefix reversals.

It remains open, however, whether there exists a polynomial-time algorithm for grouping strings defined over an alphabet of fixed size $k \geq 4$. Hurkens et al. [215], though, showed the following.

Theorem 9.23 [215] For any fixed k, there is a PTAS for grouping k-ary strings by prefix reversals.

9.2.5 Reversals of Length at Most 2

This particular case has been studied under the name of "adjacent swaps" by Chitturi et al. [112]. We refer the reader to section 3.5.2.2 for its equivalent in unsigned permutations, where the associated distance was denoted inv_2. Chitturi et al. [112] proved the following results.

Theorem 9.24 [112] Let S and T be two balanced strings of length n, built over an alphabet of size k.

- For both signed and unsigned strings, there is an $O(nk)$ algorithm for rearranging by reversals of length *equal to* 2.
- In the signed case, there is an $O(nk)$ algorithm for rearranging by reversals of length *at most* 2.

9.3 Unsigned Transpositions

9.3.1 Unit Cost Transpositions

Christie and Irving [116] generalized the transposition distance on unsigned permutations (see section 3.1) to unsigned strings.

- Introduced by Christie and Irving [116].
- Complexity: *Rearranging* by transpositions is **NP**-hard, even for binary alphabets, by a reduction of 3-PARTITION (see Radcliffe et al. [309]).

  Sorting by transpositions is polynomial for binary strings (see Christie and Irving [116]), but the complexity is open in the general case.
- Best approximation ratio: *Rearranging* can be approximated (1) within $O(\log n \log^* n)$ (see Cormode and Muthukrishnan [122] and Shapira and Storer [335]), and (2) within $O(c)$ when $occ(S) = c$ (see Kolman and Waleń [239]).

  A PTAS exists for *rearranging* dense instances built over an alphabet of fixed size k (see Radcliffe et al. [309]).
- Diameter: $\lfloor \frac{n}{2} \rfloor$ for binary strings belonging to the class $\mathcal{L}(\lfloor \frac{n}{2} \rfloor, \lceil \frac{n}{2} \rceil)$ (see Christie and Irving [116]). The general case is open.

Bounds and Diameter Just as they did for reversals, Christie and Irving [116] generalized the notion of transposition breakpoints on unsigned permutations to binary strings, the goal being to derive a lower bound on the transposition distance between binary strings.

Again, all strings are assumed to be linear extensions of the initial strings under comparison: S is extended with two additional characters, α and ω. Recall that $f_{ab}(S)$ denotes the number of times the substring ab occurs in S, where $a, b \in \{\alpha, 0, 1, \omega\}$. Then, the number of **breakpoints** between S and T is

$$b(S,T) = \sum_{a,b \in \{\alpha,0,1,\omega\}} \max(0, f_{ab}(S) - f_{ab}(T)).$$

Also let $td(S,T)$ denote the transposition *rearranging* distance between any two balanced strings S and T.

Christie and Irving [116] analyzed how a transposition affects the number of transposition breakpoints between two binary balanced strings. This led to the following result, which relates the values $b(S,T)$ and $td(S,T)$ for any two balanced binary strings S and T.

Lemma 9.3 [116] Let S and T be two balanced binary strings of length n; we have

$$td(S,T) \geq \begin{cases} \left\lceil \frac{b(S,T)}{2} \right\rceil & \text{if } S_1 = T_1 \text{ or } S_n = T_n, \\ \left\lceil \frac{b(S,T)-1}{2} \right\rceil & \text{otherwise.} \end{cases}$$

Christie and Irving [116] also obtained the following upper bound for $td(S,T)$ in the case of binary strings.

Lemma 9.4 [116] Let S and T be balanced binary strings of length n. Then

$$td(S,T) \leq \left\lfloor \frac{n}{2} \right\rfloor.$$

Let $tdD_k(n)$ denote the diameter for the unsigned transposition distance between any two balanced strings of length n built over an alphabet of size k. Recall that for any $p \geq 1$, $E_p = 0^p 1^p$ and $C_p = (01)^p$. It can be seen by lemma 9.3 that $td(E_p, C_p) \geq p$ and $td(0 \cdot E_p, 0 \cdot C_p) \geq p$; thus, the lower and upper bounds given by lemmas 9.3 and 9.4 match, and we have the following result.

Theorem 9.25 [116] For any $n \geq 1$, $tdD_2(n) = \left\lfloor \frac{n}{2} \right\rfloor$.

The only known result on strings built over alphabets of size $k \geq 3$ is the following lower bound, which nicely generalizes the result obtained in the case of permutations (theorem 3.1). Note that here, the definition of a transposition breakpoint, initially given for binary alphabets, is extended to ternary alphabets.

Theorem 9.26 [116] Let S and T be two balanced strings of length n on an alphabet of size $k \geq 3$; we have

$$td(S,T) \geq \left\lceil \frac{b(S,T)}{3} \right\rceil.$$

Rearranging: Complexity and Approximations The complexity of rearranging by transpositions has been studied by Radcliffe et al. [309]. Their first result is the following.

Theorem 9.27 [309] Rearranging unsigned balanced strings by transpositions is **NP**-hard, even for binary alphabets.

As they did for unsigned reversals and unsigned prefix reversals, Radcliffe et al. [309] studied the approximability of rearranging by unsigned transpositions for *dense* instances. They obtained the following result (recall that k is the size of the alphabet over which the two input strings are built).

Theorem 9.28 [309] For any fixed k, there exists a PTAS for rearranging dense instances by transpositions.

As far as approximation algorithms are concerned, no direct result is known. However, as explained in section 9.1.1 (see also theorem 9.1 and figure 9.1), the MCSP problem and the transposition problem have optima that differ by only a multiplicative constant. Thus, the approximation results that hold for MCSP also hold for transpositions, up to a constant multiplicative factor. More precisely, this implies that (1) there is an $O(\log n \log^* n)$ approximation algorithm for unsigned strings (see Cormode and Muthukrishnan [122] and Shapira and Storer [335]), and (2) when $occ(S) = c$, there is an $O(c)$ approximation algorithm (see Kolman and Waleń [239]).

Sorting Recall that for any binary string S, $z(S)$ denotes the number of uniform blocks made exclusively of 0s in S. For any string S, we let $tsd(S)$ denote the transposition *sorting* distance of S. Similarly to what they did for reversals, Christie and Irving [116] gave a characterization of $tsd(S)$ on binary strings, depending on the form of the input string S.

Lemma 9.5 [116] Let S be a binary string; then

$$tsd(S) = \begin{cases} z(S) - 1 & \text{if } S_1 = 0, \\ z(S) & \text{otherwise.} \end{cases}$$

Thanks to the above lemma, it is not difficult to determine $tsd(S)$ for strings built over a binary alphabet.

Theorem 9.29 [116] There is a linear-time algorithm for sorting unsigned strings by transpositions on binary alphabets.

Table 9.5
Diameter for sorting linear binary strings of length n

α	Diameter
$0 \leq \alpha < 1$	$\Theta(n)$
$\alpha = 1$	$\Theta(n \log n)$
$1 < \alpha < 2$	$\Theta(n^{\alpha})$
$\alpha \geq 2$	$\Theta(n^2)$

9.3.2 Length-Weighted Transpositions

Similarly to sorting strings by reversals (see section 9.2.3), the problem of *sorting* by length-weighted transpositions for binary strings has been studied independently by Qi [306] and Bongartz [77] (a subset of the results obtained by Qi has also been published in Chinese; see Qi et al. [307]). As was the case with sorting by length-weighted reversals, any transposition here has a cost, assigned by a function $f(l) = l^{\alpha}$, where $\alpha \geq 0$ is a given parameter and l is the length of the transposition (i.e., the total length of the two blocks involved in the transposition).

Note that the particular case $\alpha = 0$ corresponds to the classical problem of sorting by transpositions, presented in section 9.3.1. In the present section, we will assume that *all strings are binary*, unsigned, and linear.

- Introduced independently by Qi [306] and Bongartz [77].
- Complexity: Polynomial for unsigned linear binary strings when $\alpha \geq 2$. Unknown when $0 < \alpha < 2$.
- Best approximation ratio: see table 9.6.
- Diameter: see table 9.5.

Bounds and Diameter Qi [306] investigated diameter issues related to sorting by length-weighted transpositions. The results he obtained are summarized in table 9.5.

Algorithms Qi [306] also investigated the possibility of determining/approximating the total cost of sorting by length-weighted transpositions. The results he obtained are summarized in table 9.6.

9.3.3 Restricted Length-Weighted Transpositions

Amir et al. [12, 13] studied the problem of rearranging strings by moving only *one* character at a time: this is equivalent to a transposition for which at least one of the blocks is of length 1. As was the case in section 9.3.2, the unit-cost model is no longer assumed. More precisely, suppose string S is of length n; at each step, two locations $1 \leq i, j \leq n$ are chosen, and character S_i is moved to the j-th position of S.

Table 9.6
Approximation ratios for sorting linear binary strings of length n

α	Approximation ratio
$0 \leq \alpha < 1$	$O(\log n)$
$\alpha = 1$	2
$1 < \alpha < 2$	$O(1)$
$\alpha \geq 2$	Polynomial

For instance, assuming that $i < j$,

$$S = S_1\, S_2 \cdots S_{i-1}\, S_i\, S_{i+1} \cdots S_{j-1}\, S_j\, S_{j+1} \cdots S_n$$

is transformed into

$$S' = S_1\, S_2 \cdots S_{i-1}\, S_{i+1} \cdots S_{j-1}\, S_i\, S_j\, S_{j+1} \cdots S_n.$$

The cost of such a move is then defined as $|i - j|^\alpha$, where α is a given parameter. The goal is, given two balanced strings S and T, to find a sequence of such moves that transforms an S into a T that minimizes the total cost (that is, the sum of the costs of each move). Amir et al. [12, 13] called the problem the "rearrangement problem," a rather unfortunate name. In the following, we will prefer the term **character moving**. Amir et al. [12, 13] studied the character moving problem under three types of cost function.

- l_1 cost function: $\alpha = 1$ (i.e., the cost is the sum of the costs the individual characters have been moved).
- l_2 cost function: $\alpha = 2$ (i.e., the cost is the sum of the *square* of the costs the individual characters have been moved).
- l_∞ cost function: $\alpha = +\infty$ (i.e., the cost is the *maximum* of the distances the individual characters have been moved).

In each case, the rearranging by character-moving problem seeks a scenario of individual moves that minimizes the total cost, under the chosen cost model (l_1, l_2, or l_∞). The main results obtained by Amir et al. [12, 13] are summarized in the two following theorems.

Theorem 9.30 [12] For any two balanced strings of length n,

- there is an $O(n)$ algorithm for rearranging by character moving under the l_1 cost function;
- there is an $O(n)$ algorithm for rearranging by character moving under the l_2 cost function.

Theorem 9.31 [13] For any two balanced strings of length n, there is an $O(n)$ algorithm for rearranging by character moving under the l_∞ cost function.

9.3.4 Prefix Transpositions

Chitturi and Sudborough [110] studied the particular case of **prefix transpositions**, in which the transpositions have the additional constraint that one of the two substrings involved in the transposition is a prefix of S.

- Introduced by Chitturi and Sudborough [110].
- Complexity: *Rearranging* unsigned strings by prefix transpositions is **NP**-hard, even for binary alphabets, by a reduction of 3-PARTITION (see Chitturi and Sudborough [110]).
- Best approximation ratio: $O(\log n \log^* n)$ for unsigned strings (see Cormode and Muthukrishnan [122] and Shapira and Storer [335]); $O(c)$ when $occ(S) = c$ (see Kolman and Waleń [239]).
- Diameter: Upper bounded by $n - \lceil \frac{n}{k} \rceil$, where k is the size of the alphabet. The exact value is unknown.

Bounds and Diameter For any two balanced strings S and T, let $ptd(S, T)$ denote the prefix transposition distance between S and T. Recall also that the number of transposition breakpoints between S and T, as defined in section 9.3.1, is denoted by $b(S, T)$. (We note that although the definition of $b(S, T)$ was initially given for binary alphabets, it is easily extendable to any alphabet size.) Chitturi and Sudborough [110] proved the following lower and upper bounds on $ptd(S, T)$.

Lemma 9.6 [110] For any two balanced strings S and T, we have

$$ptd(S, T) \geq \left\lceil \frac{b(S, T)}{2} \right\rceil.$$

Lemma 9.7 [110] For any two balanced strings S and T of length n belonging to the class $\mathcal{L}(a_1, \ldots, a_k)$, we have

$$ptd(S, T) \leq n - \max_i a_i.$$

Let $ptdD_k(n)$ denote the diameter for the prefix transposition distance between any two balanced strings of length n defined over an alphabet of size k. The above lemma implies the following bound on $ptdD_k(n)$.

Corollary 9.2 For any $n \geq k \geq 1$, $ptdD_k(n) \leq n - \lceil \frac{n}{k} \rceil$.

Rearranging: Complexity and Approximation Chitturi and Sudborough [110] studied the complexity of *rearranging* by prefix transpositions, and obtained the following result.

Theorem 9.32 [110] Rearranging by prefix transpositions is **NP**-complete, even for strings built over a binary alphabet.

As far as approximations are concerned, it is easy to see that there is a strong relationship between transpositions and prefix transpositions (see figure 9.1). Altogether, and since the transposition problem is itself related in a similar fashion to the MCSP problem (see figure 9.1 as well), we conclude that (1) there is an $O(\log n \log^* n)$ approximation algorithm for unsigned strings (see Cormode and Muthukrishnan [122] and Shapira and Storer [335]), and (2) when $occ(S) = c$, there is an $O(c)$ approximation algorithm (see Kolman and Waleń [239]).

9.3.5 Adjacent Swaps

Chitturi et al. [112] studied the problem of rearranging balanced unsigned strings by what they called "adjacent swaps." This kind of operation can be seen as a transposition where each block is of length 1, or as a reversal of blocks of length 2. It was presented in the latter context in section 9.2.5, and we refer the reader to that section for more information.

9.4 Unsigned Block Interchanges

9.4.1 Unit-Cost Block Interchanges

Christie [115] studied the generalization of translocations in the context of unsigned strings; for this, he used the term **block interchange**, which we will also use in the following.

- Introduced by Christie [115].
- Complexity: *Rearranging* strings by block interchanges is **NP**-hard, even for binary alphabets, by a reduction of 3-PARTITION (see Christie [115]).

  The *sorting* problem is polynomial for binary strings (see Christie [115]).
- Best approximation ratio: $O(\log n \log^* n)$ (see Cormode and Muthukrishnan [122] and Shapira and Storer [335]). There also exists an $O(c)$ ratio when $occ(S) = c$ (see Kolman and Waleń [239]).
- Diameter: $\lceil (n-1)/4 \rceil$ for binary strings. Unknown in the general case.

Bounds and Diameter For any two balanced strings S and T, we denote the block-interchange distance between S and T by $bid(S, T)$. As he did for reversals and for transpositions, Christie [115] generalized the notion of block-interchange breakpoints on unsigned permutations to binary strings, the goal being to derive a lower bound on the block-interchange distance between binary strings.

Recall that strings in this case are linear extensions, so that S is extended with two additional characters, α and ω, and that $f_{ab}(S)$ denotes the number of times the substring ab occurs in S, where $a, b \in \{\alpha, 0, 1, \omega\}$. The number of breakpoints between S and T is

$$b(S,T) = \sum_{a,b \in \{\alpha,0,1,\omega\}} \max(0, f_{ab}(S) - f_{ab}(T)).$$

Thanks to this, Christie [115] obtained the following lower bound on the block-interchange distance.

Lemma 9.8 [115] Let S and T be two balanced binary strings of length n; we have $bid(S,T) \geq \lceil b(S,T)/8 \rceil$.

Christie [115] also obtained the following upper bound for $bid(S,T)$ in the case of binary strings.

Lemma 9.9 [115] Let S and T be two balanced binary strings of length n; we have $bid(S,T) \leq \lceil (n-1)/4 \rceil$.

Let $bidD_k(n)$ denote the diameter for the block-interchange problem between two strings of length n, built over an alphabet of size k. Christie [115] gave pairs of strings for which the lower and upper bounds given by lemmas 9.8 and 9.9 match; hence the following theorem.

Theorem 9.33 [115] For all $n \geq 1$, $bidD_2(n) = \lceil \frac{n-1}{4} \rceil$.

Rearranging: Complexity and Approximations Concerning the complexity of rearranging unsigned strings by block interchanges, Christie [115] proved the following.

Theorem 9.34 [115] Rearranging by block interchanges is **NP**-hard, even for binary alphabets.

As far as approximation algorithms are concerned, one can observe that the problem is closely related to the transposition problem, discussed in section 9.3.1. Again, it is not difficult to see that both optima differ by a constant multiplicative factor not exceeding 2 (see figure 9.1), and thus all approximation algorithms for one problem are approximation algorithms for the other: only the ratio may change, but only by a constant multiplicative factor. Thus, we conclude that rearranging unsigned balanced strings by block interchanges can be approximated within a ratio (1) $O(\log n \log^* n)$ (see Cormode and Muthukrishnan [122] and Shapira and Storer [335]), and (2) $O(c)$ when $occ(S) = c$ (see Kolman and Waleń [239]).

Sorting For any string S, let $bisd(S)$ be the block-interchange sorting distance of S. Recall also that for any binary string S, $z(S)$ denotes the number of blocks composed of 0s in S.

Table 9.7
Rearranging by character swaps under the l^α cost model

α	Binary strings	General strings
$\alpha = 0$	$O(n)$	**NP**-hard $O(n)$ 1.5-approximation
$0 < \alpha \leq \frac{1}{\log n}$	$O(n^3)$	$O(n)$ 3-approximation
$\frac{1}{\log n} < \alpha < 1$	$O(n^3)$	$O(n^3)$ k-approximation where k is the size of the alphabet
$\alpha = 1$	$O(n)$	$O(n)$
$1 < \alpha \leq \log 3$	$O(n)$	$O(n)$ 2-approximation
$\alpha > \log 3$	$O(n)$	$O(n)$

Lemma 9.10 [115]

$$bisd(S) \geq \begin{cases} \left\lceil \frac{z(S)-1}{2} \right\rceil & \text{if } S_1 = 0, \\ \left\lceil \frac{z(S)}{2} \right\rceil & \text{otherwise.} \end{cases}$$

Theorem 9.35 [115] There is a linear-time algorithm for sorting unsigned strings by block interchanges on binary alphabets.

9.4.2 Character Swaps

Amir et al. [14] studied the particular case of the block-interchange problem where the lengths of the blocks are always equal to 1. In other words, only pairs of characters are exchanged at each step. See section 3.5.2.1 for a description of the equivalent problem for permutations, where the associated distance is denoted *exc*. The same problem on permutations was first solved by Cayley [100], and in the same article he mentioned the problem on strings.

- Introduced by Cayley [100].
- Complexity: **NP**-complete for the unit-cost model and polynomial for the linear cost model (see Amir et al. [14]).
- Best approximation ratio: 1.5 for the unit-cost model (see Amir et al. [14]).

In the rest of this section, we will call the problem the **character swaps** problem. Amir et al. [14] studied different cost functions, depending on the value of a parameter $\alpha \geq 0$. More precisely, if during a given step, characters S_i and S_j from string S are swapped, then the associated cost will be l^α, where $l = |i - j|$.

Rearranging: Complexity and Approximations Rearranging by character swaps seeks a series of swaps that minimizes the sum of the costs of each swap. Depending on the

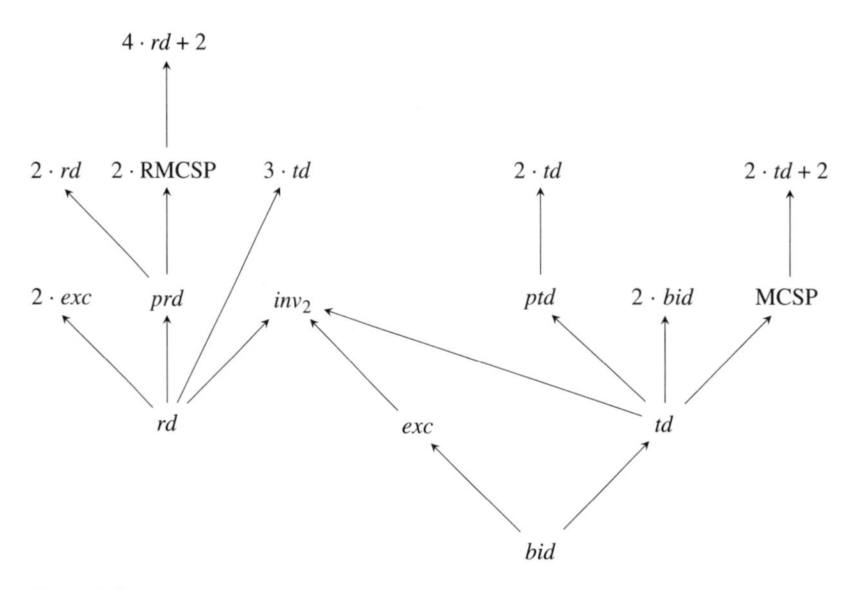

Figure 9.1
Some relations between distances on unsigned balanced strings; an arrow from distance d_1 to distance d_2 means that for all balanced strings S and T, $d_1(S,T) \leq d_2(S,T)$

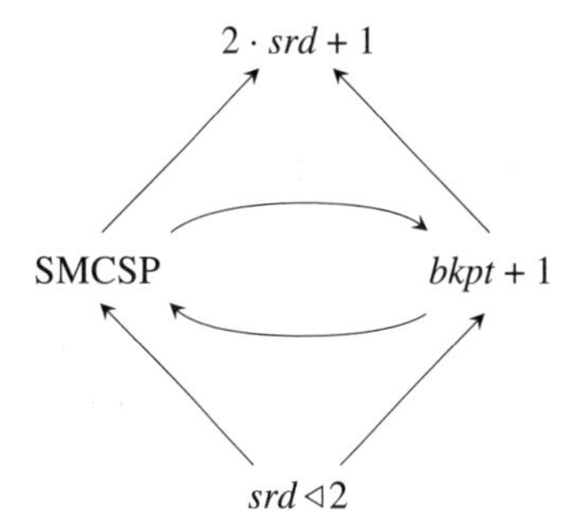

Figure 9.2
Some relations between distances on signed balanced strings; an arrow from distance d_1 to distance d_2 means that for all balanced strings S and T, $d_1(S,T) \leq d_2(S,T)$

value of α, but also on the size k of the alphabet, the results differ. They are all summarized in table 9.7.

Note that the problem is polynomial for any $\alpha \geq 0$ in the case of binary strings, and also in the case $\alpha > \log 3$ for general strings. The problem is known to be **NP**-hard for general strings only in the case $\alpha = 0$. When $0 < \alpha \leq \log 3$, the complexity is unknown for general strings.

Sorting Very few results are specific to sorting by character swaps, and they are just improvements of two of the results of table 9.7 concerning the time complexity for sorting binary strings. Those results can be summarized by the following theorem.

Theorem 9.36 [14] For any $0 < \alpha < 1$, there is an $O(n)$ algorithm for sorting binary strings of length n by character swaps under the l^{α} cost model.

9.5 Relations between Distances

As was done at the end of chapter 3, we conclude this chapter with a summary of some relations between distances on balanced strings. Those relations are summarized in figure 9.1 for unsigned strings, and in figure 9.2 for signed strings.

Proving the relations depicted in figures 9.1 and 9.2 is quite straightforward: either by simulating one operation using another, or by noting that one is a restriction of another (we leave this as an exercise for the reader). Other obvious relations, such as $2 \cdot d_1 \leq 2 \cdot d_2$ whenever $d_1 \leq d_2$, are not drawn.

III MULTICHROMOSOMAL MODELS

So far, we have examined models in which the input genomes are given as a single sequence of genes, thereby implying that the corresponding species consist of only one (linear or circular) chromosome. But this restricts the studies to a part of the domain of the living world (bacteria, archaebacteria, viruses), or to small parts of multichromosomal genomes that did not undergo interchromosomal rearrangements (which is the case for most mammalian X chromosomes, for example). In order to compare the genomes of eukaryotes (e.g., animals, plants, or fungi), we need a model that handles several chromosomes. The rearrangements that can occur in a multichromosomal genome contain those that alter single chromosomes, which we have extensively surveyed in parts I and II, plus some operations that concern several chromosomes (see section 1.1):

- **fusion**, which merge two chromosomes,
- **fission**, which split one chromosome into two, and
- **reciprocal translocations**, which exchange the arms of two different chromosomes (i.e., they exchange segments containing the telomeres in both chromosomes).

Transpositions and block interchanges may also be interchromosomal: they can move or exchange parts of a chromosome to a location in another chromosome.

Interestingly, combinatorial problems on genomes with multiple chromosomes are more than mere generalizations of the unichromosomal cases. Indeed, in unichromosomal rearrangement scenarios, all intermediary steps are most of the time required to be unichromosomal genomes. Relaxing this constraint and allowing more general objects in the scenarios sometimes leads to easier problems.

The most general framework can be adopted by defining a genome as a set of paths and cycles in a graph where genes are either vertices (if their orientation is not known), or edges (if their orientation is known). These models generalize unsigned or signed permutations, and are investigated in chapter 10.

Multichromosomal genomes can also be modeled using permutations. Indeed, the disjoint cycle decomposition (page 14) of a permutation may be used to represent several *circular* chromosomes (each cycle represents the order of the genes along a chromosome). This is the subject of chapter 11. In that model, first used by Meidanis and Dias [264], a lot of operations can be taken into account, including intrachromosomal rearrangements (e.g., reversals, transpositions, block interchanges) and interchromosomal rearrangements (fusions, fissions).

The above frameworks model multichromosomal genomes in the case where the order of genes along chromosomes matters. If we wish to disregard this information, we can instead use *set systems*: in that model, each chromosome consists of a subset of $\{1, 2, \ldots, n\}$. There can therefore be multiple copies of the same gene distributed among several chromosomes, but each gene may appear only once in each chromosome. Only the interchromosomal fusions, fissions, and translocations are taken into account in this model; indeed, since the order of genes along chromosomes is not taken into account, it makes no sense to use reversals, for instance. This model will be discussed in chapter 12.

10 Paths and Cycles

A genome with several chromosomes can be naturally modeled by a set of paths and cycles in a graph whose vertex set represents the genes. This model captures all the information about the order of the genes and their partition into chromosomes. Every path or cycle may be written as a linear or circular string, by choosing for each path an arbitrary starting point and for each cycle an arbitrary direction, and enumerating the genes along the path or cycle. Two strings are then considered as equivalent when they are obtained from one another by changing the starting point or the direction. This models the chromosomes, for which the two telomeres or the two reading directions are indeed undifferentiated.

Many combinatorial problems can be considered, depending on whether the genomes are signed or not, and on the operations one wants to take into account. The main result here is the general polynomial-time algorithm of Hannenhalli and Pevzner [196] for transforming one signed genome into another, using reversals, fissions, fusions, and reciprocal translocations. Later, Yancopoulos et al. [375] introduced the double cut-and-join (or 2-break) operation, which encompasses all these operations and led to a linear-time algorithm for computing a standard genomic distance (see Bergeron et al. [52]).

10.1 Genomes

In this chapter, multichromosomal genomes are assumed to contain no duplicates, so that the length (denoted n, as usual) of such a genome is the same as the size of the alphabet (denoted $\{1, 2, \ldots, n\}$ as usual).

An **unsigned gene** is considered as a vertex. A **signed gene** is considered as a pair of vertices: the **tail** and the **head**, linked by an edge. The tail and the head are called the **extremities** of the gene.

An **unsigned genome** is a graph whose vertex set consists of unsigned genes and additional **telomeric markers**, all denoted by T, such that each gene has degree 2 and each telomeric marker has degree 1. Loops and multiple edges are allowed. Edges of this graph are called **adjacencies**. Genes adjacent to telomeric markers are called **telomeres**. Edges incident to a telomeric marker are called **telomeric adjacencies**.

A **signed genome** is a graph whose vertex set consists of the extremities of the signed genes and additional telomeric markers, all denoted by T. Its edge set contains the edges of the genes, joining the tail and head of each gene, plus additional edges called the **adjacencies**, such that each gene extremity has degree 2 and each telomeric marker has degree 1. Multiple edges are allowed. Gene extremities adjacent to telomeric markers are called **telomeres**. An edge incident to a telomeric marker is called a **telomeric adjacency**.

Signed and unsigned genomes are thus a collection of elementary paths and cycles. Connected components are called **chromosomes**. A chromosome is **circular** if it is a cycle, and it is **linear** if it is a path. A genome is called **linear** if it contains only linear chromosomes. Linear genomes are the most realistic model for the nuclear genome of animals or plants.

A linear genome can be written by choosing, for each linear chromosome, one of the two telomeres as a starting point, and enumerating the genes along the chromosome. For signed genomes, every gene is given a positive sign if it is read from its tail to its head, and a negative sign if it is read from its head to its tail. This model of genomes generalizes permutations, signed or unsigned, linear or circular. Permutations can then be seen as genomes with only one chromosome.

An example of an unsigned genome on eight genes with three chromosomes in that model would be

$$\Pi = \{\underbrace{(3\ 1\ 5)}_{C_1}, \underbrace{(2\ 6)}_{C_2}, \underbrace{(8\ 7\ 4)}_{C_3}\}.$$

Chromosomes are equivalent up to reversals, which means that this example might be rewritten, for example, as $\{(5\ 1\ 3), (2\ 6), (4\ 7\ 8)\}$ and still correspond to the same genome. An example of a signed multichromosomal genome is given in figure 10.1.

10.2 Breakpoints

Just as in the case of permutations (page 20), a breakpoint distance between two genomes can be computed under the models described above. The breakpoint distance is not really a rearrangement distance, but it provides a first measure of dissimilarity between genomes.

Not many discussions have been published on the breakpoint distance for multichromosomal genomes. No definition is even reported for unsigned genomes. For signed genomes, Pevzner and Tesler [295] propose definitions for *internal* or *external*

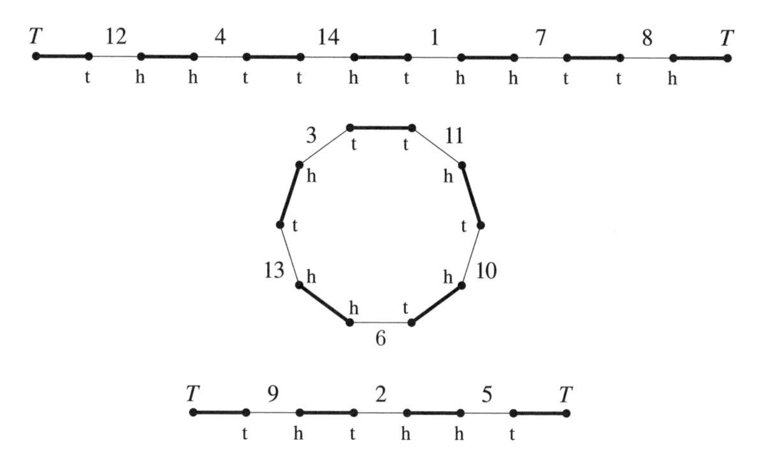

Figure 10.1
A multichromosomal genome with linear and circular chromosomes given by $C_1 = \{T12_t, 12_h4_h, 4_t14_t, 14_h1_t, 1_h7_h, 7_t8_t, 8_hT\}$; $C_2 = \{3_t11_t, 11_h10_t, 10_h6_t, 6_h13_h, 13_t3_h\}$; and $C_3 = \{T9_t, 9_h2_t, 2_h5_h, 5_tT\}$. Adjacencies are represented by thick lines

breakpoints, according to whether or not the breakpoint involves a telomere, and count them equally. The following definition, which weights internal and external breakpoints differently, is due to Tannier et al. [353].

Definition 10.1 For any two genomes Π and Γ on a gene set of size n, let a be the number of common nontelomeric adjacencies between Π and Γ, and e be the number of common telomeres. Then the **breakpoint distance** between Π and Γ is

$$bd(\Pi, \Gamma) = n - a - \frac{e}{2}.$$

Its linear-time computation is immediate from the definition. Note that the breakpoint distance does not need to be an integer, and for unichromosomal genomes it does not correspond exactly to the definition of the breakpoint distance for permutations given in section 2.6.1. This definition has the advantage that fusions and fissions account for one breakpoint, whereas reversals and translocations account for two.

10.3 Intervals

It is also possible to adapt the various notions of intervals and their variants introduced in the case of unsigned permutations. A subset of the union of genes and telomeric markers (telomeric markers are not distinguishable) I is an **interval** of a genome Π if the subgraph of Π induced by the extremities of the genes in I is connected. If this connected subgraph is a path, the extremities of this path are the

extremities of the interval I. A **common interval** of two genomes Π and Γ is an interval of both genomes. A **conserved interval** of two genomes Π and Γ is a common interval that has the same extremities in Π and Γ. Single genes are **trivial** conserved intervals. A **component** of two genomes Π and Γ is a nontrivial conserved interval of Π and Γ that is not the union of other conserved intervals. A component is **minimal** if it does not contain any other nontrivial component.

For example, if the signed genomes are $\Pi = \{(2\ 1\ 3\ 5\ 4), (6\ 7\ -11\ -9\ -10\ -8\ 12)\}$ and $\Gamma = \{(1\ 2\ 3\ 4\ 5), (6\ 7\ 8\ 9\ 10\ 11\ 12)\}$, then the components of Π and Γ are $\{T, 2, 1, 3\}$, $\{3, 5, 4, T\}$, $\{T, 6\}$, $\{6, 7\}$, $\{8, 9, 10, 11\}$, $\{7, 8, 9, 10, 11, 12\}$ and $\{12, T\}$.

These concepts are analogous to those introduced in the context of permutations (see pages 21 and 63), adding the possibility to have telomeric markers in intervals and components. This terminology is widely used to compute distances, but no similarity or distance measure based on common intervals has been studied in the specific context of multichromosomal genomes.

10.4 Translocation Distance

In this section on translocations, genomes are assumed to be linear, in order for the translocation rearrangement to be defined on all chromosomes.

- Introduced by Kececioglu and Ravi [233].
- Complexity:

— For unsigned genomes, the problem is **NP**-hard and not approximable within a factor of 1.00017 (or 5717/5716) (see Zhu and Wang [384]).
— For signed genomes, the problem is solvable in polynomial time, according to Hannenhalli [195], with further corrections by Bergeron et al. [51].

- Algorithms: Computing the translocation distance between signed genomes is linear in the number of genes (see Li et al. [246]), while giving a scenario realizing the distance can be done in $O(n^{3/2}\sqrt{\log n})$ time (see Ozery-Flato and Shamir [286]).
- Best approximation ratio: $1.5 + \varepsilon$ for unsigned genomes (see Cui et al. [129]).
- Diameter: unknown.

Definition 10.2 A **translocation** removes two adjacencies ab and cd from two different chromosomes of a genome (some of a, b, c, and d may be telomeric markers T), and replaces them either with adjacencies ac and bd or with adjacencies ad and bc. A translocation is **reciprocal** if a, b, c, and d are all different from T.

Figure 10.2 shows an example of a reciprocal translocation. Chromosome fusions and fissions are particular cases of (nonreciprocal) translocations:

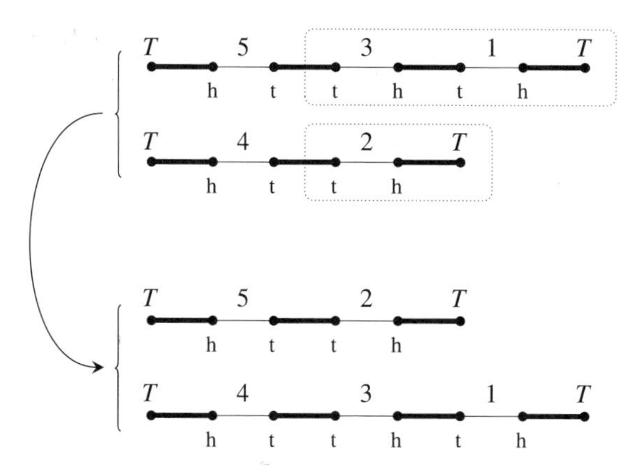

Figure 10.2
A reciprocal translocation in a signed genome

Definition 10.3 A chromosome **fusion** is a translocation involving telomeric adjacencies of two different chromosomes and resulting in an adjacency TT. A chromosome **fission** is a translocation involving a nontelomeric adjacency and TT.

If the genome is written as a set of strings on the set of genes, the starting points of each chromosome can be chosen in such a way that a translocation exchanges the prefix of one chromosome with the prefix of another. It is reciprocal if the two prefixes are not empty. Using "suffix" instead of "prefix" is equivalent, because of the choice of the starting point. A fusion concatenates two strings, and a fission splits one string into two. Here is an example of a reciprocal translocation:

$\{(1\ 2\ \boxed{3\ 4}), (5\ 6\ \boxed{7\ 8})\} \rightarrow \{(1\ 2\ 7\ 8), (5\ 6\ 3\ 4)\}.$

The equality between strings 5 6 7 8 and -8 -7 -6 -5 in a signed genome implies that a reciprocal translocation may also result in chromosomes 1 2 -6 -5 and -8 -7 3 4.

Definition 10.4 Given two genomes Π and Γ, the **translocation distance** $tld(\Pi, \Gamma)$ is the minimum number of reciprocal translocations necessary to transform Π into Γ.

The translocation distance problem has received much attention, and is the subject of a survey by Wang [364]. Maybe the main property to notice is that if a translocation exchanges a prefix of chromosome X with a prefix of chromosome Y, then it corresponds to a reversal on the string composed by the concatenation of X and Y (reversed). For example, the reciprocal translocation

$\{(1\ 2\ \boxed{3\ 4}), (5\ 6\ \boxed{7\ 8})\} \longrightarrow \{(1\ 2\ 7\ 8), (5\ 6\ 3\ 4)\}$

can be seen as the reversal of the interval $\{3, 4, 7, 8\}$ in the permutation $(1\ 2\ 3\ 4\ -8\ -7\ -6\ -5)$, concatenation of the two chromosomes, which, when split again at the same point, gives the two chromosomes $\{(1\ 2\ 7\ 8), (5\ 6\ 3\ 4)\}$.

This explains why, in most cases, the complexity of computing the translocation distance between genomes inherits hardness results from that of computing the reversal distance on permutations: both are **APX**-hard for the unsigned case, and polynomial for the signed case, with a linear algorithm for the distance computation and a subquadratic algorithm for computing an optimal scenario. But solutions to those problems differ in their details, and the distance formulas are different. Close relations between the two problems are completely investigated by Ozery-Flato and Shamir [288].

10.4.1 Feasibility

Two arbitrary genomes may not always be transformed into one another by reciprocal translocations. The definition, and in particular the non-emptiness of the exchanged prefixes, forces both genomes to share the same set of genes, to contain the same number of chromosomes (since this number is maintained by a reciprocal translocation), and to share the same set of telomeres (since this set is also maintained by a reciprocal translocation).

These three conditions are sufficient (see Feng et al. [172]). Because of the constraints on the number of chromosomes and the telomeres, we can define a *sorting* problem as in the case of permutations: given a genome Π, the **identity genome** Id_Π is the one with the same extremities as Π, and every other adjacency is of the form $i \cdot i+1$ for unsigned genomes, and $i_h \cdot (i+1)_t$ for signed genomes. This "identity genome" is not canonical, since it depends on Π. Therefore, transforming any genome Π into a genome Γ if they satisfy the required properties is a problem equivalent to transforming a genome Π into Id_Π, as was the case for permutations thanks to left-invariance: just relabel the genes of Γ to obtain Id_Π, and label the genes in Π accordingly. That is why we use the "sorting" terminology and the distance of one genome Π as well as the distance between two genomes.

Definition 10.5 Given a genome Π, the **translocation distance** $tld(\Pi)$ is the minimum number of reciprocal translocations necessary to transform Π into Id_Π.

10.4.2 Unsigned Genomes

The proofs of the complexity results obtained by Zhu and Wang [384] are similar to those obtained for unsigned reversals on permutations. The authors reduce the ALTERNATING CYCLE DECOMPOSITION problem, also used to prove **NP**-hardness and inapproximability results for sorting unsigned permutations by reversals (see section 3.3).

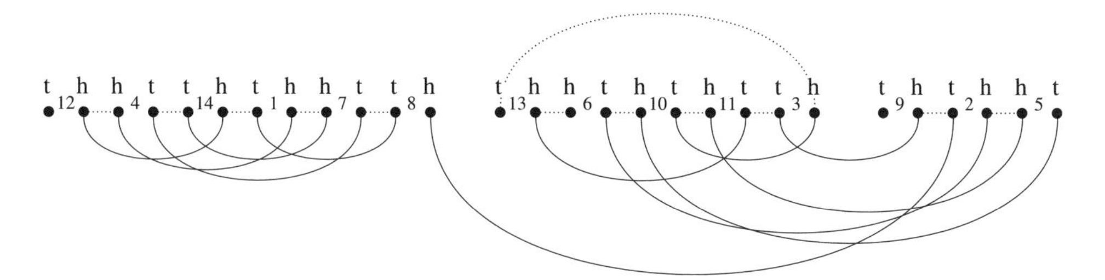

Figure 10.3
The breakpoint graph of the genomes Π given by $C_1 = \{T12_t, 12_h4_h, 4_t14_t, 14_h1_t, 1_h7_h, 7_t8_t, 8_hT\}$; $C_2 = \{3_t11_t, 11_h10_t, 10_h6_t, 6_h13_h, 13_t3_h\}$; and $C_3 = \{T9_t, 9_h2_t, 2_h5_h, 5_tT\}$, and Γ given by $C_1 = \{T12_t, 12_h14_h, 14_t7_h, 7_t4_t, 4_h1_h, 1_t8_t, 8_h2_t, 2_h6_t, 6_hT\}$ and $C_2 = \{T9_t, 9_h3_t, 3_h10_t, 10_h5_t, 5_h11_h, 11_t13_h, 13_tT\}$. Π edges are dotted lines, and Γ edges are solid lines

The original 2-approximation for sorting by translocations, obtained by Kececioglu and Ravi [233] and based on a 2-approximation by Kececioglu and Sankoff [232] for sorting permutations by unsigned reversals, has been improved by Cui et al. [128], who obtained a 1.75-approximation, and by Cui et al. [129], who obtained a $(1.5 + \varepsilon)$-approximation, still based on approximation algorithms for the ALTERNATING CYCLE DECOMPOSITION problem.

10.4.3 Signed Genomes

Hannenhalli [195] first proved a formula for computing the signed reciprocal translocation distance between two genomes.

Definition 10.6 The **breakpoint graph** $BG(\Pi, \Gamma)$ of two multichromosomal signed genomes Π and Γ is the graph whose vertex set contains all gene extremities, and whose edges are the nontelomeric adjacencies of Π (the Π-edges) and Γ (the Γ-edges).

Breakpoint graphs consist of degree vertex-disjoint elementary paths and cycles, alternating with Π-edges and Γ-edges. An example of a breakpoint graph is given in figure 10.3. The number of cycles in $BG(\Pi, \Gamma)$ is denoted by $c(BG(\Pi, \Gamma))$. When genomes satisfy the conditions for being comparable using the translocation distance, all paths of the corresponding breakpoint graph are degree 0 vertices.

Theorem 10.1 [195] For an arbitrary genome Π with n genes and N chromosomes, we have

$$tld(\Pi) = n - N - c(BG(\Pi, Id_\Pi)) + s(\Pi) + o + 2b,$$

where $s(\Pi)$ is the number of minimal components of Π, o is 1 if $s(\Pi)$ is odd and 0 otherwise, and b is 1 if

1. all minimal components of Π reside on a single chromosome,
2. $s(\Pi)$ is even, and
3. all minimal components are contained within a single component,

and 0 otherwise.

This formula yields an $O(n^3)$ algorithm for both computing the distance and solving the translocation rearrangement problem between two signed genomes. Li et al. [246] devised an $O(n)$ algorithm for computing the distance, and Bergeron et al. [51] exhibited an error in Hannenhalli's algorithm for the rearrangement scenario (the distance formula was, however, correct), yielding a corrected $O(n^3)$ algorithm. Following the sorting by reversals history, several improvements of the running time were achieved: by Zhu and Ma [383] ($O(n^2 \log n)$, then $O(n^2)$); by Wang et al. [365] ($O(n^2)$); and eventually by Ozery-Flato and Shamir [286], who adapted the algorithm of Tannier et al. [352] for sorting signed permutations by reversals and achieved a time complexity of $O(n^{3/2}\sqrt{\log(n)})$. All quadratic algorithms for sorting by translocations were also adapted to sorting by *reciprocal* translocations, yielding a crop of $O(n^2)$ time algorithms (see Ozery-Flato and Shamir [288]).

10.4.4 Translocations Preserving Centromeres

A **centromere** is a region of a chromosome that plays a vital role during DNA replication. In a linear chromosome of a signed genome, the centromere is a (possibly telomeric) adjacency of this chromosome, showing that the centromere is located between the two elements forming the adjacency, or at the extremity of a chromosome. Every linear chromosome has a centromere, so reciprocal translocations that yield chromosomes with no centromeres (and, as a by-product, chromosomes with several centromeres) are very unlikely. This motivates the study of genomic distances such that each rearrangement preserves centromeres in every chromosome. This problem has been investigated only in the case of signed genomes.

- Introduced by Kececioglu and Ravi [233].
- Complexity: polynomial for signed genomes (see Ozery-Flato and Shamir [287]).
- Best algorithm: $O(Nn + n^{3/2}\sqrt{\log n})$ for signed genomes, where n is the number of genes and N is the number of chromosomes in the compared genomes (see Ozery-Flato and Shamir [287]).

Kececioglu and Ravi [233] noted that some translocations are not valid, according to the position of the centromere in the chromosomes. This led to an *oriented* version of rearrangement by translocations in unsigned genomes that have *acrocentric* chromosomes (i.e., their centromeres are located at one telomere). In this case, only segments of chromosomes not containing the centromere may be translocated. Kececioglu and Ravi [233] devised a 2-approximation algorithm for this problem.

The general problem of rearranging by translocations was given a solution by Ozery-Flato and Shamir [287] in the context of signed genomes where centromeres are arbitrarily positioned. In a signed linear genome, a centromere is a label on an adjacency. A chromosome is said to be **legal** if exactly one of its adjacencies is labeled as a centromere. A genome is legal if all its chromosomes are legal. Given a legal genome, a rearrangement is said to be **legal** if the resulting genome is legal. Reversals are always legal, whereas fusions of chromosomes never are. A fission of a chromosome is legal if the broken adjacency contains the centromere. For translocations, some are legal and some are not.

A first interesting remark is that two genomes Π and Γ can be transformed into one another by legal reciprocal translocations only if the following condition is satisfied, which is more stringent than the condition of section 10.4.1. If either genome contains a chromosome $C = x_1 x_2 \cdots x_k$, whose centromere is on the adjacency (x_i, x_{i+1}), then we say that the **elements** of the chromosome are $\{x_1, \ldots, x_i, -x_{i+1}, \ldots, -x_k\}$. The elements of a genome are the elements of all its chromosomes. One genome can be transformed into another by legal reciprocal translocations if and only if, in addition to the conditions of section 10.4.1, they have the same elements. Ozery-Flato and Shamir [287] devised a polynomial-time algorithm for transforming one signed linear genome into another by legal reciprocal translocations, under the assumption that these genomes satisfy the above condition.

10.4.5 Variants and Special Cases

In the case where the exchanged prefixes or suffixes are required to have equal lengths, both the signed and unsigned versions are polynomial, with a linear algorithm by Kececioglu and Ravi [233].

It seems that every variant of sorting signed permutations by reversals can be transformed into a variant of sorting by translocations. Following the work of El-Mabrouk [156], Qi et al. [305] devised exact polynomial-time algorithms for sorting by translocations and deletions, assuming that the set of genes present in Γ is a subset of the set of strings present in Π (the same problem was handled for reversals by El-Mabrouk [154]).

The similarity between the reversal and translocation problems has given the idea of mixing the two operations, but surprisingly no further study has been reported since the introduction of the problem by Kececioglu and Ravi [233]. They give a $\frac{3}{2}$-approximation algorithm for sorting signed genomes by translocations and reversals, and a 2-approximation algorithm for the same problem on unsigned genomes.

The relatively small amount of work on this variant is probably due to the success of more general models, which include other rearrangements and which we survey in the next section.

10.5 Double Cut-and-Joins (2-Break Rearrangement)

- Introduced by Yancopoulos et al. [375].
- Complexity: polynomial (see Yancopoulos et al. [375]).
- Best algorithm: $O(n)$, both for computing the distance and for rearranging (see Bergeron et al. [52]).

Double cut-and-join (DCJ) operations, or 2-break rearrangements (see section 4.5), are the most general and the simplest rearrangement framework to have been developed to date. We have already defined them on permutations (section 4.5), but they are better suited to multichromosomal models, since intermediary genomes may be multichromosomal, which cannot happen in the permutation model. DCJ operations are defined only on signed genomes.

Definition 10.7 A **DCJ operation** is an operation acting on two adjacencies ab and cd of a genome Π, which transforms them either into the two adjacencies ac and bd or into the two adjacencies ad and bc. Some of a, b, c, and d may be telomeric markers, and an adjacency may even be TT. We say the DCJ **cuts** the adjacencies ab and cd, and **joins** ac and bd (or ad and bc).

The result is of course a genome. This operation generalizes the operation defined in section 4.5 on permutations. DCJ operations may be reversals, translocations, fusions, or fissions, and two consecutive DCJ operations mimic transpositions and block interchanges.

The linear genomes Π and Γ under comparison can have different numbers of chromosomes and different extremities. Therefore, the "sorting" vocabulary no longer makes sense, and what is defined is just the distance between two genomes.

Definition 10.8 Given two genomes Π and Γ, the **double cut-and-join distance** between Π and Γ, denoted by $dcj(\Pi, \Gamma)$, is the minimum number of DCJ operations needed to transform Π into Γ.

Examples of two DCJ operations are illustrated in figure 10.4.

This is a very general genomic distance that encompasses many kinds of rearrangements and tends to be universally adopted by the community. The formula for the DCJ distance is very simple.

Theorem 10.2 [375] Given two genomes Π and Γ on n genes,

$$dcj(\Pi, \Gamma) = n - \left(c(BG(\Pi, \Gamma)) + \frac{p_e(BG(\Pi, \Gamma))}{2} \right),$$

where p_e is the number of paths of even length in the breakpoint graph.

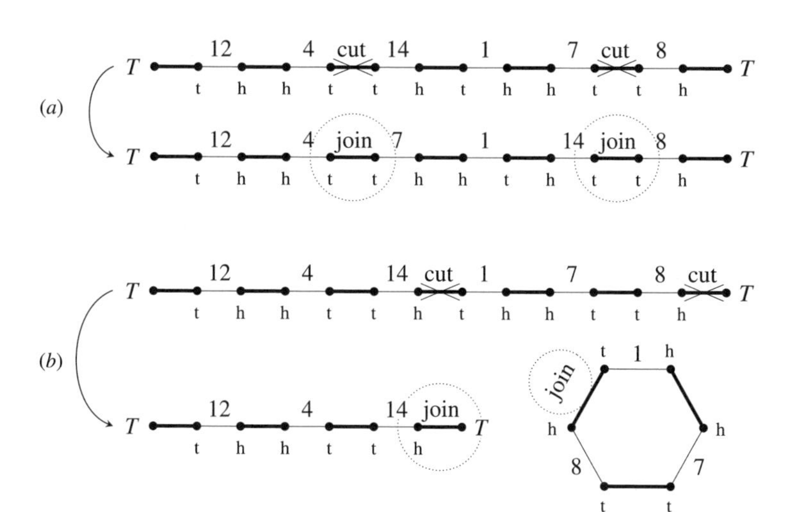

Figure 10.4
Two examples of DCJ operations: (*a*) the DCJ cuts 4_t14_t and 7_t8_t and joins 4_t7_t and 14_t8_t (it is a reversal); (*b*) the DCJ cuts 14_h1_t and 8_hT and joins 14_hT and 8_h1_t. This operation produces a circular chromosome

Note the similarity of this formula to the breakpoint distance formula (see section 10.2), which relies on the following correspondence between the three parameters of the two distances: n is the same number, $c \geq a$ because any common adjacency is a cycle of the breakpoint graph, and $p_e \geq e$ because every common telomere is an even path (with no edge) of the breakpoint graph. A linear-time algorithm for computing the DCJ distance is immediate. Yancopoulos et al. [375] first gave a quadratic algorithm to compute DCJ scenarios, and Bergeron et al. [52] gave a linear-time algorithm for the same task. It is deduced from the following lemma.

Lemma 10.1 For two genomes Π and Γ, if a DCJ operation on Π results in a genome Π' containing an adjacency that is present in Γ but not in Π, then $dcj(\Pi, \Gamma) = dcj(\Pi', \Gamma) + 1$.

10.6 *k*-Break Rearrangement

The DCJ problem has been generalized by Alekseyev and Pevzner [8] to *k*-break rearrangements.

- Introduced by Alekseyev and Pevzner [8].
- Complexity: polynomial for fixed *k* and circular genomes (see Alekseyev and Pevzner [10]); unknown for linear genomes.
- Best algorithm: $O(n^{k-2})$ (see Alekseyev and Pevzner [10]).

A k-**break rearrangement** cuts k adjacencies, and forms k new adjacencies by joining the $2k$ extremities according to an arbitrary matching. For two genomes Π and Γ, the k-**break distance** $kbd(\Pi, \Gamma)$ is the minimum number of k-break rearrangements necessary to transform Π into Γ.

Alekseyev and Pevzner [10] give a general formula for the k-break rearrangement distance on circular genomes. A subset of cycles in the breakpoint graph of circular genomes Π and Γ is called **breakable** if the total number of reality edges in these cycles equals 1 (mod $k-1$). Let $sk(\Pi, \Gamma)$ be the maximum number of disjoint breakable subsets in $BG(\Pi, \Gamma)$.

Theorem 10.3 [10] For two circular genomes Π and Γ, the k-break rearrangement distance $kbd(\Pi, \Gamma)$ is

$$kbd(\Pi, \Gamma) = \frac{bd(\Pi, \Gamma) - sk(\Pi, \Gamma)}{k-1}.$$

Alekseyev and Pevzner [10] give a polynomial-time algorithm for computing the k-break distance for fixed values of k, and a general framework for computing it in time $O(n^{k-2})$, when k is arbitrary. Alekseyev [7] attempted to extend the solution to linear genomes, but provided only bounds; no exact formula is known yet.

10.7 Fusions, Fissions, Translocations, and Reversals

- Introduced by Hannenhalli and Pevzner [196].
- Complexity: polynomial (see Hannenhalli and Pevzner [196], with subsequent corrections by Tesler [355], Ozery-Flato and Shamir [285], and Jean and Nikolski [220]).
- Best algorithm: $O(n^2)$ (see Tesler [355], Bergeron et al. [53]).

Although a distance taking fusions, fissions, translocations, and reversals into account had been defined a long time before the DCJ distance, it can be expressed in terms of a particular case of the DCJ framework that is adapted to linear genomes. Rearrangements by fusions, fissions, translocations, and reversals have been addressed only for signed genomes.

Given a linear signed genome Π, a DCJ operation on Π that does not create any circular chromosome is called **linear**. Linear DCJ operations are fusions, fissions, translocations (not necessarily reciprocal), and reversals. Since fusions and fissions are special cases of translocations, we simply write this distance RT.

Definition 10.9 Given two genomes Π and Γ, the **RT-distance** between Π and Γ, denoted by $rtd(\Pi, \Gamma)$, is the minimum number of linear DCJ operations required to transform Π into Γ.

A formula for the distance that is computable in polynomial time (more specifically in $O(n^4)$ time) was first announced by Hannenhalli and Pevzner [196]. Tesler [355] found several errors and gaps in the description of Hannenhalli and Pevzner [196] while trying to implement their algorithm. He finally achieved an $O(n^2)$ algorithm for the rearrangement scenario, and an $O(n)$ algorithm for computing the distance, both based on the theory of sorting by reversals. Later, Ozery-Flato and Shamir [285] pointed out one case in which the duality theorem for genomic distance of Hannenhalli and Pevzner [196] is incorrect, thus implying errors in Tesler's algorithm as well. Ozery-Flato and Shamir [285] revised the duality theorem and the corresponding algorithms. Their formula is correct, but their algorithm still had a flaw, eventually corrected by Jean and Nikolski [220]. The formula of Ozery-Flato and Shamir [285] still involves a few tricky parameters, and Bergeron et al. [53] made an effort to explain the formula in a simple way, on which we will base our presentation.

Recall the definitions of intervals, common intervals, conserved intervals, and components given at the beginning of this chapter (page 164). Here, an adjacency of a genome Π is said to **belong** to a component C of two genomes Π and Γ if both extremities of the adjacency involve genes that are in C but in no other component included in C. The component C of two genomes Π and Γ is said to be **oriented** if there is a linear DCJ operation on Π cutting two adjacencies that belong to C and decreasing the DCJ distance between Π and Γ by 1. It corresponds to the definition of components (oriented, unoriented) defined on permutations, with additional possible components including telomeres (see page 65).

Property 10.1 [53] Two different components of two genomes are either disjoint, nested with different extremities, or overlapping on one element.

This property allows us to give a PQ-tree structure to the set of components of a genome, just as in the case of permutations (see page 65).

Definition 10.10 A sequence of components $C_1, \ldots, C_k$ of two genomes, such that two consecutive components overlap, is called a **chain of components**. A chain C is **maximal** if there is no chain containing all the components of C and other components.

Definition 10.11 The **PQ-tree of components** of genomes Π and Γ is defined as follows:

- the root is a P-node, containing all maximal components;
- each unoriented component is represented by a P-node;
- each maximal chain of unoriented components (possibly a single component) is represented by a Q-node whose (ordered) children are the P-nodes that represent the components of this chain;

• a Q-node is the child of the minimal component that contains the chain it represents.

For two genomes Π and Γ, let $\mathcal{T}$ be the minimal subtree of the PQ-tree of components of Π and Γ containing all the unoriented components of Π and Γ. A **cover** of $\mathcal{T}$ is a collection of paths in $\mathcal{T}$ joining all nodes corresponding to unoriented components, such that an extremity of a path is not covered by another path. The **cost** of a cover of $\mathcal{T}$ is the sum of the costs of the paths it contains. A path containing only one unoriented component, or two unoriented components that both contain a telomere, costs 1, and any other path costs 2. The following theorem gives a formula for the RT-distance.

Theorem 10.4 [53] For two genomes Π and Γ, let $\mathcal{T}$ be the minimal subtree of the PQ-tree of components of Π and Γ, and t be the minimum cost of a cover of $\mathcal{T}$. Then

$$rtd(\Pi, \Gamma) = dcj(\Pi, \Gamma) + t.$$

Therefore, restricting the problem to linear genomes increases the distance by a factor t that depends on the structure of unoriented components (if there are none, the two distances are equal). Bergeron et al. [53] give an algorithm to compute t efficiently via seemingly tricky parameters.

10.8 Rearrangements with Partially Ordered Chromosomes

Genomic maps often do not specify the order within some groups of two or more markers. The synthesis of several maps from divergent or incomplete sources introduces additional order ambiguity due to markers missing from some sources. Zheng and Sankoff [379] represent chromosomes using directed acyclic graphs (DAGs), to account for poor resolution and missing data. The genome rearrangement problem in that setting is then to compute a minimum number of translocations and reversals for transforming a set of linear extensions (one for each chromosome) into a set of linear extensions (one for each chromosome) of another species.

- Introduced by Zheng and Sankoff [379].
- Complexity: Unknown.

Zheng and Sankoff [379] proposed a heuristic that, as in section 5.4, consists in embedding the set of all possible linear extensions of a poset into a directed graph by appropriately augmenting the arc set of the DAG naturally associated with the poset. These two sets of directed graphs that represent two genomes are then used to produce a single large bicolored graph from which a maximal decomposition into alternating cycles is extracted.

11 Cycles of a Permutation

In the first part of this book, we have seen how permutations are used to represent the relative order of markers along one chromosome, compared to another. Alternative ways have been proposed by Meidanis and Dias [264] and others after them (see, e.g., Lin et al. [254, 256]; Mira and Meidanis [269]; Lu et al. [259]). These representations provide other views of some genome rearrangement problems, and make it possible to retrieve the structure of the cycle graph or of the breakpoint graph of a permutation (signed or unsigned). Some results proved using other models have been re-proved with these techniques, which have occasionally led to new results.

11.1 A Model for Multichromosomal Circular Genomes

If we want to handle genomes containing several circular chromosomes that can exchange genes, such that each gene appears only once in a given genome, then each chromosome can be represented by a cyclic ordering of its genes, and the genome is represented by the permutation whose disjoint cycle decomposition is the set of chromosomes. An example is given below.

$$\underbrace{(1,9,12,20,3,10,17)}_{\text{chromosome 1}}\underbrace{(2,5,14,22,24,7,6,4,13,15,8,18)}_{\text{chromosome 2}}\underbrace{(11,19,23,16,21)}_{\text{chromosome 3}}$$

This disjoint cycle representation corresponds to the following permutation in the standard representation:

$\pi = (9\ 5\ 10\ 13\ 14\ 4\ 6\ 18\ 12\ 17\ 19\ 20\ 15\ 22\ 8\ 21\ 1\ 2\ 23\ 3\ 11\ 24\ 16\ 7)$.

Genome rearrangement problems are often proved to be equivalent for circular and linear chromosomes when there is only one chromosome (see, e.g., Hartman and Shamir [202]; Lin et al. [254]), but this correspondence does not always hold for multichromosomal genomes (see, for example, chapter 10 for the breakpoint median problem, which is polynomial if circular chromosomes are allowed, and **NP**-complete otherwise).

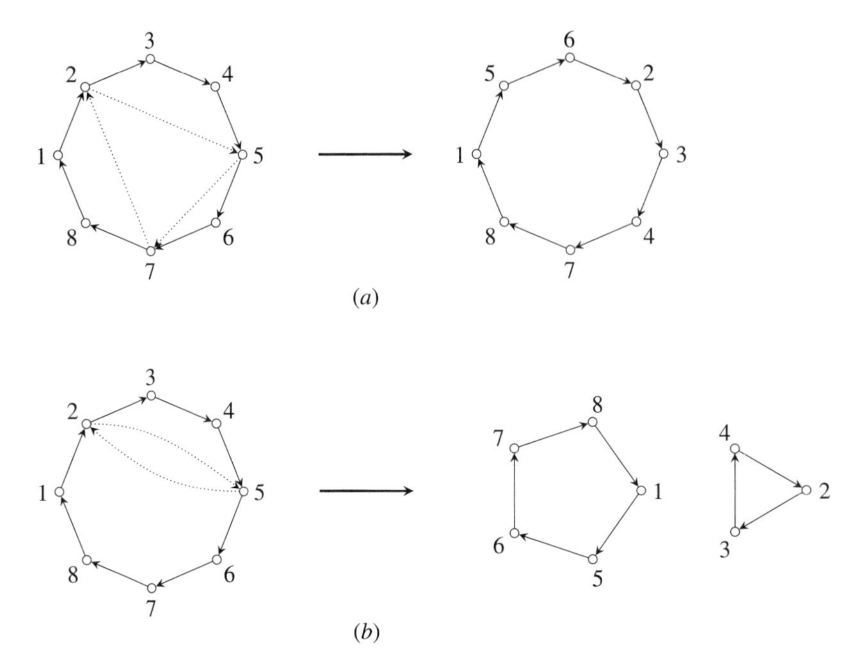

Figure 11.1
Rearrangements on unsigned circular chromosomes: (*a*) a transposition, (*b*) a fission whose inverse is a fusion

In this model, fusions and fissions are 2-cycle permutations (i.e., they consist of a single nontrivial cycle of length 2): a fusion acts on two elements belonging to distinct cycles, whereas a fission acts on two elements of the same cycle. Moreover, 3-cycles of the form (π_i, π_j, π_k), where $i < j < k$ and π_i, π_j, and π_k belong to the same chromosome, represent transpositions. To see this, simply denote one of these 2- or 3-cycles by ρ, and perform the composition $\rho \circ \pi$, where π is an arbitrary permutation. These operations are illustrated on the permutation $(1, 2, 3, 4, 5, 6, 7, 8)$ in figure 11.1.

Therefore, a minimum-length factorization into 2-cycles of a permutation $\sigma^{-1}\pi$ for two permutations π, σ in S_n may be seen as an optimal rearrangement scenario transforming π into σ using fusions and fissions, which is easily computed in linear time since it corresponds to the Cayley distance $exc(\pi)$ (see section 3.5.2).

Dias and Meidanis [137] consider a rearrangement problem using fissions, fusions, and transpositions as expressed above, where transpositions weigh twice as much as fusions and fissions. They prove that the problem essentially remains equivalent to computing the Cayley distance, and is therefore solvable in $O(n)$ time for computing the distance and in $O(n^2)$ time for giving a rearrangement scenario. They observe

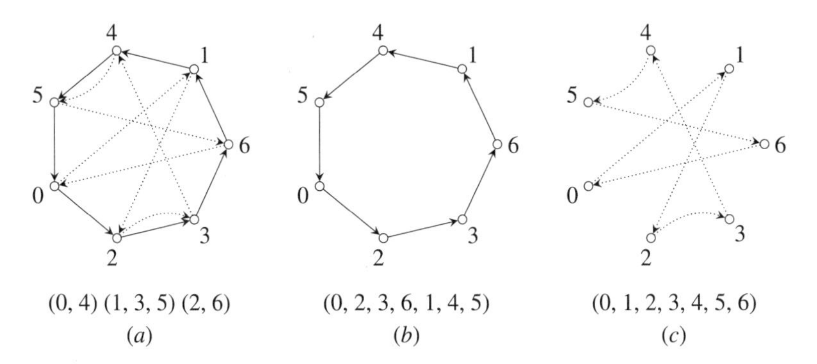

Figure 11.2
(*a*) The circular cycle graph of (5 4 1 6 3 2) and the corresponding permutation $\overline{(5\ 4\ 1\ 6\ 3\ 2)}$; (*b*) and (*c*) are the two cycles formed, respectively, by the black edges and by the gray edges, with the corresponding permutations

that their solution remains valid when assigning an arbitrary larger weight to transpositions.

It is sometimes possible to recover some solutions developed with combinatorial objects in the first part of this book with this alternative formalism. Formulas are very similar, though the objects are defined independently. This is because the manipulated objects are usually the same, and therefore it is possible to obtain the usual combinatorial structure with the disjoint cycle representation.

For example, Doignon and Labarre [147] proposed an alternative representation of the cycle graph of a permutation (see definition 3.3) that simply consists in representing the cycle graph $G(\pi)$ of a permutation π in S_n, using the disjoint cycle decomposition of the following permutation:

$$\bar{\pi} = (0, \pi_n, \pi_{n-1}, \ldots, \pi_1) \circ (0, 1, 2, \ldots, n). \tag{11.1}$$

This model is based on a circular representation of $G(\pi)$ in which vertices 0 and $n+1$ are identified (clearly, this does not affect the alternating cycle decomposition). The cycle $(0, \pi_n, \pi_{n-1}, \ldots, \pi_1)$ represents the cycle formed by the black edges, and the cycle $(0, 1, 2, \ldots, n)$ represents the cycle formed by the gray edges; the cycles in the disjoint cycle decomposition of $\bar{\pi}$ are exactly the alternating cycles of $G(\pi)$. Figure 11.2 illustrates this by showing the circular version of the cycle graph of figure 3.1 and the corresponding permutations (i.e., $\bar{\pi}$, $(0, \pi_n, \pi_{n-1}, \ldots, \pi_1)$ and $(0, 1, 2, \ldots, n)$). It can be easily checked that $\bar{\pi} = (0, 2, 3, 6, 1, 4, 5) \circ (0, 1, 2, 3, 4, 5, 6) = (0, 4)(1, 3, 5)(2, 6)$; the cycles of $\bar{\pi}$ list the vertices that are encountered in each alternating cycle of $G(\pi)$ after following a gray edge, then a black edge.

A first use of this model is the construction of a bijection between cycle graphs and factorizations of $(0, 1, 2, \ldots, n)$ into the product of an $(n+1)$-cycle and a permutation that decomposes into k cycles, easily deduced from equation (11.1). Thanks to this bijection, the problem of enumerating permutations in S_n with $c(G(\pi)) = k$ is solved, and this fully characterizes the distribution of the block-interchange distance (see section 3.5.1).

A second use of this model is the reformulation of *any* rearrangement problem in terms of minimum-length factorizations of $\bar{\pi}$ into the products of particular permutations. Recall that sorting a permutation π using a set X of given allowed operations is equivalent, if the inverse of an allowed operation belongs to X as well, to finding a minimum-length factorization of π into the product of elements of X (see section 2.4). The following result reformulates that problem in terms of factorizations of $\bar{\pi}$.

Theorem 11.1 [241] Let $X = \{s_1, s_2, \ldots\}$ be a subset of S_n whose elements are mapped onto $X' = \{\overline{s_1}, \overline{s_2}, \ldots\} \subseteq A_{n+1}$, using equation (11.1). Moreover, let $\mathcal{C}$ be the union of the conjugacy classes (of S_{n+1}) that intersect with X'; then for any π in S_n, any factorization of π into t elements of X yields a factorization of $\bar{\pi}$ into t elements of $\mathcal{C}$.

Clearly, theorem 11.1 directly provides a way to obtain lower bounds on our rearrangement problems. Previous results such as lower bounds on sorting by transpositions (theorem 3.4) and by block interchanges (theorem 3.21) can be easily recovered, and new results such as the lower bound of theorem 3.14 on sorting by prefix transpositions can be obtained (see Labarre [241]).

11.2 A Generalization to Signed Genomes

11.2.1 A Different Kind of Signed Permutation

Meidanis and Dias [264] generalized their model to signed genomes by representing both strands of DNA molecules in the disjoint cycle decomposition. That is, the permutation π has base set $\{-n, \ldots, n\} \backslash \{0\}$, and verifies $\pi_i = -\pi^{-1}_{-i}$. Note that this constraint is different from the usual constraint on signed permutations, so that the permutation group is different from the hyperoctahedral group, where a signed permutation verifies $\pi_{-i} = -\pi_i$ (see section 2.3). This implies that a cycle of the permutation cannot contain both a number and its opposite. The disjoint cycles of this permutation model the gene order on the circular chromosomes of a genome, but each chromosome is represented by two cycles, modeling its two strands.

An example is given below.

$$\underbrace{(1,-9,3)(9,-1,-3)}_{\text{chromosome 1}}\;\underbrace{(2,-7,6,-4,8)(-2,-8,4,-6,7)}_{\text{chromosome 2}}\;\underbrace{(10,-5)(5,-10)}_{\text{chromosome 3}}$$

This disjoint cycle representation corresponds to the following permutation in the standard two-row notation:

$$\begin{pmatrix} -10 & -9 & -8 & -7 & -6 & -5 & -4 & -3 & -2 & -1 & 1 & 2 & 3 & 4 & 5 & 6 & 7 & 8 & 9 & 10 \\ 5 & 3 & 4 & 6 & 7 & 10 & 8 & 9 & -8 & -3 & -9 & -7 & 1 & -6 & -10 & -4 & -2 & 2 & -1 & -5 \end{pmatrix}.$$

Let π and σ be two signed multichromosomal circular genomes (or permutations of $\{-n, \ldots, n\} \backslash \{0\}$). Let γ be the permutation $(-0, +0)(-1, +1)\ (-2, +2) \cdots (-n, +n)$. Then both $\gamma \circ (0\ \pi_1\ \cdots\ \pi_n)$ and $\gamma \circ (0\ \sigma_1\ \cdots\ \sigma_n)$ are permutations that decompose into 2-cycles, each of which corresponds to the following:

1. a π-edge of the breakpoint graph of the two genomes π and σ, in the case of $\gamma \circ \pi$, and
2. a σ-edge of the breakpoint graph of the two genomes π and σ, in the case of $\gamma \circ \sigma$.

The permutation $\gamma \circ \pi \circ \gamma \circ \sigma$ gives a "square" decomposition of the breakpoint graph of π (with respect to σ) into alternating cycles, which means that each alternating k-cycle in the breakpoint graph of π with respect to σ corresponds to two k-cycles in the disjoint cycle decomposition of $\gamma \circ \pi \circ \gamma \circ \sigma$.

11.2.2 The Operations

In this representation, we have to redefine all rearrangements we have considered so far or, more accurately, to model them differently. If u and v are distinct elements of the same cycle of π, then

1. the permutation $(u, \gamma \circ \pi(v))(v, \gamma \circ \pi(u))$ is the (signed) reversal of the segment with extremities u and v,
2. an unsigned reversal is defined as $(u, v)(\gamma \circ \pi(u), \gamma \circ \pi(v))$,
3. a transposition is the permutation $(u, v, w)(\gamma \circ \pi(w), \gamma \circ \pi(v), \gamma \circ \pi(u))$,
4. a block interchange is the permutation $(u, w)(\gamma \circ \pi(w), \gamma \circ \pi(u))(v, x)(\gamma \circ \pi(x), \gamma \circ \pi(v))$.

Additionally, a fusion is a permutation $(u, \gamma \circ \pi(v))(v, \gamma \circ \pi(u))$ where u and v are not in the same cycle, and a fission is a permutation $(u, v)(\gamma \circ \pi(v), \gamma \circ \pi(u))$.

11.2.3 Some Results

One of Meidanis and Dias's [264] goals in developing these models was to prove in an algebraic way some previous results that were proved using graph-theoretic arguments. This approach is nice because a lot of results such as sorting by block interchanges, or reversals and block interchanges, and in general k-break rearrangements, can be expressed in this way.

These techniques are less powerful in some cases (for example, the result of Hannenhalli and Pevzner [199] on sorting signed permutations by reversals (page 64) has no equivalent, due to the impossibility of reformulating hurdles). But this formulation allows us to solve ad-hoc variants as sorting by fusions, fissions, and transpositions (see, for example, Dias and Meidanis [137]). Some results are proved again in this formalism, such as sorting by DCJ rearrangements (Meidanis and Dias [269]; Lu et al. [259]), providing an alternative proof of the result of Yancopoulos et al. [375].

The same formalism has been adopted by Lin et al. [256, 254], Lu et al. [259], and Lin and Xue [252], who prove the same results again, and work on minimizing the number of block interchanges in a scenario sorting a signed permutation by reversals and block interchanges.

12 Set Systems and the Syntenic Distance

We have seen that some models assume that the order of genes on chromosomes is either known (chapter 2 and subsequent chapters) or partially known (chapter 5). We will now examine the case in which the order of the genes is not known at all. This assumption makes no sense for unichromosomal genomes, but a distance between multichromosomal genomes can be studied, which takes into account only the information about which gene belongs to which chromosome. This distance is known as the *syntenic distance*. The model presented in this chapter disregards the order of genes: the only relevant information is how genes are scattered among chromosomes.

- Introduced by Ferretti et al. [174].
- Complexity: **NP**-hard (see DasGupta et al. [131]). If the distance is bounded by d, then it can be computed exactly in $O(hk + 2^{O(d \log d)})$ time, where h and k are the number of chromosomes of the two input genomes (see Liben-Nowell [249]).
- Best approximation ratio: 2. Such approximations have been proposed by DasGupta et al. [131], Ferretti et al. [174], and Liben-Nowell [248]; for all of them, the ratio is tight.
- Diameter: For two genomes containing, respectively, h and k chromosomes (with $h \geq k \geq 4$), the diameter is equal to $h + k - 4$ (see Kleinberg and Liben-Nowell [236] or Pisanti and Sagot [300]).

12.1 Introduction

Throughout this chapter, a genome Π is a partition of the set of genes into h non-empty chromosomes, that is, $\Pi = \{C_1, C_2, \ldots, C_h\}$. For completeness, we note that the order of the chromosomes in the set $\{C_1, C_2, \ldots, C_h\}$ is not important for the syntenic distance problem (as we will see, the rearrangement events considered in this problem do not take the order of chromosomes into account). We will refer to the number h of chromosomes in Π as its **size**.

One can see that, in that model, any intrachromosomal event in any chromosome C_i does not modify C_i. Hence, only interchromosomal events are to be considered; the operations that will be taken into account are fusions, fissions, and translocations (see page 159).

Adapted to the syntenic model of genomes, let C_i, C_j, C_q, and C_l be four sets of genes (chromosomes), such that (1) $C_i \cup C_j = C_q \cup C_l$, and (2) at most one of C_i, C_j, C_q, C_l is empty. Then the three aforementioned moves can be summarized as

$$\{C_i, C_j\} \to \{C_q, C_l\},$$

where

- a **fusion** corresponds to the case where either $C_q = \emptyset$ or $C_l = \emptyset$,
- a **fission** corresponds to the case where either $C_i = \emptyset$ or $C_j = \emptyset$, and
- a **reciprocal translocation** corresponds to the case where none of the four sets is empty and where $\{C_i, C_j\} \neq \{C_q, C_l\}$.

Here translocations always mean reciprocal translocations.

Definition 12.1 The **syntenic distance** between two genomes Π_1 and Π_2, denoted by $\text{syd}(\Pi_1, \Pi_2)$, is the minimum number of fusions, fissions, and translocations needed to transform Π_1 into Π_2.

Note that the term *syntenic* comes from the fact that two genes are said to be **in synteny** if they belong to the same chromosome.

12.2 Structural Properties

The syntenic distance problem was introduced by Ferretti et al. [174], but the first paper that studies the problem from a formal algorithmic and combinatorial point of view is that of DasGupta et al. [131], who in particular prove that the syntenic distance satisfies the three axioms of definition 1.1.

12.2.1 Compact Representation

The *compact representation* of a syntenic instance was first described by Ferretti et al. [174], then formalized by DasGupta et al. [131]. This is in analogy with the left-invariance property in the permutation model, where computing a given distance is equivalent to a sorting problem (see definition 2.11); in the synteny model there is no order, but the compact representation allows for a similar concept.

Definition 12.2 The **compact representation** of a genome Π_1 with respect to a genome Π_2 is obtained by replacing each C'_i in Π_2 with its index i, and each C_j in Π_1 with the set of indices of the chromosomes on which each gene of C_j lies in Π_2.

As an example, consider the following two genomes:

$$\Pi_1 = \{\overbrace{\{a,b,e,f,h\}}^{C_1}, \overbrace{\{c,g\}}^{C_2}, \overbrace{\{d,i,j\}}^{C_3}\},$$

$$\Pi_2 = \{\underbrace{\{a,b,c,d\}}_{C_1'}, \underbrace{\{e,f\}}_{C_2'}, \underbrace{\{g,h,i\}}_{C_3'}, \underbrace{\{j\}}_{C_4'}\}.$$

Then the compact representation of Π_1 with respect to Π_2 consists of $\Pi_1' = \{\{1,2,3\},\{1,3\},\{1,3,4\}\}$ and $\Pi_2' = \{\{1\},\{2\},\{3\},\{4\}\}$.

Note that, in the compact representation, fissions or translocations can possibly create duplicates, which could be problematic since duplicates are not allowed in the definition of a genome in this model. For instance, consider the following fission in the original instance: $\{a,b,e,f,h\} \to \{\{a,e,f\},\{b,h\}\}$. In the compact representation, this fission becomes $\{1,2,3\} \to \{\{1,2\},\{1,3\}\}$, and in that case, element 1 appears twice in the output. Similarly, the following translocation in the original instance, $\{\{a,b,e,f,h\},\{d,i,j\}\} \to \{\{a,e,h,i,j\},\{b,d,f\}\}$, becomes $\{\{1,2,3\},\{1,3,4\}\} \to \{\{1,2,3,4\},\{1,2\}\}$ in the compact representation.

However, it is easy to see, as mentioned by DasGupta et al. [131], that in any optimal move sequence in the compact representation, *no element is present in both output sets of a translocation or a fission* (this was formally proved later by Liben-Nowell [248], where such a move was called a *redundant* move). Thus, as long as we are concerned with computing the distance (that is, the length of a shortest move sequence), fissions and translocations do not create duplicates, and the problem remains well defined. Since we are interested here in such optimal moves, we can always assume that, after any move, no element is duplicated. Under this assumption, we are now able to restate the synteny problem in the compact representation, as follows.

Definition 12.3 Given a collection $\mathcal{S}(h,k)$ of k (not necessarily distinct) subsets $S_1,\ldots,S_k$ of $\{1,\ldots,h\}$, the **synteny problem** seeks to compute the minimum number of translocations, fusions, and fissions required to transform $\mathcal{S}(h,k)$ into the collection $\{\{1\},\{2\},\ldots,\{h\}\}$. This number is denoted by $syd(\mathcal{S}(h,k))$ and is called the **syntenic distance** of $\mathcal{S}(h,k)$.

According to definition 12.2, there are two ways to construct the compact representation of two genomes Π_1 and Π_2 of respective sizes h and k ($h \geq k$):

- either Π_1 becomes $\Pi_1' = \{\{1\},\{2\},\ldots,\{h\}\}$ and Π_2 is modified accordingly, or
- Π_2 becomes $\Pi_2' = \{\{1\},\{2\},\ldots,\{k\}\}$ and Π_1 is modified accordingly.

The first possibility is called the **dual** of the second, and vice versa. One might wonder whether the resulting compact instances are equivalent; DasGupta et al. [131] proved the following result.

Proposition 12.1 [131] Let $\mathcal{S}(h,k)$ be a synteny instance and $\mathcal{S}'(h,k)$ be its dual. Then $syd(\mathcal{S}(h,k)) = syd(\mathcal{S}'(h,k))$.

Consequently, in the following, starting with two genomes Π_1 and Π_2 of respective sizes k and h ($h \geq k$), we will focus only on the compact representation $\mathcal{S}(h,k)$, where we wish to obtain $\{\{1\},\{2\},\ldots,\{h\}\}$ using the minimum number $syd(\mathcal{S}(h,k))$ of moves.

Liben-Nowell [248] proved that the syntenic distance is *monotonic* in the following sense.

Proposition 12.2 [248] Let $\mathcal{S}(h,k) = \{S_1, S_2, \ldots, S_k\}$ and $\mathcal{T}(h',k) = \{T_1, T_2, \ldots, T_k\}$ be two instances of the synteny problem such that, for all $1 \leq i \leq k$, $S_i \subseteq T_i$. Then $syd(\mathcal{T}(h',k)) \geq syd(\mathcal{S}(h,k))$.

Finally, DasGupta et al. [131] showed the existence of a *canonical form* for the syntenic distance.

Lemma 12.1 [131] There always exists an optimal sequence of moves between two sets in which all fusions occur first, then all translocations, and finally all fissions.

12.3 Lower Bounds

Just as for other rearrangement problems described so far, a graph-theoretic model was introduced (by DasGupta et al. [131] in this case) to attack the problem at hand and to successfully obtain bounds and approximation algorithms.

Definition 12.4 The **synteny graph** of an instance $\mathcal{S}(h,k)$, denoted by $SG(\mathcal{S}(h,k))$, contains a vertex for each set of $\mathcal{S}(h,k)$, and two distinct vertices are joined by an edge if the corresponding sets have a nonempty intersection.

Proposition 12.3 [131] For any synteny instance $\mathcal{S}(h,k)$, we have

$$syd(\mathcal{S}(h,k)) \geq h - p,$$

where p is the number of connected components in $SG(\mathcal{S}(h,k))$.

Indeed, let $\mathcal{S}(h) = \{\{1\},\{2\},\ldots,\{h\}\}$, and let $SG(\mathcal{S}(h))$ be its synteny graph. Clearly, $SG(\mathcal{S}(h))$ is a collection of h isolated vertices (i.e., it has h connected components). Moreover, after any given move (fusion, fission, or translocation), the number of components in the synteny graph cannot increase by more than 1. Therefore, starting with p components, it takes at least $h - p$ moves to reach h components.

Another lower bound is given by Liben-Nowell [248]. Let $\mathcal{S}(h,k)$ be an instance, and for any gene g, let $count(g)$ denote the number of sets of $\mathcal{S}(h,k)$ in which g appears.

Theorem 12.1 [248] For any instance of synteny $\mathcal{S}(h,k)$, we have

$$syd(\mathcal{S}(h,k)) + \log_{4/3}(syd(\mathcal{S}(h,k))) \geq h - 1 + \max_{1 \leq c \leq k-1} \{c - |\{g \mid count(g) \leq c+1\}|\}.$$

12.4 Diameter

Kleinberg and Liben-Nowell [236], as well as Pisanti and Sagot [300], studied the maximal value of the syntenic distance, which we will refer to as the *syntenic diameter*.

Theorem 12.2 [236, 300] For any $h \geq k \geq 4$, the syntenic diameter is $h + k - 4$.

It is relatively easy to see that the diameter is reached, for example, by $\mathcal{S}(h,k) = \{\{1,2,\ldots,h\},\{1,2,\ldots,h\},\ldots,\{1,2,\ldots,h\}\}$. Liben-Nowell [249] showed an interesting relationship between the syntenic distance problem and the so-called *gossiping* problem (see Hedetniemi et al. [211] for a survey on the gossiping problem). This relationship was extensively used by Chauve and Fertin [103] to characterize all instances $\mathcal{S}(h,k)$ that reach the diameter, for any $h \geq k$. Their result also implies the existence of a polynomial-time algorithm that determines whether a given instance $\mathcal{S}(h,k)$ reaches the diameter.

12.5 Algorithmic Results

12.5.1 Syntenic Distance

DasGupta et al. [131] proved that computing the syntenic distance is an **NP**-hard problem, by a transformation from LARGEST BALANCED INDEPENDENT SET.

Proposition 12.4 [131] There exists a polynomial-time approximation algorithm $\mathcal{H}$ for computing $syd(\mathcal{S}(h,k))$ whose ratio equals 2. Its running time is $O(hkA^{-1}(hk,k))$, where A^{-1} is the inverse of Ackermann's function.

For more information about Ackermann's function, see, for instance, Cormen et al. [121]. Algorithm $\mathcal{H}$ works as follows. Take the synteny graph $SG(\mathcal{S}(h,k))$ and let CC_i be its p connected components, for $1 \leq i \leq p$. Let k_i be the number of vertices of CC_i, and remember that $\sum_{i=1}^{p} k_i = k$. For each $1 \leq i \leq p$, operate $k_i - 1$ fusions within CC_i. We then end up with p isolated vertices in the synteny graph, which represent p sets. Now, operate $h - p$ fissions on those p sets to obtain the h singletons $\{1\},\{2\}\ldots\{h\}$. Altogether, we have used $\sum_{i=1}^{p}(k_i - 1) = k - p$ fusions and $h - p$ fissions, and thus $h + k - 2p$ moves. However, by proposition 12.3 we know that $syd(\mathcal{S}(h,k)) \geq h - p$. Since $h \geq k$ by hypothesis, we conclude that algorithm $\mathcal{H}$ is a 2-approximation algorithm.

Earlier, Ferretti et al. [174] also provided a heuristic algorithm, $\mathcal{F}$, without analyzing its approximation ratio. Liben-Nowell [248] undertook this study and proved that (1) $\mathcal{F}$ always outperforms $\mathcal{H}$, and (2) $\mathcal{F}$ has ratio 2. Liben-Nowell also showed the following surprising result.

Theorem 12.3 [248] Any algorithm that works only within connected components of the synteny graph cannot have an approximation ratio better than 2.

Indeed, it can be easily seen that solving the following instance using only intracomponent moves requires at least $2h-4$ moves ($h-2$ fusions and $h-2$ fissions): $\mathcal{S}(h,h) = \{\{1,2,\ldots,h-1\},\{h\},\{h\},\ldots,\{h\}\}$. However, the above instance can be solved in $h-1$ moves (more precisely, $h-1$ translocations), thus yielding a ratio of $2-\varepsilon$, for any $\varepsilon > 0$.

Since both $\mathcal{H}$ and $\mathcal{F}$ use only intracomponent moves, we conclude by theorem 12.3 that their approximation ratio 2 is tight. Liben-Nowell [249] gives a fixed-parameter algorithm for determining $syd(\mathcal{S}(h,k))$, where the parameter d is an upper bound on $syd(\mathcal{S}(h,k))$.

Theorem 12.4 [249] Whenever $syd(\mathcal{S}(h,k)) \leq d$, there exists an $O(hk + 2^{O(d \log d)})$ algorithm that determines the exact value of $syd(\mathcal{S}(h,k))$.

The above theorem improves the result of DasGupta et al. [131], where an $O(hk + 2^{O(d^2)})$ algorithm was given. Note that since $syd(\mathcal{S}(h,k)) = O(h)$ (indeed, $syd(\mathcal{S}(h,k)) \leq h+k-4$; see theorem 12.2), the above result implies an $O(2^{O(h \log h)})$ exact algorithm for computing the syntenic distance of any instance $\mathcal{S}(h,k)$.

12.5.2 Easy Cases

Though many structural properties and some algorithmic results are known for the syntenic distance problem, few positive results exist. Hence, some authors have undertaken the study of restricted versions of synteny. In this section, we summarize the main results that have been obtained for linear synteny, nested synteny (in their linear and exact restrictions), uncovering synteny, and synteny using only fusions and fissions.

12.5.2.1 Linear Synteny DasGupta et al. [131] introduced a restricted version of the problem, called **linear synteny**. Linear synteny differs from "general" synteny in that the set of moves is predefined, and each of them must appear in a specific order. More precisely:

1. the first $k-1$ moves must be either fusions or translocations that produce a singleton that does not appear in any other set, and
2. all following moves must be fissions that create singletons.

The associated distance of an instance $\mathcal{S}(h,k)$ under these restrictions is denoted by $lsyd(\mathcal{S}(h,k))$.

DasGupta et al. [131] proved that the linear synteny problem is **NP**-hard as well; moreover, algorithm $\mathcal{H}$ (see proposition 12.4) yields a 2-approximation for that problem. Here are some other results. The first one, theorem 12.5, motivates the study of linear synteny.

Theorem 12.5 [131] If linear synteny can be approximated within ratio c in polynomial time, then for any $\varepsilon > 0$, general synteny can be approximated within ratio $c + \varepsilon$ in polynomial time.

Later, Liben-Nowell and Kleinberg [250] studied structural properties of linear synteny, and obtained results similar to those in the general case: monotonicity, duality, and canonicity. Upper and lower bounds on $lsyd(\mathcal{S}(h,k))$ have also been given, as stated below.

Theorem 12.6 [248] For any instance of synteny $\mathcal{S}(h,k)$,

$$lsyd(\mathcal{S}(h,k)) \geq h - 1 + \max_{1 \leq c \leq k-1} \{c - |\{g \mid count(g) \leq c + 1\}|\}.$$

The following result relates the linear syntenic distance to the general syntenic distance.

Lemma 12.2 [131] Let $\mathcal{S}(h,k)$ be an instance of synteny. Then

$$lsyd(\mathcal{S}(h,k)) \leq syd(\mathcal{S}(h,k)) + \log_{4/3} syd(\mathcal{S}(h,k)).$$

Lemma 12.2 and theorem 12.6 were used to obtain the lower bound on the general syntenic distance given in theorem 12.1. Finally, Liben-Nowell [248] studied the diameter of the linear syntenic distance in the case $h = k$, and obtained the following result.

Theorem 12.7 [248] For any $h \geq 4$, the maximal linear syntenic distance between two h-sized genomes is $2h - 3$.

12.5.2.2 Linear Nested Synteny

Definition 12.5 An instance $\mathcal{S}(h,k) = \{S_1, S_2, \ldots, S_k\}$ is said to be **nested** if, for all $1 \leq i \neq j \leq k$, one of the three following cases occurs:

- $S_i \cap S_j = \emptyset$,
- $S_i \subseteq S_j$, or
- $S_j \subseteq S_i$.

The problem of determining $lsyd(\mathcal{S}(h,k))$ in the specific case of *nested* instances is solvable in polynomial time. This result was initially proved by Liben-Nowell and Kleinberg [250], who gave an algorithm with time complexity $O(hk^2 + pk^2 \log k)$ (recall that p is the number of components of the synteny graph). Later, Pisanti and Sagot [300] gave a more efficient algorithm, running in $O(hk + k^2 + pk \log^2 k)$ time.

12.5.2.3 Exact Nested Synteny

Definition 12.6 An instance $\mathcal{S}(h,k) = \{S_1, S_2, \ldots, S_k\}$ is said to be **exact** if $syd(\mathcal{S}(h,k)) = h + k - 1$. The exact synteny problem is thus the problem of determining whether a general synteny instance is exact.

Though the exact synteny problem is **NP**-hard (a direct consequence of DasGupta et al.'s [131] result for the general synteny problem), the problem restricted to *nested* instances turns out to be solvable in polynomial time. Pisanti and Sagot [300] give an $O(hk + k^2 + k \log k)$ algorithm for solving the exact nested synteny problem.

12.5.2.4 Uncovering Synteny More recently, Ting and Yong [358] studied the **uncovering synteny** problem.

Definition 12.7 An instance $\mathcal{S}(h,k) = \{S_1, S_2, \ldots, S_k\}$ (with $h \geq k \geq 4$) is said to be **uncovering** if there exists an ordering of the sets S_j, for $1 \leq j \leq k$, such that $S_i \backslash (S_{i+1} \cup S_{i+2} \cup \cdots \cup S_k) \neq \emptyset$ for all $1 \leq i \leq k-3$.

Ting and Yong [358] note that determining whether a syntenic instance is uncovering can be achieved in polynomial time. Their main result is the following: for any uncovering instance, there exists an $O(hk)$ algorithm for computing $syd(\mathcal{S}(h,k))$. Actually, for any uncovering instance, they show that either $syd(\mathcal{S}(h,k)) = h - p$ or $syd(\mathcal{S}(h,k)) = h - p + 1$, where p is the number of connected components in the synteny graph.

12.5.2.5 Fusions and Fissions Only Dias and Meidanis [139] studied another restricted case of synteny, in which only fusions and fissions are allowed. Let $syd_{ff}(\mathcal{S}(h,k))$ be the distance under this restriction. Dias and Meidanis [139] proved the following two results.

Theorem 12.8 [139] For any $h \geq k$, $syd_{ff}(\mathcal{S}(h,k)) = h + k - 2p$, where p is the number of connected components of the synteny graph $SG(\mathcal{S}(h,k))$.

As in proposition 12.4, let $A^{-1}(p,q)$ denote the inverse of Ackermann's function for any two integers p, q.

Theorem 12.9 [139] For any $h \geq k$ and any syntenic instance $\mathcal{S}(h,k)$, there exists an $O(h^2 + hkA^{-1}(hk, h))$ algorithm that finds an optimal sequence of $syd_{ff}(\mathcal{S}(h,k)) = h + k - 2p$ fusions and fissions to solve this instance.

12.6 Conjectures and Open Problems

The main open problem regarding general synteny is to obtain an approximation algorithm with a better ratio.

Problem 12.1 [131] Devise an approximation algorithm with ratio strictly smaller than 2 for computing the syntenic distance, or show that this ratio cannot be improved.

Other problems are related to the variant in which only translocations are allowed. Clearly, this variant can apply only to *square instances* (i.e., instances with $h = k$). This problem is mentioned and considered by Liben-Nowell [249], as well as Pisanti and Sagot [300], but there exist few results apart from those concerning the diameter.

Problem 12.2 What is the complexity of the syntenic distance problem using only translocations?

We stress that this is not the same problem as that of sorting by unsigned translocations discussed in section 10.4.2, since the underlying objects on which translocations are performed differ. Besides, Liben-Nowell [249] made the following interesting remark: there are instances for which fusions and fissions are needed to reach optimality. For instance, he gave the following square instance:

$$\mathcal{S}(h,k) = \{\{1,2,3,4\},\{1,2,3,4\},\{1,2,3,4\},\{5,6,7\},\{5,6,7\},\{5,6,7\},\{5,6,7\}\},$$

for which eight translocations are needed, but only seven moves suffice: one fusion, five translocations, and one fission. This raises the following question.

Problem 12.3 [103] Is it possible to characterize those syntenic instances that can be optimally solved by translocations only?

Another variant could also be considered, in which genes could be duplicated along the genome. To our knowledge, there exists no result on the subject.

Problem 12.4 What can be said about the variant where chromosomes do not form a partition of the set of genes (i.e., genes can appear in several chromosomes of a genome)?

It should be noted that in this case, the compact representation is no longer suitable. Take, for instance, $\Pi_1 = \Pi_2 = \{\{a,b\},\{a,c\}\}$. The compact representation yields $\Pi_1' = \{\{1,2\},\{1,2\}\}$ and $\Pi_2' = \{\{1\},\{2\}\}$, since gene a belongs to both chromosomes. The syntenic distance between Π_1 and Π_2 is clearly 0, but the compact representation does not reflect it anymore.

IV MULTIGENOMIC MODELS

A natural generalization of genome rearrangement problems, if not their ultimate goal, is the reconstruction of the evolutionary events that explain the differences and the relations between more than two species. This leads to the inference of common ancestor configurations and eventually of *phylogenies*, which are trees depicting the kin relationships between organisms or between species (examples will appear in figures 14.1 and 14.2).

Many techniques can be used to reconstruct a phylogeny from gene order data; for instance, *distance-based* phylogenetic reconstruction methods compute all pairwise distances between the input genomes (according to any method we have mentioned so far, although sequence alignment has come to be the de facto standard), and then use the resulting distance matrix to infer a tree. A lot of methods (whether they are distance-based or not) are available for reconstructing trees and assessing their accuracy, but reviewing them is far beyond the scope and point of this book. We refer the reader to Felsenstein [170] for more information on phylogenetic inference.

In this part, we are interested in methods for inferring evolutionary events, given a phylogenetic tree, or for inferring both the events and the tree, minimizing the number of evolutionary events at the same time. In both cases, we obtain the configuration of ancestral genomes as a by-product of the reconstruction of rearrangement scenarios, which is not possible when the comparison is limited to two genomes.

The computational complexity of many combinatorial optimization problems usually increases when the number of objects that are taken into account increases from two to three, and genome rearrangement problems are no exception: as soon as there are three genomes to compare, most problems we have examined so far become **NP**-hard, even when the rearrangement problem on two genomes is trivial. Therefore, chapter 13 first focuses on the so-called *median problem*, which consists in reconstructing an ancestral configuration of two genomes, given a third as an out-group. Still, the three genomes are used in a symmetrical way, and the out-group has no special status in the instance. The median problem can be used as a hint for more general problems with more than three genomes: either for reconstructing ancestral configurations, given a phylogenetic tree (a problem known as the "small parsimony" problem), or for inferring the phylogenetic tree, the ancestral genomes, and the evolutionary events at the same time (the "large parsimony" problem). Some variants of these ancestral genome reconstruction techniques rely on the assumption that a genome in the instance has undergone a whole genome duplication, which means that it contains two copies of each gene. We survey those "halving" problems at the end of chapter 13.

Finally, chapter 14 reviews all problems that deal with more than three genomes. It contains the most general formulations of the genome rearrangement problems we have investigated, and is probably the most relevant part of this book for readers interested in biological applications. Few theoretical results are specific to this chapter: it uses all results we have previously reviewed, including those presented in chapter 13, to design heuristics for inferring phylogenies.

13 Median and Halving Problems

Given a rearrangement distance d and three genomes Π_1, Π_2, and Π_3, the **median problem** consists in finding a fourth genome Π_M, the **median genome**, such that the sum of the distances between Π_M and each other genome is minimized. Given a genome Π_M, the quantity to minimize is therefore

$$M = d(\Pi_1, \Pi_M) + d(\Pi_2, \Pi_M) + d(\Pi_3, \Pi_M),$$

which is called the **median score** of Π_M.

Genome median problems are used to infer ancestral configurations (see Bourque and Pevzner [81]), or as a hint for phylogenetic inference methods (see chapter 14). They can also be used to infer statistics on the rearrangement rates on different lineages (see Bourque et al. [84]). Solutions to median problems have also been used to model a consensus genome when different sources are inconsistent (see Jackson et al. [218]).

As in pairwise genome rearrangement problems, a number of choices can be made as to how genomes are modeled and which distance is chosen, spawning just as many different combinatorial optimization problems. However, a universal lower bound on the optimal median score can be computed, regardless of which distance is chosen and how genomes are modeled. It will prove useful in the subsequent sections, when the median problem is hard but the distance computations can be achieved fast.

Lemma 13.1 Given any three genomes Π_1, Π_2, and Π_3, and a rearrangement distance d, let M be the median score of a genome Π_M. We then have

$$M \geq \frac{d(\Pi_1, \Pi_2) + d(\Pi_1, \Pi_3) + d(\Pi_2, \Pi_3)}{2}.$$

The models for genomes used in this chapter are the permutations (part I), signed multichromosomal genomes in the "paths and cycles" configuration (chapter 10), and multichromosomal genomes in the "set system" configuration (chapter 12). It seems that median problems have not been studied in the case of strings, posets, or other multichromosomal models described in chapters 11 and 12. In consequence, we always refer to paths and cycles when we speak about multichromosomal models in this chapter, except in section 13.3.5, on genome halving, which uses set systems.

13.1 Breakpoint Median

As for the comparisons of two permutations, we begin the exploration of median problems with the simplest problem, the breakpoint median. Its conceptual simplicity and the fact that the distance on which it is based is trivial to compute for every definition of a genome make it a good starting point for our review of median problems.

Here, multichromosomal genomes are paths and cycles as defined in chapter 10. Therefore, a genome with n genes (i.e., of length n) is built over the alphabet $\mathcal{A} = \{1, 2, \ldots, n\}$.

- Introduced by Sankoff and Blanchette [317] and Blanchette et al. [67].
- Complexity: **NP**-hard for signed and unsigned permutations, whether they are circular (Pe'er and Shamir [292]) or linear (Bryant [88]). In the signed multichromosomal case, the problem is **NP**-hard for linear genomes, but polynomial with an $O(n^3)$ time solution for signed genomes with circular chromosomes allowed (see Tannier et al. [353]).
- Best approximation ratio: 7/6 for signed permutations (see Pe'er and Shamir [293]) and 5/3 for unsigned permutations (see Caprara [96]).
- Exact algorithms: Using a reduction to the TRAVELING SALESMAN PROBLEM by Sankoff and Blanchette [317].

Recall that for two signed multichromosomal genomes Π and Γ, the breakpoint distance between two genomes is $bd(\Pi, \Gamma) = n - (a + e/2)$ (see section 10.2), where n is the number of genes, a is the number of common adjacencies, and e is the number of common telomeres. Note that this formula does not apply directly to linear permutations, where the breakpoint distance has been defined as $n - a + 1$, where n and a are computed on the linear extension of the permutations. This is due to the model of chromosomes we use here, which can be read indifferently in one direction or the other, which is not necessarily the case for permutations.

13.1.1 Complexity

Pe'er and Shamir [292] and Bryant [88] independently proved that finding an optimal breakpoint median of three (signed or unsigned) permutations is **NP**-hard, by a reduction of HAMILTONIAN PATH (Pe'er and Shamir [292] use the directed version, whereas Bryant [88] uses the undirected version). Bryant studied the complexity of various related constrained problems: for example, he proved the following:

• The problem remains **NP**-hard if the median genome is required to contain no adjacencies not present in one of the input genomes.

• Determining whether a solution to the problem on unsigned permutations is unique, and whether a given adjacency belongs to all median genomes, are **NP**-complete problems.

Since permutations are particular strings, these **NP**-completeness results on the breakpoint median problem on permutations imply analogous results for the same problem on strings.

Surprisingly, the problem turns out to be easier to solve in the more general case of signed multichromosomal genomes; indeed, in this case, medians are not required to be permutations, which actually makes the problem less constrained. In the most general case, where both linear and circular genomes are allowed, the problem reduces to a maximum perfect matching problem, which yields an algorithm running in time $O(n^3)$, where n is the number of genes (see Tannier et al. [353]). However, if genomes are linear and if the median is required to be linear as well, then the problem is **NP**-hard again, which Tannier et al. [353] prove by a reduction of the same problem on circular signed permutations.

13.1.2 Algorithms

13.1.2.1 Exact Algorithms

Permutations Sankoff and Blanchette [317, 318] proposed the following reduction of the breakpoint median problem on unsigned permutations to the TRAVELING SALESMAN problem, which can be extended to signed permutations. Let Π_1, Π_2, Π_3 be three unsigned circular permutations that form an instance of the breakpoint median problem, and let G be the complete undirected graph with vertex set $\{1, 2, \ldots, n\}$. Each edge $\{x, y\}$ of G has weight $3 - u(x, y)$, where $u(x, y)$ is the number of genomes among Π_1, Π_2, Π_3 for which xy or yx is a point. Then, a minimum weight solution to the TRAVELING SALESMAN problem on G yields the optimal solution to the median problem. The TRAVELING SALESMAN problem can be solved to optimality in time $O(n^2 2^n)$ by dynamic programming (see Applegate et al. [20]).

Gramm and Niedermeier [189] give an exact algorithm with time complexity $O(2.15^b n)$ for computing the breakpoint median of three signed permutations, where b is the maximum number of breakpoints between any two input permutations. The result is given for the median of $k \geq 3$ permutations, which gives a complexity $O(2.15^b kn)$.

Multichromosomal Genomes Tannier et al. [353] proposed the following approach for handling the breakpoint median problem on multichromosomal signed genomes, which yields a polynomial-time algorithm for solving the problem to optimality. Let

Π_1, Π_2, and Π_3 be three genomes (possibly with circular chromosomes) on a gene set $\mathcal{A}$ of size n, and let G be the complete undirected graph whose vertex set consists of all the extremities of the genes in $\mathcal{A}$ and of an additional vertex t_x for each gene extremity x. Weights are then added to edges as follows:

- For every pair of gene extremities $\{x, y\}$, assign a weight to edge $\{x, y\}$ equal to the number of genomes, among Π_1, Π_2, and Π_3, in which xy is an adjacency (each edge in G joining two gene extremities therefore has weight 0, 1, 2, or 3).
- For every vertex v (which is the extremity of a gene), assign a weight to edge $\{v, t_v\}$ equal to half the number of genomes, among Π_1, Π_2, and Π_3, in which v is a telomere (each edge $\{v, t_v\}$ in G therefore has weight 0, $\frac{1}{2}$, 1, or $\frac{3}{2}$).
- Finally, assign weight 0 to every other edge in G.

Theorem 13.1 [353] Let $\mathcal{M}$ be a perfect matching in G. Clearly, edges between gene extremities in G define the adjacencies of a genome $\Pi_{\mathcal{M}}$, and the weight of $\mathcal{M}$ in G is $3n - (bd(\Pi_1, \Pi_{\mathcal{M}}) + bd(\Pi_2, \Pi_{\mathcal{M}}) + bd(\Pi_3, \Pi_{\mathcal{M}}))$.

Therefore, the breakpoint median problem of three multichromosomal genomes can be solved by means of finding a maximum-weight perfect matching in G, which can be achieved in time $O(n^3)$ (see Edmonds [153]).

13.1.2.2 Lower Bounds Using a transformation of the breakpoint median problem into the TRAVELING SALESMAN problem, Bryant [90] gives a lower bounding technique for the breakpoint median problem on signed permutations. If π^1, π^2, and π^3 are three permutations of $\{1, \ldots, n\}$ that form an instance of the breakpoint median problem (seeking a permutation for median), and $g \in \{1, \ldots, n\}$, let $d_g(\pi^i, \pi^j)$ equal 0 if g has the same successor in π^i and π^j (i.e., if $\pi^i_{(\pi^i)^{-1}_g+1} = \pi^i_{(\pi^i)^{-1}_g+1}$), and 1 otherwise. The formula is computed on the linear extensions of the permutations, so that 0 and $n+1$ are valid indices. For any permutation π^M, let $\delta_g(\pi^M) = d_g(\pi^M, \pi^1) + d_g(\pi^M, \pi^2) + d_g(\pi^M, \pi^3)$.

Theorem 13.2 [90] If M is the minimum score of a median permutation with respect to the breakpoint distance, then

$$M \geq \sum_{g \in \{1,\ldots,n\}} \min_{\pi^M} \delta_g(\pi^M).$$

Bryant [90] shows how to compute this bound in polynomial time, and improves it using Lagrange multiplier techniques.

13.1.2.3 Approximation Pe'er and Shamir [293] give a 7/6-approximation for the breakpoint median problem on signed permutations. Using a classical approach for the TRAVELING SALESMAN problem, Caprara [96] provides a 5/3-approximation for the breakpoint median problem on unsigned permutations by using a primal-dual integer programming formulation.

13.2 Reversal and DCJ Median

Two other median problems that have received much attention are the reversal median problem and the DCJ median problem (see sections 4.5 and 10.5 for a definition of DCJ distances). They have been investigated only in the signed case; therefore, all genomes and permutations in this section are signed, and we omit that adjective in the following.

Here, multichromosomal genomes are paths and cycles as defined in chapter 10.

- Introduced by Sankoff et al. [327].
- Complexity: The reversal and the DCJ median problems are **NP**-hard, and even **APX**-hard for signed permutations (see Caprara [97]) and multichromosomal signed genomes (see Tannier et al. [353]).
- Approximation: $\frac{4}{3}$-approximation by Caprara [92] for the reversal and DCJ median for permutations.

Several studies (see, e.g., Moret et al. [273, 275]) argue, based on experimental tests, that phylogenies based on solutions to the reversal median problem are more accurate than phylogenies based on solutions to the breakpoint median problem.

13.2.1 Complexity

Caprara [97] proved the **NP**-hardness of the reversal median problem by a transformation from BREAKPOINT GRAPH DECOMPOSITION, the same problem that was used to prove the **NP**-hardness of sorting by unsigned reversals (see section 3.3.4). Note that the **NP**-hardness result of Caprara [97] contains (and even uses) **NP**-hardness for the median with the DCJ rearrangement distance on permutations. The reversal median problem is even **APX**-hard, which can be shown using the result of Berman and Karpinski [57] on approximating the BREAKPOINT GRAPH DECOMPOSITION problem. Tannier et al. [353] also use a reduction of the BREAKPOINT GRAPH DECOMPOSITION problem to prove the **NP**-hardness of the DCJ median problem on multichromosomal genomes.

13.2.2 Algorithms

The potential practical importance of the reversal median problem motivated the design of several heuristics and exact exponential-time algorithms, and experimental results are also available (see, e.g., Siepel and Moret [341] or Caprara [97]).

13.2.2.1 Exact Algorithms Caprara [92, 97] proposed an integer programming formulation of the reversal median problem for permutations, and an exact solver based on a branch-and-bound technique that is efficient for small instances (i.e., when $n \leq 50$), implemented in GRAPPA (a software described in section 15.2.4). This exact algorithm, which requires the use of a linear programming solver, combines bounds from the relaxation of the integer programming formulation with combinatorial bounds, one of which is derived from lemma 13.1. Note that the reversal distance is always at least as large as the DCJ distance, so bounds for the DCJ distance may also be used for the reversal median problem.

13.2.2.2 Heuristic Algorithms Beyond $n = 50$, no exact algorithm is able to find the solution in a reasonable amount of time. Many heuristic principles have been published; most of them are based on a greedy strategy that consists in using one genome as a starting point and finding rearrangements that may bring it closer to the median (usually the hint is that they bring it closer to the other genomes). This strategy, which varies depending on which events should be considered, has been implemented by various authors (see, e.g., Bourque and Pevzner [81], Mira et al. [270], Interian and Durrett [216], or Arndt and Tang [21]). Local searches based on the same principles have been proposed by Interian and Durrett [217] and Lenne et al. [245] for the 2-break median problem. Other methods are mentioned by Wu and Gu [373, 374].

13.2.3 Variants

13.2.3.1 Perfect Reversal Median The reversal median with some "perfectness" constraints has been studied by Bernt et al. [60, 62, 63]. Recall (section 4.3.1) that a scenario of reversals transforming one genome into another is **perfect** if it does not break any common interval. The perfect reversal median problem consists in finding a median with minimal score, with the additional constraint that there exists a perfect sequence of reversals for transforming each of the input genomes into the median (or conversely). A parallel algorithm for solving the median problem with preservation of common intervals was proposed by Bernt et al. [61]. A slight relaxation of the perfectness contraint led Bernt et al. [60] to study the reversal median problem with preservation of "conserved intervals" (see page 64) instead of common intervals. They proposed a heuristic to solve this problem approximately.

The general case is solved by a branch-and-bound technique, but there is an interesting particular case where the problem is solvable in polynomial time (since this is quite rare in this part, it deserves to be mentioned). This occurs when the PQ-tree structure of common intervals (defined on page 23) satisfies a particular nontrivial property.

Theorem 13.3 [62] If the PQ-tree of the common intervals of three given permutations has only linear nodes, then the perfect reversal median problem can be solved in linear time.

Indeed, the only allowed operations here are reversals of strong intervals, that is, common intervals that do not overlap other common intervals. The problem reduces to a simple median problem on binary vectors compared by a Hamming distance. It is easy to see that the latter problem can be solved in linear time.

13.2.3.2 Reversals Around an Origin of Replication Ohlebusch et al. [283] solve, in linear time, a variant of the median problem in which the only allowed operations are reversals that are symmetric around an origin or a terminus of replication.

13.3 Duplicated Genomes

Whereas genome median problems aim at reconstructing ancestral configurations from comparisons of several species, **genome halving** problems use the intrinsic information about a genome that has undergone a whole duplication to reconstruct its ancestral configuration. *Guided* halving problems combine approaches inspired by both median problems and genome halving problems.

Beyond the duplication of individual genes, it is possible for the complete genome of a species to be fully duplicated, resulting in a genome where each chromosome exists in two copies. Although this event is usually lethal, in rare cases a duplicated genome can stabilize after a series of rearrangements. Evidence supporting the occurrence of whole genome duplications has been adduced in numerous plant genomes [2, 184], as well as in vertebrate genomes [260, 185, 219]. A particularly convincing example of whole genome duplication is found in the yeast genome [372, 243, 234]; however, alternative views do exist [267, 119].

From an algorithmic point of view, duplicated genomes are a particular case of genomes with duplicates: each gene family occurs exactly twice. However, the goal here is not to compute the distance between two given genomes, but to reconstruct the ancestral genome of a given genome. Except in section 13.3.5, multichromosomal genomes are paths and cycles as defined in chapter 10.

13.3.1 The Double Distance

- Introduced by Alekseyev and Pevzner [9].
- Complexity: $O(n^3)$ for the breakpoint distance, **NP**-hard for the DCJ distance for multichromosomal signed genomes (circular chromosomes allowed) (see Tannier et al. [353]).

We first give a formalization of a genome that has undergone a whole genome duplication. Recall that a signed gene g is defined as an oriented sequence of DNA identified by its tail g_t and its head g_h, linked by an edge.

Definition 13.1 A **duplicated gene** g is a pair of signed genes $\{g1_t, g1_h\}$ and $\{g2_t, g2_h\}$ that are biologically identified as homologous.

A genome is a set of paths and cycles on a set of genes, and a **duplicated genome** is a genome on a set of duplicated genes. This should not be confused with the following concept.

Definition 13.2 For a genome Π on a gene set $\mathcal{A}$, a **doubled genome** $\Pi \oplus \Pi$ is a duplicated genome on the set of duplicated genes from $\mathcal{A}$ such that if $g_x h_y$ is an adjacency of Π (where $x, y \in \{t, h\}$), then either both $g1_x h1_y$ and $g2_x h2_y$, or both $g2_x h1_y$ and $g1_x h2_y$, are adjacencies in $\Pi \oplus \Pi$.

Figure 13.1 illustrates the two concepts. Note the difference between a multichromosomal duplicated genome and the special case of a doubled genome: the former has two copies of each gene, whereas in the latter these copies are organized in such a way that there are two identical copies of each chromosome (when we ignore the 1's and 2's in the $g1_x$'s and $g2_x$'s): it has two linear copies of each linear chromosome and, for each circular chromosome, either two circular copies or one circular chromosome containing the two successive copies. Note also that for a genome Π, there is an exponential number of possible doubled genomes $\Pi \oplus \Pi$ (exactly 2 to the power of the number of nontelomeric adjacencies).

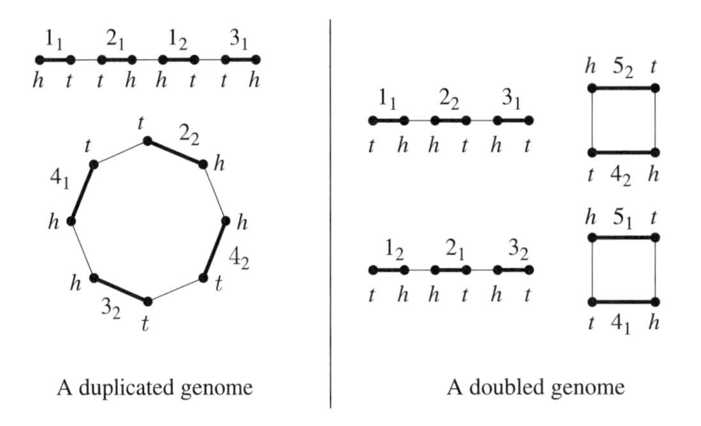

Figure 13.1
A duplicated genome and a doubled genome

Definition 13.3 For a duplicated genome Θ, an ordinary genome Π, and a distance d between genomes, the **double distance** between Π and Θ is

$$dd(\Pi, \Theta) = \min_{\Pi \oplus \Pi} d(\Pi \oplus \Pi, \Theta).$$

The problem of computing the double distance was mentioned as being open by Alekseyev and Pevzner [9]. It can be solved in polynomial time for multichromosomal genomes when d is the breakpoint distance, but it is **NP**-complete when d is the DCJ distance (see Tannier et al. [353]). The problem remains open for unichromosomal genomes.

13.3.2 Genome Halving

Given a duplicated genome Θ and a distance d between genomes, the **genome halving** problem consists in finding an ordinary genome Π that minimizes $d(\Theta, \Pi)$.

- Introduced by El-Mabrouk et al. [159].
- Complexity: Linear for multichromosomal genomes (El-Mabrouk and Sankoff [158]) with the fusion/fission/translocation/reversal distance (RT, defined on page 172), and quadratic for unichromosomal circular genomes with the reversal and DCJ distance (see Alekseyev and Pevzner [10], who also claim that there is a flaw in the solution of El-Mabrouk and Sankoff [158]).

Although computing the double distance is **NP**-hard for the DCJ distance, El-Mabrouk [155], El-Mabrouk and Sankoff [158], and El-Mabrouk et al. [159] showed that it can be computed in polynomial time for linear genomes when d is the RT-distance (see definition 10.9). A simple solution has been proposed when d is the DCJ distance, in the case of multichromosomal genomes (see Warren and Sankoff [368]; Mixtacki [271]) and in the case of signed unichromosomal genomes (see Alekseyev and Pevzner [9]). It can be solved in polynomial time when d is the breakpoint distance on multichromosomal signed genomes (see Tannier et al. [353]), but it is open in the case where genomes are linear.

The polynomial solutions are based on the **contracted breakpoint graph** for duplicated genomes, which is implicit in the work of El-Mabrouk and Sankoff [158] and named by Alekseyev and Pevzner [11]. The vertex set of the contracted breakpoint graph is composed of the set of gene extremities, in which two heads or two tails of one duplicated gene are contracted in a single vertex. Then there is an edge between A_x and B_y (for $x, y \in \{h, t\}$) if at least one of $A1_xB1_y$, $A1_xB2_y$, $A2_xB1_y$, and $A2_xB2_y$ is an adjacency in the genome.

The connected components of this graph are paths and cycles. The double distance can be computed as follows. For the breakpoint distance, construct the weighted complete undirected graph as on page 196, and compute a maximum-weight perfect matching (see Tannier et al. [353]). For the DCJ distance, choose half of the edges of each even cycle or path, alternating so that the result is a perfect matching on this cycle or path. For odd cycles, join the cycles by pair, choose an alternating matching on both, and join the unmatched vertices. For odd paths, take one edge over two, starting from the extremities. This forms a matching of the graph, and thus a genome of the set of genes, and it is an optimal solution to the DCJ halving problem (see Warren and Sankoff [368]; Mixtacki [271]).

This solution allows several chromosomes and possibly circular ones. Sharp verifications have been done by El-Mabrouk and Sankoff [158] to construct only linear chromosomes, and to prove the solution is also valid for the RT-distance. Alekseyev and Pevzner [11] have claimed that the solution of El-Mabrouk and Sankoff [158] contains an error in the unichromosomal case.

13.3.3 Solving Tetraploidy

Tetraploidy is a variant of genome halving that is based on the assumption that among the two copies of the duplicated genome, it is known which copy belongs to which copy of the ancestral genome. In other words, each gene in the duplicated genome belongs either to gene set A or to gene set B, and the goal is to construct a doubled genome where all chromosomes contain only genes from one set. This variant was introduced by El-Mabrouk and Sankoff [157], who give a heuristic and some special cases where this heuristic yields the exact solution.

13.3.4 Guided Halving

- Introduced by Zheng et al. [381].
- Complexity: Polynomial for the breakpoint distance and multichromosomal genomes with circular chromosomes allowed, **NP**-hard for breakpoint distance and linear genomes or permutations; also **NP**-hard for the DCJ distance and linear genomes or permutations (see Tannier et al. [353]; Caprara [97]).

Seoighe and Wolfe [331] first observed the extreme non-uniqueness associated with the solutions to the genome halving problem and suggested that this inherent difficulty could be attenuated through the use of a reference genome, or *out-group*. Given a duplicated genome Θ and an ordinary genome Π, the **guided halving** problem consists in finding an ordinary genome Π_M that minimizes $d(\Theta, \Pi_M) + d(\Pi_M, \Pi)$, where d is any distance on genomes. The guided halving problem is similar to the genome halving problem for Θ, but it takes into account the ordinary genome Π of an organism that is presumed to share a common ancestor with Π_M, the reconstructed undoubled ancestor of Θ.

The guided halving problem is known to be polynomial only for the breakpoint distance on general genomes (see Tannier et al. [353]). It is **NP**-hard for the breakpoint distance on linear genomes, and for the DCJ distance on general genomes (see Tannier et al. [353]; Zheng et al. [381]). Other variants are open.

Zheng et al. [381, 382] have proposed heuristics and used them, for instance, to infer the ancestor of the maize genome with the rice and sorghum genomes as out-groups. Another variant of the guided halving problem, introduced by Zheng et al. [381], seeks a genome Π_M that minimizes $d(\Pi_M, \Pi)$, given $d(\Theta, \Pi_M) = \min_H d(\Theta, H)$. Since computing $\min_H d(\Theta, H)$ is polynomial-time solvable for every distance and every definition of a genome (it is the halving problem), this variant may have different complexities than the first one. The computation problems associated with this variant, according to the distance and definition of genome, are all open.

13.3.5 Genome Halving with Unordered Chromosomes

Genome halving aims at finding a doubled genome that minimizes a distance to a duplicated genome. The problem has been studied in the case where chromosomes are considered as sets of genes, that is, the *synteny model* (see chapter 12).

- Introduced by El-Mabrouk et al. [159].
- Complexity: Unknown. Heuristics have been proposed by El-Mabrouk et al. [159] and Yin and Hartemink [377].

As in chapter 12, here a **genome** Π is a collection of N subsets $C_1, C_2, \ldots, C_N$ of a set $B = \{b_1, b_2, \ldots, b_h\}$ such that each gene in B appears exactly once among these subsets C_i. A **duplicated genome** is a collection of N subsets $C_1, C_2, \ldots, C_N$ of a set $B = \{b_1, b_2, \ldots, b_h\}$ such that each gene in B appears exactly twice among these subsets C_i. In a duplicated genome, if a gene occurs twice in the same set (or chromosome) C_i, it is called a 2-**gene**. The genome Π is called a **doubled genome** if it is made up of two identical copies of $N/2$ chromosomes.

Operations on chromosomes are defined as for the syntenic distance (see page 182). The distance between two genomes is the syntenic distance, and the goal is to find a doubled genome minimizing the distance to a given duplicated genome Π. The distance to this closest doubled genome is denoted as $hvd(\Pi)$.

At the heart of the analysis of the genome halving problem is the **synteny graph** (see also section 12.3) induced by a genome Π. The synteny graph induced by Π is the graph $\Omega(\Pi) = (V, E)$ defined by $V = \{C_1, C_2, \ldots, C_N\}$ and $E = E_1 \cup E_2$, where $E_1 = \{\{C_i, C_j\} : i \neq j \wedge C_i \cap C_j \neq \emptyset\}$ and $E_2 = \{\{C_i, C_i\} : C_i \text{ contains a 2-gene}\}$. Note that $\Omega(\Pi)$ is not necessarily simple (i.e., it may contain loops). From a graph-theoretic point of view, our goal becomes transforming the graph $\Omega(\Pi)$ into a matching bipartite graph by eliminating (and occasionally adding) appropriate edges through translocations. See figure 13.2 for an illustration.

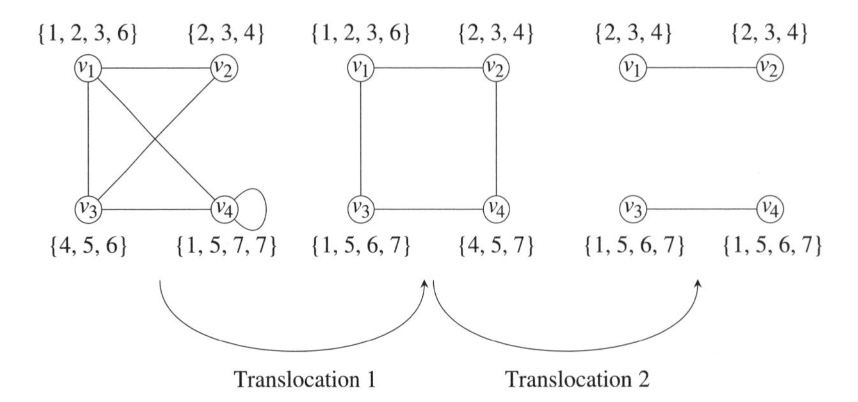

Figure 13.2
Two translocations are enough to transform the synteny graph on the left, representing a genome with four chromosomes, into a perfect matching graph. During the first translocation, v_3 exchanges $\{4\}$ with $\{1, 7\}$ from v_4. In the second translocation, v_1 exchanges $\{1, 6\}$ with $\{4\}$ from v_4

13.3.5.1 Lower and Upper Bounds The genome halving problem with unordered chromosomes requires the computation of the minimum number $hvd(\Pi)$ of fusions, fissions, and reciprocal translocations that are sufficient to transform a given genome Π into an ancestral duplicated genome Π' containing two identical sets of chromosomes. Clearly, for any genome with N chromosomes, we have $hvd(\Pi) \leq N$, since a trivial duplicated genome can be obtained through $N - 1$ fusions followed by a single fission. Yin and Hartemink [377] prove more accurate lower and upper bounds.

Theorem 13.4 [377] Let Π be a genome with N chromosomes and $\Omega(\Pi)$ be the corresponding synteny graph. Then

$$\lceil \Delta/2 \rceil \leq hvd(\Pi) \leq N - 2 + \ell - \min\{r, (N-4)/2\},$$

where Δ is the maximum degree of $\Omega(\Pi)$, $\ell = 1$ if $\Omega(\Pi)$ has at least $N - 1$ loops and $\ell = 0$ otherwise, and r is the number of nonloop vertices of degree 1 in $\Omega(\Pi)$.

The upper bound of theorem 13.4 can be refined using the following concept.

Definition 13.4 A subset $W \subset V$ of vertices without loops in a graph $G = (V, E)$ is **well-separated** if it satisfies the following two conditions:

1. for all $u, v \in W$, $\{u, v\} \notin E$, and u and v share no common neighbor if $\deg(u) > 1$ and $\deg(v) > 1$;
2. $\sum_{u \in W} \deg(u) \leq (N-4)/2$.

Yin and Hartemink [377] prove that $hvd(\Pi) \leq N - 2 + \ell - |W^*|$, where ℓ is defined as in theorem 13.4 and W^* is a maximum-cardinality well-separated vertex set in $\Omega(\Pi)$.

13.3.5.2 Diameter

The **genome halving diameter** for genomes with N chromosomes is defined by

$$hvdD(N) = \max\{hvd(\Pi) : \Pi \text{ is a genome with } N \text{ chromosomes, } N \text{ even}\}.$$

The genome halving diameter problem is thus to find $hvdD(N)$ for even N; Yin and Hartemink [377] prove lower and upper bounds on $D(N)$.

Theorem 13.5 [377] For all values of N we have

$$N - u \leq hvdD(N) \leq N - 1,$$

where u is the largest integer such that $u2^u \leq N$. In the case of nonloopy genomes, the upper bound reduces to $N - 2$.

13.4 Other Variants, Generalizations, and Discussion

13.4.1 Other Operations

Even though finding an optimal reversal or 2-break median are **NP**-hard problems, the underlying operations have been well studied in the pairwise genome rearrangement version, so they may be handled with theoretical studies. However, in the case of the translocation distance, with genomes modeled as set systems (see chapter 12), computing even the pairwise distance is **NP**-complete. This result does not trivially imply **NP**-completeness of the median problem, which is proved by DasGupta et al. [131]. Ferretti et al. [174] present a heuristic with tests to predict the number of chromosomes in the mammalian ancestor.

The transposition median problem is even more complicated since no complexity result is known even for the pairwise distance (see section 3.1). Yue et al. [378] report on a heuristic and experiments on simulated data. A study by Eriksen [163], cited again below for discussion, reports some statistical work and experiments on reversal and transposition medians.

13.4.2 More Permutations in the Input

A median permutation may be searched for more than three input permutations. Although that generalization does not make sense in the context of evolution (the Steiner tree problem, studied extensively in chapter 14, is more relevant), the techniques are similar and have been used, for example, by Jackson et al. [218] to design consensus arrangements when the input is the data from the same genome, but for some experimental or polymorphism reasons, all sources are different.

13.4.3 Medians and Centers

As seen in the study of Popov [302], the combinatorics of genome rearrangement literature goes far beyond actual genomic rearrangements, in the sense that an increasing number of operations are studied under this name without being really connected to the biological reality. Popov [302] defines an alternative to the median, which is no longer motivated by biology, but still belongs to the field due to the similarity of objects, questions, and methods.

Definition 13.5 A **center** permutation for a group of input permutations Π and a distance d is a permutation π that minimizes

$$\max_{\sigma \in \Pi} d(\pi, \sigma).$$

This center has been investigated, for instance, for the element interchange distance (see page 50) by Popov [302]. Klein [235] proves the fixed parameter tractability of the reversal median problem, the reversal center problem, the breakpoint center problem, the induced breakpoint median problem, and the exemplar breakpoint median problem.

13.4.4 Discussion

Though the reversal median problem has received a lot of attention from researchers and developers, and although there have been several attempts to reconstruct phylogenies and ancestral genome configurations using such techniques (see chapter 14), this approach is now being criticized (see Eriksen [163] or Adam and Sankoff [1]) because of the large number of optimal or near-optimal reversal medians that can exist for a single instance. Moreover, several optimal medians may be as distant as two genomes in the instance, making unlikely even the hope that an optimal solution will be close to reality. Moreover, results obtained using these techniques are often far from well-established biological results (see Froenicke et al. [179]), leading us to think that the model still needs to be improved for handling real data.

14 Rearrangement Phylogenies

14.1 The Large Parsimony Problem

The comparison of three or more genomes based on their gene orders faces many of the classical difficulties inherent in the comparison of species based on their nucleotide sequences. Some particular additional difficulties include the following:

- Rearrangement distances are sometimes difficult to compute even for genomes without duplicates (e.g., the transposition distance, discussed in section 3.1).
- Possible rearrangement events are numerous, and their relative importance is not always precisely known.
- Gene orders, and therefore inferred rearrangements, depend on the quality of the homology inference between genes.

Though these arguments should prevent us from trying to reconstruct phylogenies from gene orders, research in this direction is motivated both by the fact that gene orders are much less sensitive to mutations (character insertion, character deletion, character replacement) than nucleotide sequences, and by the fact that they represent the evolution process in a deeper way.

The *large parsimony problem* was noticeably used by Sankoff et al. [324] to simultaneously produce a multiple sequence alignment and a phylogeny based on nucleotide sequences. It appears as a particular case of a more general problem, called the STEINER TREE PROBLEM, which we define below in a graph-theoretic formulation (which is the most general one if we do not restrict ourselves to finite graphs).

To this end, given a tree $\mathcal{T} = (V(\mathcal{T}), E(\mathcal{T}))$ and a weight function w on its edges, write

$$W_w(\mathcal{T}) = \sum_{\{x,y\} \in E(\mathcal{T})} w(x, y).$$

The Steiner tree problem is then defined as a decision problem, as follows.

STEINER TREE PROBLEM

INSTANCE: Graph $G = (V, E)$ with weight function w, which is a metric; $\Pi \subseteq V$; positive integer K.

QUESTION: Is there a tree $\mathcal{T}$ within G that spans all vertices in Π and such that

$W_w(\mathcal{T}) \leq K$?

The tree $\mathcal{T}$ is called a **Steiner tree** with **Steiner set** Π. When the Steiner set Π is a subset of taxa from a set V, w is a distance between pairs of taxa in V, and G is a complete graph, the Steiner tree problem seeks a phylogenetic tree explaining the evolution of taxa in Π following the maximum parsimony criterion. From a phylogenetic point of view, however, the taxa in Π are often considered as extant taxa that must appear as leaves of the phylogenetic tree, contrary to ancestral taxa that correspond to internal nodes.

Definition 14.1 A Steiner tree $\mathcal{T}$ for a set Π is **full** (or **terminal**) if the elements of Π label the leaves of $\mathcal{T}$.

The full Steiner tree problem is the variant of the Steiner tree problem in which a solution is further required to be a full Steiner tree. The specific problem at hand in the phylogenetic context is the following one (as defined by Sankoff et al. [324]).

LARGE PARSIMONY PROBLEM

INSTANCE: A set $\Pi = \{\Pi_1, \Pi_2, \ldots, \Pi_r\}$ of taxa from a set V, a distance d between taxa, a positive integer K.

QUESTION: Is there a full Steiner tree $\mathcal{T} = (V(\mathcal{T}), E(\mathcal{T}))$ for Π with $V(\mathcal{T}) \subseteq V$ such that

$W_d(\mathcal{T}) \leq K$?

The Steiner tree problem is **APX**-hard even when the edge weights are only 1 or 2 (see Bern and Plassmann [59]), but admits a 1.55-approximation algorithm (see Robins and Zelikovsky [313]). On the other hand, the full Steiner tree problem is also **APX**-hard (see Lin and Xue [253]) and admits a 2.52-approximation algorithm (see Viduani Martinez et al. [362]). Although from a formal point of view the large parsimony problem is a particular case of the full Steiner tree problem, serious problems arise when one tries to approximate the former using an algorithm for the latter. Indeed, approximation algorithms for the full Steiner tree problem use the whole input of the problem (i.e., the entire input graph). For the large parsimony

PROBLEM, the vertex set of this graph is usually very large; for instance, if taxa are represented by signed permutations of n elements, then V is the set of all such permutations, and therefore contains $2n(2n-2)(2n-4)\ldots 2 = 2^n n!$ vertices.

The LARGE PARSIMONY PROBLEM must therefore be considered as a specific problem for which specific solutions have to be found. It remains **NP**-complete in very particular cases—for instance, when the taxa are nucleotide sequences and d is the Hamming distance (see Day [133]). In the next section, we discuss this problem in the context of rearrangement-based distances, which is our main point in this chapter.

14.2 The Large Parsimony Problem with Gene Orders

The jump from sequence comparison to genome comparison, both combined with the reconstruction of the evolutionary history of a given set of taxa, is the starting point of a new era in phylogenetic reconstruction. However, the formal problem to solve (in the hope that it will allow us to obtain good phylogenetic solutions) remains the same: the LARGE PARSIMONY PROBLEM.

The first insights in the area are due to Hannenhalli et al. [200] and Sankoff et al. [327], whose approaches to multiple genome comparison are already based on parsimony. Subsequent experimental researches by Rice and Warnow [312] and Tang et al. [350] showed that parsimony is usually more accurate than other reconstruction methods. However, it is worth pointing out that although the LARGE PARSIMONY PROBLEM is supported by these results as a good framework for approaching phylogenetic reconstruction, its relevance becomes real only when the representation of the genomes and the definition of the distance d become concrete.

14.2.1 Breakpoint and Reversal Phylogenies on Permutations

The representation of genomes as signed permutations, thus avoiding duplicates, is the most natural one for a first investigation of the problem on genomes with only one chromosome, since several important distances are polynomial in this case (e.g., breakpoint distance and signed reversal distance; see chapter 4). The distance d is then chosen from among those distances that can be computed in (a low-order) polynomial time, which explains why the breakpoint and reversal distances are candidates of choice.

- Introduced by Blanchette et al. [67] and Sankoff and Blanchette [318] (breakpoint phylogeny); Moret et al. [274] and Bourque and Pevzner [81] (reversal phylogeny).

- Complexity: Computing a breakpoint phylogeny is **NP**-complete (see Pe'er and Shamir [292]), as well as computing a reversal phylogeny that is both **NP**-complete and **APX**-hard (see Caprara [92]).

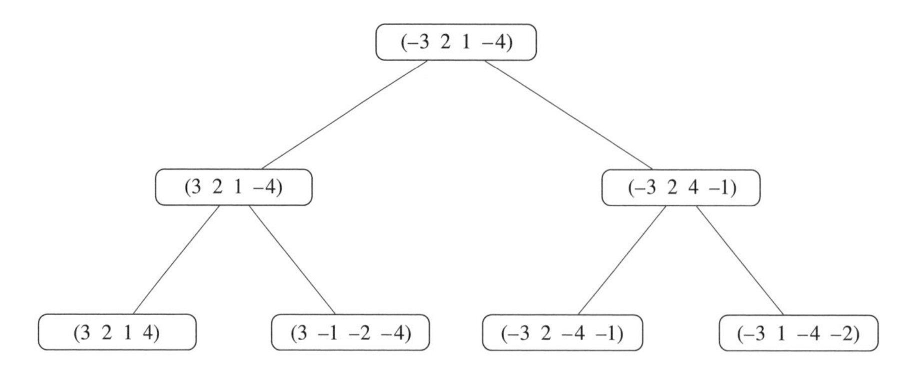

Figure 14.1
Breakpoint phylogeny for the Steiner set {(3 2 1 4), (3 −1 −2 −4), (−3 2 −4 −1), (−3 1 −4 −2)}. All edges have weight 2. The edges of the reversal phylogeny with the same labels are all of weight 1

The need to strongly simplify the LARGE PARSIMONY PROBLEM in order to avoid unnecessary and unfruitful difficulties was first expressed by Blanchette et al. [67] and Sankoff and Blanchette [318]. Their rationale for the use of the breakpoint distance instead of some edit distance (such as the reversal distance or the transposition distance) is the simplicity of computing the breakpoint distance and the fact that breakpoints are not associated with any rearrangement event.

Definition 14.2 A **breakpoint phylogeny** for a set Π of signed permutations is a full Steiner tree with Steiner set Π and edges that are weighted by the breakpoint distance between the permutations associated with their end points.

Notice here that the internal nodes of the tree are implicitly labeled by signed permutations. Figure 14.1 shows an example of breakpoint phylogeny where all edges have weight 2. In this context, the LARGE PARSIMONY PROBLEM can be formulated as follows:

BREAKPOINT PHYLOGENY PROBLEM

INSTANCE: Set $\Pi = \{\Pi_1, \Pi_2, \ldots, \Pi_r\}$ of signed permutations of n elements, a positive integer K.

QUESTION: Is there a breakpoint phylogeny $\mathcal{T}$ for Π such that

$W_{bd}(\mathcal{T}) \leq K$?

Moret et al. [274] and Bourque and Pevzner [81] note several drawbacks of the breakpoint analysis, including the difficulty of adapting it to multichromosomal genomes. They propose to use the signed reversal distance instead of the breakpoint distance. The resulting problem is called the REVERSAL PHYLOGENY PROBLEM, and the associated desired phylogeny is a **reversal phylogeny**. The breakpoint phylogeny shown in figure 14.1 can be seen as a reversal phylogeny where all edges have weight 1. Note that the STEINER TREE PROBLEM (but not the FULL variant) associated with the reversal distance had previously been introduced by Caprara [92].

Not surprisingly, both the BREAKPOINT PHYLOGENY PROBLEM and the REVERSAL PHYLOGENY PROBLEM inherit the hardness results from the corresponding median problems. All algorithms devised for the breakpoint or reversal phylogeny problem are heuristics, and will be presented in section 14.3.

14.2.2 Variants

Breakpoint and reversal phylogenies have been extensively studied, leaving very little room for other variants of the LARGE PARSIMONY PROBLEM. Earnest-DeYoung et al. [152] studied a variant where constrained block insertions and block deletions are added to reversals, and Yue et al. [378] studied another variant where the TRANSPOSITION PHYLOGENY PROBLEM is approached. The simplicity of computing the distance between two genomes is a strongly required feature and is the main reason (when compared to the transposition distance, for instance) for the omnipresence of the breakpoint and reversal distances in the current studies.

The multichromosomal case faces the difficulty of appropriately redefining the classical distances (e.g., breakpoint distance or reversal distance) in this context. However, adaptations were initiated by Pevzner and Tesler [295], and Tannier et al. [353] showed that when genomes are represented using the "paths and cycles" model in chapter 10, the median breakpoint problem is polynomial for general genomes (i.e., allowing both circular and linear chromosomes; see also chapter 13). Consequently, in this case the SMALL PHYLOGENY PROBLEM and the LARGE PHYLOGENY PROBLEM do not inherit from the complexity result on the corresponding median problem, and are open.

14.3 Heuristics for the Breakpoint/Reversal Phylogeny Problem

Breakpoint and reversal phylogenies are not necessarily binary trees, but the search for solutions to the BREAKPOINT PHYLOGENY PROBLEM and the REVERSAL PHYLOGENY PROBLEM may be restricted to binary trees, assuming that the internal nodes of binary solutions are not required to have distinct labels. Indeed, if the optimal tree is not a binary tree, then its weight is identical to that of a binary tree obtained by appropriately splitting some internal nodes and appropriately joining the resulting new nodes using 0-weight edges.

Consequently, unless explicitly stated, all phylogenies discussed in the rest of this chapter are considered to be unrooted binary trees whose vertices are labeled using signed permutations of n elements. The exceptions to this rule are typically *genomes with unequal gene content* (i.e., genomes represented by signed permutations of possibly different lengths and of different sets of elements).

14.3.1 Tree Steinerization

The approach presented here, initiated by Blanchette et al. [67] and Sankoff and Blanchette [318] in the context of breakpoint phylogenies, consists of two steps: generate all tree topologies with r leaves labeled $\Pi_1, \Pi_2, \ldots, \Pi_r$, and label each internal node with a signed permutation in order to minimize the weight of the resulting tree. Here, a *tree topology* is simply an unrooted binary tree, uniquely identified by the adjacencies between nodes and independent from any embedding of it in a planar space. Figure 14.2 shows two different embeddings of the same tree topology with the same leaf labels.

When the two steps are considered as independent parts of the algorithm, the first one is classical from an algorithmic point of view, and the second one is a specific problem called the SMALL PARSIMONY PROBLEM (see, for instance, Sankoff et al. [327]). However, these two steps are not completely independent. Roughly estimating the optimal weight of each topology based on a preliminary partial labeling allows us

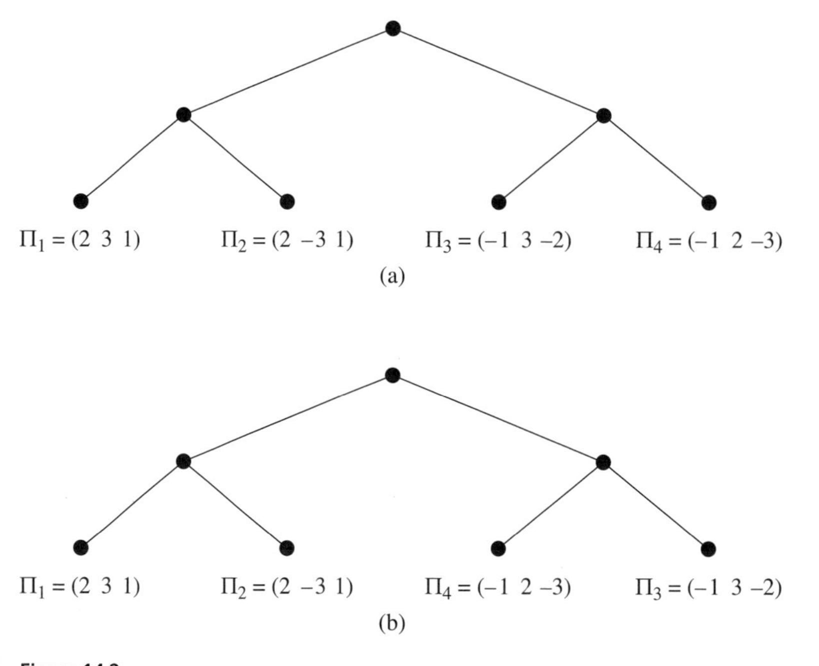

Figure 14.2
A tree topology $\mathcal{T}$ with $r = 4$ leaves labeled Π_1, Π_2, Π_3, Π_4 embedded in a plane in two different ways. The circular permutation associated with the leaves is $\pi^\circ = [1\,2\,3\,4]$ in (*a*) and $\sigma^\circ = [1\,2\,4\,3]$ in (*b*), yielding $C_{bd,\pi^\circ}(\mathcal{T}) = 5$ and $C_{bd,\sigma^\circ}(\mathcal{T}) = 5.5$, where *bd* is the breakpoint distance

to drop, without completely labeling, the obviously bad topologies (i.e., those whose best possible labeling has a weight exceeding some appropriate threshold). Only a small number of candidates is thus kept for the computationally hard task of topology labeling.

The resulting method is given in algorithm 14.1. In this algorithm, steps 5 to 13 (called the Steinerization algorithm by Sankoff and El-Mabrouk [320]) iteratively improve the labeling of the internal nodes, using a method based on local optimization initially proposed by Sankoff et al. [325] and used by Sankoff et al. [327] for the SMALL PARSIMONY PROBLEM. The stabilization expected in step 13 may take very long to reach, so the heuristic remains exponential even if good filters are used in steps 2 and 3 to reduce the number of topologies to label.

Several steps crucially affect the running time, and thus the upper bound on the size of the entry (number r of genomes, their size n) that may be handled by this method: the *weight estimation* at line 2, the *initialization procedure* at line 4, and the *median solver* at line 8. These key parts of the algorithm have been extensively studied, and several variants exist for each of them. We do not intend to present them all here, but only to provide a few outlines for each of those three steps in the following sections.

Remark 14.1 The reader should keep in mind, while reading the following, that two main implementations of the Steinerization-based procedure exist. The first one,

Algorithm 14.1
Steinerization-based heuristic

Input: signed permutations $\Pi_1, \Pi_2, \ldots, \Pi_r$
Output: minimum weight phylogeny with leaves $\Pi_1, \Pi_2, \ldots, \Pi_r$
1: **for** each tree topology $\mathcal{T}$ with leaves $\Pi_1, \Pi_2, \ldots, \Pi_r$ **do**
2: estimate the optimal weight of $\mathcal{T}$;
3: **if** the estimated weight does not exceed the current threshold **then**
4: find an initial genome for each internal node of $\mathcal{T}$;
5: **repeat**
6: **for** each internal node v of $\mathcal{T}$ with the label denoted Π^v **do**
7: let Π^a, Π^b, Π^c be the labels of the three neighbors of v;
8: compute a median Π_M of Π^a, Π^b, Π^c;
9: **if** Π_M achieves smaller total weight with Π^a, Π^b, Π^c than Π^v, **then**
10: replace Π^v with Π_M;
11: **end if**
12: **end for**
13: **until** no more genome replacement is performed;
14: compute the weight of $\mathcal{T}$;
15: **end if**
16: **end for**
17: **return** the tree with minimum weight.

---

called BPAnalysis, solves the BREAKPOINT PHYLOGENY PROBLEM only, and was proposed by Sankoff's group (see Blanchette et al. [67] and Sankoff and Blanchette [318]). The second one, called GRAPPA, solves both the BREAKPOINT PHYLOGENY PROBLEM and the REVERSAL PHYLOGENY PROBLEM, and was proposed by Moret's group in a series of papers [274, 273, 276]. Both implementations are presented in chapter 15.

14.3.1.1 Weight Estimation In a context where the heuristic attempts to find a *minimum* weight tree, the role of weight estimation is to evaluate as accurately and as quickly as possible a tight lower bound on the optimal weight of a tree topology $\mathcal{T}$, assuming its leaves are already labeled $\Pi_1, \Pi_2, \ldots, \Pi_r$. If this lower bound exceeds the threshold, then no labeling of $\mathcal{T}$ can provide a competitive weight and $\mathcal{T}$ can therefore be safely discarded.

Remark 14.2 Although the threshold in line 3 of algorithm 14.1 is not emphasized as a main optimization goal in our presentation, its value is of noticeable importance in the algorithm. The trivial approach of updating it each time a tree weight is computed or evaluated is overtaken by a good upper bound on the minimum weight of the tree. An example of such an upper bound is proposed by Moret et al. [273].

Since the only data available with the tree topology consist of genomes $\Pi_1, \Pi_2, \ldots, \Pi_r$, a simple idea suggested by Sankoff et al. [327] and Moret et al. [276] relies on using these genomes and the triangle inequality (when the distance between genomes is really a metric) to obtain a lower bound. The following result is an easy consequence of the triangle inequality.

Lemma 14.1 [276] Let $\mathcal{T}$ be a tree topology with leaves labeled $\Pi_1, \Pi_2, \ldots, \Pi_r$, and let π° be the circular permutation of r elements describing a circular ordering of the leaves in $\mathcal{T}$ under some planar embedding of $\mathcal{T}$. Then $W_d(\mathcal{T})$, for an arbitrary metric d between genomes, is lower-bounded by

$$C_{d,\pi^\circ}(\mathcal{T}) = \frac{1}{2}\sum_{i=1}^{r} d(\Pi_{\pi_i^\circ}, \Pi_{\pi_{i+1}^\circ}),$$

called the **circular lower bound of $\mathcal{T}$ with parameters d and π°.**

Note that the same tree topology yields different circular orderings for different planar embeddings. In other words, swapping the left and the right children of an internal node will change neither the topology nor its minimum weight, but will change the permutation π° and, consequently, the value of the circular ordering lower bound. (See figure 14.2.) The best lower bound obtained in this way will then be the maximum of all circular ordering lower bounds, which can be computed in $O(r^3)$ time (see Bachrach et al. [25]); however, heuristics are preferred in the implemented software.

Another lower bound for arbitrary metrics is computed, using linear programming, by Tang and Moret [348] and Bachrach et al. [25]. Bryant [90] proposes a lower bound for genomes with possibly unequal content in the case of breakpoint phylogenies, which he improves using Lagrange multipliers. When limited to signed permutations of the same length, this lower bound is proved tighter than the circular ordering one, for a normalized breakpoint distance.

14.3.1.2 Initialization Procedure For a given tree topology $\mathcal{T}$, the convergence of the tree Steinerization process strongly depends on the initial labeling of $\mathcal{T}$. For breakpoint phylogenies, Blanchette et al. [67] and Sankoff and Blanchette [318] propose three initialization procedures, based on solving the TRAVELING SALESMAN PROBLEM (TSP) at each internal node of $\mathcal{T}$, for different data. Among them, one is identified as better for low-divergence data and another shows better for high-divergence data. Moret et al. [274] test the efficiency of these initialization procedures, identifying one of them as the best compromise of accuracy and speed, although they further propose six other initialization procedures.

14.3.1.3 Median Solver The median solver is eventually called many times by the Steinerization-based heuristic, and the quality of the solution it returns is an important factor affecting the number of calls. Therefore, it must be carefully chosen from among the numerous heuristics and exact algorithms available for either the breakpoint or the reversal median (see a complete description in chapter 13).

Blanchette et al. [67] and Sankoff and Blanchette [318] use a reduction of the median breakpoint problem to the TSP and a branch-and-bound algorithm to exactly solve the resulting instance of TSP. Moret et al. [274] give greater place to efficiency in solving the median breakpoint problem by proposing the use of approximate TSP solvers. Heuristics aiming at directly computing a breakpoint median were proposed by Sankoff et al. [328] for a normalized breakpoint distance.

As far as reversal median solvers are concerned, two of them experimentally proved their efficiency. Caprara [95] combined branch-and-bound and divide-and-conquer strategies on a generalization of the breakpoint graph, whereas Siepel and Moret [341] searched the space of genome rearrangements, using branch-and-bound based on the triangle inequality. The latter algorithm has the advantage of being extensible to other metrics, as shown by Tang et al. [350], who added insertions and deletions to reversals, in order to handle genomes with unequal content. Other reversal median solvers are available (see, e.g., Sankoff et al. [327]; Siepel [339]; Wu and Gu [374]; Bernt et al. [60, 63]).

Algorithm 14.2
Sequential addition-based heuristic

Input: signed permutations $\Pi_1, \Pi_2, \ldots, \Pi_r$
Output: minimum weight phylogeny with leaves $\Pi_1, \Pi_2, \ldots, \Pi_r$
1: solve the median problem for Π_1, Π_2, Π_3, and call $\mathcal{T}$ the resulting tree;
2: let $\sigma = (\sigma_4\sigma_5 \ldots \sigma_r)$ be a permutation of $4, 5, \ldots, r$;
3: **for** $l := 4$ to r **do**
4: **for** each edge $\{u, v\}$ in $\mathcal{T}$ with labels Π^u, Π^v for u, v **do**
5: compute a median Π_M^{uv} of Π^u, Π^v, Π_{σ_l};
6: $C(u, v) := d(\Pi^u, \Pi_M^{uv}) + d(\Pi^v, \Pi_M^{uv}) + d(\Pi_{\sigma_l}, \Pi_M^{uv}) - d(\Pi^u, \Pi^v)$
7: **end for**
8: let $C(u_0, v_0) = min\{C(u, v) \mid \{u, v\} \in E(\mathcal{T})\}$;
9: remove edge $\{u_0, v_0\}$ from $\mathcal{T}$;
10: add vertices x, y with respective labels $\Pi_M^{u_0v_0}$, Π_{σ_l} to $\mathcal{T}$;
11: add edges $\{x, u_0\}$, $\{x, v_0\}$, $\{x, y\}$ to $\mathcal{T}$;
12: **end for**
13: **return** the tree $\mathcal{T}$.

14.3.2 Sequential Addition

Rather than generating all topologies and labeling them, the methods in this subsection attempt to directly build one or several trees with small weight, from among which the best tree is then selected. A tree is built by sequentially adding new permutations, among $\Pi_1, \Pi_2, \ldots, \Pi_r$, to an already partially built tree. Algorithm 14.2 formalizes this idea.

Steps 4 to 11 in this heuristic search for the best edge (u_0, v_0) (called a *split edge* by Bourque and Pevzner [81]) to be replaced with a 3-star whose leaves are u_0, v_0 and a new vertex labeled Π_l. Then the insertion of Π_l into $\mathcal{T}$ results into a splitting of the original edge (u_0, v_0), which minimally affects the weight of $\mathcal{T}$.

14.3.2.1 Seeking a Unique Tree The sequential addition-based heuristic allows us to build one tree, with hopefully small weight, based on local optimization. One particular advantage of this method is that it can start with an arbitrarily large $\mathcal{T}$ (to replace the one obtained in step 1), assuming another method was used to build trees with a smaller number of vertices. This is the approach used by Bourque and Pevzner [81].

Again, a good median solver is needed; Bourque and Pevzner [81] focus on the reversal distance, which they consider to be more accurate than the breakpoint distance, from a biological point of view. Furthermore, the proposed median solver (a heuristic) relies strongly on the fact that the reversal distance is an edit distance, in that it enumerates the intermediate genomes one has to build in order to transform one genome into the other. On the one hand, this allows us to identify paths between the three genomes, and genomes at the intersection of these paths that are good candidates for the median genome (a similar idea was used by Sankoff et al. [327]). On the other hand, the generalization to other edit distances, and more particularly to distances between multichromosomal genomes, is easy, as Bourque and Pevzner [81] show, using reversals, translocations, fusions, and fissions. The resulting implementation, including many improvements to this general approach and called MGR, is presented in section 15.2.2.

14.3.2.2 Searching for a Set of Trees The sequential addition-based heuristic has an obvious drawback—its results strongly depend on the ordering σ of the input genomes—and a less obvious drawback—fixing one median in step 5 of algorithm 14.2 permanently affects the weight of the tree and the subsequent choices. Bernt et al. [64] propose to abandon the definitive choices and to allow the algorithm (1) to choose among several candidates for each median problem, and (2) to select the best order σ depending on the context (i.e., on the partial tree in construction). To reach this goal, three main modifications are performed:

1. Steps 2 and 4 of algorithm 14.2 are merged into a single loop whose counter takes its values in the set of triples (Π^u, Π^v, Π_h) where $\{u, v\}$ is an edge and Π_h, $4 \leq h \leq r$, is a genome not yet introduced into the tree.
2. Step 5 of algorithm 14.2 generates a *set* $M(u, v, h)$ of solutions to the median problem for (Π^u, Π^v, Π_h), instead of a single median.
3. For each of the most promising triples $(\Pi^{u_0}, \Pi^{v_0}, \Pi_{h_0})$, selected on the basis of an estimation of their score, and for each solution in $M(u_0, v_0, h_0)$, one performs step 9 and step 10 of algorithm 14.2, followed by a recursive call of the modified algorithm (the latter step is used only if the current tree has fewer than r leaves).

With these changes, the resulting algorithm attempts to generate a large number of trees, but a collection of appropriate bounds and selection methods ensures a limited number of "fully generated trees," from among which the best trees are then output.

14.3.3 Character Encodings

The definition of the maximum parsimony based on gene orders followed many studies performed on sequence-based maximum parsimony. Methods designed specifically for the gene-order variant result from a natural attempt to solve a particular problem with particularly appropriate methods, but they do not contradict the idea that sequence-based methods could help solve the gene-order variant. The heuristic presented in algorithm 14.3 shows the outlines of the current approaches that try to reuse the ordinary sequence parsimony software.

The difficult steps in this heuristic are the *genome-to-sequence* procedure in step 1 and mainly the *sequence-to-genome* transformation in step 4. The distance *sd* in step

Algorithm 14.3
Character encodings-based heuristic

Input: signed permutations $\Pi_1, \Pi_2, \ldots, \Pi_r$
Output: minimum weight phylogeny with leaves $\Pi_1, \Pi_2, \ldots, \Pi_r$
1. encode each genome Π_i as an appropriate sequence U_i, $i = 1, 2, \ldots, r$;
2. consider a distance *sd* between sequences;
3. compute a maximum parsimony phylogeny P, given U_i $(i = 1, 2, \ldots, r)$ and *sd*;
4. relabel each node of P with a genome.

---

2 is usually the Hamming distance. The maximum parsimony phylogeny P in step 3 is a solution to the SEQUENCE-BASED MAXIMUM PARSIMONY PROBLEM, which is a variant of the LARGE PARSIMONY PROBLEM where species are represented by sequences of the same length and the distance between species is *sd*. Since this problem is **NP**-complete (see Foulds and Graham [178]) even for the simple Hamming distance on binary sequences, step 3 is solved using one of the several available fast heuristics.

14.3.3.1 From Genomes to Sequences The ordered pair "(set of sequences, *sd* distance)" should encode very accurately the ordered pair "(set of genomes, gene order-based distance)," so as to ensure a simultaneous optimization of the tree weight in both spaces. Both encodings we present here satisfy the property that the Hamming distance between two resulting sequences is closely related to the breakpoint distance between the two initial genomes. There is, however, a gap between the error rate obtained in those two spaces, since the relabeling in step 4 deteriorates the quality of the solution in the space of genomes.

MPBE In an early work, Cosner [124] developed an encoding technique for which Cosner et al. [125] proposed a simplified approach called *maximum parsimony on binary encodings* (MPBE). In this encoding, each position in the resulting binary sequence U_i corresponds to an ordered pair of signed genes a, b (here, (a, b) and $(-b, -a)$ are considered equivalent) that are adjacent in at least one of the genomes $\Pi_1, \Pi_2, \ldots, \Pi_r$. Given a genome Π_i, this position is valued 1 in the sequence U_i if either a, b or $-b$, $-a$ appears consecutively in Π_i, and 0 otherwise. See figure 14.3.

MPME Initially introduced by Bryant [90] (inspired by Sankoff and Blanchette [319]) under the name *SB encoding*, this encoding was further used by Wang et al. [366] and Tang and Wang [349] under the name MPME (*maximum parsimony on multistate encodings*). Bryant [90] noted that every MPME encoding can be converted into an MPBE encoding by using a Hamming distance-preserving mapping, and showed that the minimum weight of a tree under the MPME encoding better approximates the breakpoint weight of the tree than its minimum weight under the MPBE encoding.

$$\begin{array}{lrrrrr} \Pi_1: & 3 & 1 & -4 & 2 & 5 \\ \Pi_2: & 4 & -1 & -3 & 2 & -5 \\ \Pi_3: & -3 & 2 & 5 & 1 & -4 \end{array}$$

MPBE

	(3, 1)	(1, −4)	(−4, 2)	(2, 5)	(−3, 2)	(2, −5)	(5, 1)
U_1:	1	1	1	1	0	0	0
U_2:	1	1	0	0	1	1	0
U_3:	0	1	0	1	1	0	1

MPME

	1	2	3	4	5	6	7	8	9	10
U_1:	−4	5	1	−1	3	−3	4	−5	2	−2
U_2:	−4	−5	1	−1	−2	−3	3	2	5	4
U_3:	−4	5	4	−1	1	−5	3	2	−3	−2

Figure 14.3
Examples of MPBE and MPME encodings of genomes Π_1, Π_2, and Π_3

In the MPME encoding, the resulting sequences are of length $2n$, where n is the length of the genome, and are built on the alphabet formed by all signed genes, that is, $\{-n, \ldots, -2, -1, 1, 2, \ldots, n\}$. Position a $(1 \leq a \leq n)$ in the MPME encoding corresponds to gene a (that is, the positive member of the gene family $a \in \mathcal{A}$), and its value is the signed gene immediately following gene a in the genome (in a circular way). Position $n + a$ corresponds to gene $-a$, and its value is defined similarly. By convention, if the signed genes a, b appear consecutively and in this order in the genome (in a circular way), then a is followed by b and $-b$ is followed by $-a$. See figure 14.3.

14.3.3.2 From Sequences to Genomes The result of step 3 in algorithm 14.3 is a tree with a sequence in each internal node. Though sequences at the leaves do correspond to genome encodings, sequences in the internal nodes are simply inferred during the construction of the parsimony tree with respect to the *sd* distance, and therefore do not necessarily correspond to genomes.

Two solutions were proposed to deal with the transformation of the resulting sequence-based tree into a gene-order-based tree, which hopefully would approximate well the most parsimonious gene-order-based tree. The first one, due to Cosner et al. [125], ignores the sequences at the internal nodes, and uses the tree topology only to infer, using a Steinerization algorithm, a best labeling with respect to any available distance (Cosner et al. [125] use breakpoints only, reversals only, and reversals/transpositions/transreversals).

The second one, proposed by Tang and Wang [349], attempts to transform the sequences at the internal nodes into genomes, and thus strongly depends on the chosen encoding and the distance *sd*. Focusing on the MPBE and MPME encodings with the Hamming distance, Tang and Wang [349] claim that finding a gene order minimizing the Hamming distance between its MPBE encoding and a given MPBE-like sequence is **NP**-complete, whereas the similar MPME problem is open.

14.4 Variants

The methods presented in this chapter handle signed permutations of the same set of elements. Many authors emphasize the need to enrich these methods in order to take into account genomes with different sets of genes and/or genomes with duplicates and/or multichromosomal genomes. The current approaches attempt either to reduce the problem to the signed permutation version by simply discarding the duplicates (see, for instance, Sankoff et al. [328]), or to make appropriate hypothesis on the input genomes (e.g., they are balanced), which would possibly allow the use of existing models for comparing two genomes with unequal content and/or duplicates (see, for instance, Tang and Moret [347]).

V MISCELLANEOUS

15 Software

We present here a selection of available software implementing some of the algorithms that we have mentioned or discussed in this book. We start in section 15.1 with tools that analyze rearrangements between two genomes using unichromosomal or multichromosomal models, then consider software that can also deal with more than two genomes in section 15.2.

15.1 Pairwise Rearrangements

15.1.1 Unichromosomal Models

15.1.1.1 DERANGE The first software dedicated to the reconstruction of evolutionary scenarios by genome rearrangements is probably DERANGE, written by Sankoff, Leduc, and Rand in 1992. DERANGE sorts unsigned circular permutations by reversals and by transpositions using branch-and-bound.

- Written by Sankoff, Leduc, and Rand (see Sankoff et al. [326]).
- Platform: Macintosh.
- Home page: software available from the authors.

15.1.1.2 DERANGE II Four years after DERANGE, Blanchette et al. [66] released an enhanced version of DERANGE, called DERANGE II, which allowed the user to assign a weight to each operation. It handles circular and linear permutations, signed or not.

- Written by Blanchette et al. [66].
- Platform: Unix.
- Home page: ftp://ftp.ebi.ac.uk/pub/software/unix/derange2.tar.Z.

DERANGE II uses exhaustive search with look-ahead, but relies on heuristics to speed up the process. It is therefore not guaranteed to provide an optimal solution, but the authors claim that it is "unlikely to produce a solution containing more than one or two extra moves." It takes permutations as input.

15.1.1.3 CREx CREx (**c**ommon interval **r**earrangement **ex**plorer) handles reversals, transpositions, reverse transpositions, and tandem-duplication-random-loss events.

- Written by Bernt et al. [65].
- Platform: Web.
- Home page: http://pacosy.informatik.uni-leipzig.de/crex.

Given two genomes, CREx builds their strong interval tree (see page 22), then tries to detect patterns in the tree that correspond to genome rearrangement operations. CREx takes as input a file in FASTA format, then computes a distance matrix (based on reversals, breakpoints, or common intervals). It then displays the distance matrix, and a possible scenario.

15.1.1.4 ROBIN ROBIN (**r**earrangement **o**f **b**lock **in**terchanges) analyzes linear and circular genomes using block interchanges, based on the algorithms by Lin et al. [254]. It takes FASTA files as input.

- Written by Lu et al. [258].
- Platform: Web.
- Home page: http://genome.life.nctu.edu.tw/ROBIN.

15.1.1.5 SPRING SPRING (**s**orting **p**ermutations by **r**eversals and block **in**terchan**g**es) handles block interchanges and signed reversals, and gives an optimal rearrangement scenario along with the corresponding distance on linear and circular genomes.

- Written by Lin et al. [255].
- Platform: Web.
- Home page: http://algorithm.cs.nthu.edu.tw/tools/SPRING.

SPRING can be set to rearrange genomes using only reversals or only block interchanges, or both rearrangement operations; in the latter case, it assigns weight 1 to reversals and weight 2 to block interchanges. It also computes the breakpoint distance between the two input permutations. SPRING relies on the algorithms designed by Kaplan et al. [228] and by Lin et al. [254].

15.1.1.6 PSbR PSbR is a Java implementation of the algorithm of Sagot and Tannier [314] for finding perfect scenarios of reversals between signed permutations (which constitute its input).

- Written by Diekmann, based on the algorithm of Sagot and Tannier [314].
- Platform: all (Java applet).
- Home page: http://biomserv.univ-lyon1.fr/~tannier/PSbR.

15.1.1.7 baobabLUNA baobabLuna is a set of tools written by Braga to perform various tasks on signed permutations, such as building their breakpoint graph, computing their reversal distance, and representing the set of all optimal scenarios of reversals.

- Written by Braga.
- Platform: all (Java applet).
- Home page: http://www.geocities.com/mdvbraga/baobabLuna.html.

baobabLuna takes signed permutations as input, and performs its operations based on algorithms by Hannenhalli and Pevzner [199, 196] and Braga et al. [87].

15.1.2 Multichromosomal Models

15.1.2.1 GRIMM GRIMM (**g**enome **r**earrangements **i**n **m**an and **m**ouse) handles linear, circular, and multichromosomal genomes, signed or unsigned. It uses translocations, reversals, fusions, and fissions; computes the corresponding distance; and proposes a corresponding optimal scenario. All operations are treated equally, meaning that they are all assigned the same weight.

- Written by Tesler [356].
- Platform: Web.
- Home page: http://nbcr.sdsc.edu/GRIMM/grimm.cgi.

Contrary to most other available software, GRIMM handles genomes directly as permutations (or partitioned permutations). It is based on algorithms by Tesler [355] and by Hannenhalli and Pevzner [196, 198, 199]. GRIMM has been used to infer the number of rearrangements between the human and mouse genomes (see Pevzner and Tesler [297]), as well as to try to prove that some regions in the genome are more fragile, and broken more often by reversals, than others. Both the conclusion and the evidences of this theory are the subject of controversy (see, e.g., Sankoff and Trinh [323] and Peng et al. [294] for diverging points of view on the "fragile breakage versus random breakage" controversy).

15.1.2.2 CTRD CTRD implements Zhu and Ma's algorithm [383] for computing the signed translocation distance between two genomes, entered as sequences of integers.

- Written by Feng et al. [172].
- Platform: Web.
- Home page: http://www.cs.cityu.edu.hk/~lwang/software/Translocation/index.html.

15.2 Phylogeny Reconstruction and Medians

15.2.1 BPAnalysis

BPAnalysis was written by Blanchette, and performs breakpoint analysis based on algorithms by Sankoff and Blanchette [318]. It reconstructs a tree from a set of gene orders, using the breakpoint distance.

- Written by Blanchette, based on algorithms by Sankoff and Blanchette [318].
- Platform: Microsoft DOS/Windows, GNU/Linux.
- Home page: http://www.mcb.mcgill.ca/~blanchem/software.html.

15.2.2 MGR

MGR (**m**ultiple **g**enome **r**earrangements) is an extension of GRIMM that aims at handling multiple genomes, with an input format similar to that of GRIMM. It computes distances and phylogenetic trees for a set of multichromosomal genomes, using an exact (thus of exponential time complexity) algorithm.

- Written by Bourque, based on an algorithm by Bourque and Pevzner [81].
- Platform: Web.
- Home page: http://nbcr.sdsc.edu/GRIMM/mgr.cgi.

MGR is widely used to infer rearrangements in mammalian species whose genomes have been sequenced, and to infer the genome of a common ancestor of these species (see, e.g., Bourque et al. [82] or Murphy et al. [277]). This software has been the subject of a controversy between cytogenetic methods and bioinformatics methods to infer the genomes of common ancestors (see, e.g., Bourque et al. [85] and Froenicke et al. [179] for discussions).

15.2.3 GRIL

Motivated by the fact that genome rearrangements may not always correspond to gene boundaries, which means that some rearrangements may act on a segment of a gene instead of on the whole gene, Darling et al. [130] proposed a software called GRIL (**g**enome **r**earrangement and **i**nversion **l**ocator), which aims at identifying *locally collinear blocks* (i.e., collinear sequence regions shared by all genomes under consideration). GRIL takes sequence files (e.g., FASTA) as input.

- Written by Darling et al. [130].
- Platform: Microsoft DOS/Windows, GNU/Linux.
- Home page: http://asap.ahabs.wisc.edu/software/gril.

15.2.4 GRAPPA

GRAPPA stands for **g**enome **r**earrangements **a**nalysis under **p**arsimony and other **p**hylogenetic **a**lgorithms.

- Written by many contributors in Moret's team, and maintained by Moret and Tang.
- Platform: Unix.
- Home page: http://www.cs.unm.edu/~moret/GRAPPA.

GRAPPA takes signed permutations as input, and implements several combinatorial algorithms related to genome rearrangements. Among them, the linear algorithm of Bader et al. [26] for computing the reversal distance between two signed permutations and an extension has been proposed by Yue et al. [378] to handle transpositions. The package also contains an exact exponential algorithm by Tang and Moret [348] to compute the reversal median of three signed circular permutations, and the computation of phylogenetic trees from rearrangement distances (see Moret et al. [274] and Wang et al. [367]). A heuristic for the reversal median of three permutations has been added by Arndt and Tang [21].

15.2.5 MedRbyLS

MedRbyLS stands for **med**ian **r**eversal **by** **l**ocal **s**earch, and implements a local search algorithm for finding a reversal median of three genomes represented by signed permutations.

- Written by Interian, based on algorithms by Interian and Durrett [217, 216].
- Platform: GNU/Linux.
- Home page: http://www.cam.cornell.edu/~interian/codes.html#MedRbyLS.

15.2.6 rEvoluzer and amGRP

rEvoluzer and amGRP were written by Bernt et al. [60, 61, 64], and achieve the heuristic computation of reversal medians, with scenarios preserving framed common intervals. amGRP is based on sequential addition (see section 14.3.2) and takes permutations as input.

- Written by Bernt et al. [60, 61, 64].
- Platform: GNU/Linux.
- Available on request from the authors.

15.2.7 GENESIS

GENESIS (**gen**ome **e**volution **s**cenar**i**o**s**) was written by Gog et al. [187], and implements several algorithms for sorting unichromosomal genomes by weighted reversals and transpositions or multichromosomal genomes, either by reversals, translocations, fusions and fissions, or by weighted reversals, translocations, fusions, fissions, and transpositions.

- Written by Gog et al. [187].
- Platform: Web.
- Home page: http://www.uni-ulm.de/in/theo/research/genesis.html.

GENESIS uses the algorithms of Bader and Ohlebusch [28], Hannenhalli and Pevzner [196], Tesler [355], Ozery-Flato and Shamir [285], and Yancopoulos et al. [375]. It takes signed permutations as input.

16 Open Problems

The number of variants of genome rearrangement problems is huge, and is increasing every year. Of course, it is possible to design many other variants that have not yet been studied, but among the most studied and interesting existing variants, open problems are still numerous. They are disseminated all through the pages of this book. However, we mention again the most striking ones in this chapter and add a few that were not cited in the previous chapters. This should represent a gold mine for researchers in the mathematics of theoretical computer science.

16.1 Complexity Issues

Some problems are open regarding complexity: Is a given rearrangement problem or distance computation problem solvable in polynomial time? And if not, is it approximable? Is there an **FPT** algorithm?

16.1.1 Hardness

The complexity status of the following problems is not known:

1. Computing the transposition distance and sorting unsigned permutations by transpositions (section 3.1).

2. Computing the transposition median of three permutations (section 3.1 and chapter 13).

3. The SMALL PARSIMONY PROBLEM and LARGE PARSIMONY PROBLEM under the breakpoint distance is open regarding multichromosomal signed genomes when linear and circular chromosomes are allowed (chapters 13 and 14).

4. Computing the prefix reversal distance and sorting unsigned permutations by prefix reversals (section 3.4).

5. Computing the prefix transposition distance and sorting unsigned permutations by prefix transpositions (section 3.2).

6. Deciding whether a solution to the local complementation problem on directed graphs exists (section 6.3).

7. Counting the number of solutions to sorting signed permutations by reversals (section 4.2).

8. The double distance and genome halving problems (section 13.3) are open under all distances for unsigned genomes, whereas El-Mabrouk et al. [159] conjecture that "gene order alone, without transcription direction, would likely not suffice to permit polynomial-time exact algorithms" for genome halving.

9. Genome halving is open regarding linear genomes, both for the breakpoint, HP, and DCJ distances. Tannier et al. [353] conjecture it to be polynomial for the breakpoint distance (section 13.3).

10. Computing the minimum possibly reversed common partition between two balanced strings (problem RMCSP) (section 9.1.3).

11. Computing the reversal *sorting* distance between balanced unsigned strings, for alphabets of size $k \geq 4$ (section 9.2.1). Note that this is polynomial for $k = 2$ [116] and $k = 3$ [309] (section 9.2.1).

12. Computing the reversal *sorting* distance between balanced signed strings, for alphabets of size $k \geq 2$ (section 9.2.2).

13. Computing the transposition *sorting* distance between balanced unsigned strings, for alphabets of size $k \geq 3$ (section 9.3.1). Note that this is polynomial for $k = 2$ [116] (section 9.3).

14. Rearranging a multichromosomal genome into another (with linear and circular chromosomes) with a minimum number of k-break rearrangements (section 10.6).

15. Is it true that computing a block edit distance on strings becomes **NP**-complete as soon as its block edit collection contains transpositions (section 8.2)?

16. Computing the syntenic distance when only reciprocal translocations are allowed (section 12.5.1).

16.1.2 Approximability

A number of problems have been shown to be **NP**-hard, and polynomial-time approximation algorithms have been designed. A first problem is of course to find approximation algorithms with a smaller ratio. It is the case for almost all **NP**-hard problems cited in this book, so we will mention only the main ones here. On the other hand, it has been shown that some problems cannot be approximated within a given ratio, but there is still room for improvement since there is a gap between the best known ratio and the "tractability barrier" set by this negative result. This is the case for the following:

1. The unsigned reversal distance, which cannot be approximated within a ratio of 1.0008 and for which the best approximation to date has the ratio 1.375 (section 3.3.5);

2. Is there a **PTAS** for sorting unsigned permutations by reversals for dense instances (section 9.6)?;

3. The syntenic distance, for which the best ratio is 2, but for which no result exists about whether it is possible to lower this ratio (see section 12.5.1);

4. The transposition distance on unsigned permutations, for which there is a 1.375-approximation algorithm, whereas the complexity of this problem is not even known.

16.1.3 Polynomial Complexity

Some problems have a known polynomial-time complexity, but are believed to admit faster solution than what has been achieved to date:

1. The best time complexity for the sorting of signed permutations by reversals is $O(n^{3/2})$ (section 4.2). It is an open problem to devise an algorithm with better time complexity.

2. The best time complexity for the median and guided halving problems under the breakpoint distance on multichromosomal genomes (with circular chromosomes allowed) is $O(n^3)$ (section 13.1), using a reduction to the maximum weight perfect matching problem. It is an open problem to devise an ad-hoc algorithm with better complexity.

16.2 Diameter

Problem 16.1 (diameter) Given a distance d on a set S, determine the maximal value d can reach, that is, the value

$$\max_{s,t \in S} d(s,t).$$

The diameter problem has not been studied for many variants, but it is really an open question for the following distances:

1. The transposition distance between unsigned permutations (section 3.1.6), as well as between unsigned balanced strings for alphabets of size $k \geq 3$ (section 9.3.1);

2. The prefix reversal distance and the prefix transposition distance between unsigned permutations (sections 3.4 and 3.2, respectively);

3. The prefix reversal distance between unsigned balanced strings, for alphabets of size $k \geq 3$ (section 9.2.4);

4. The block interchange distance between unsigned balanced strings, for alphabets of size $k \geq 3$ (section 9.4.1);

5. The genome halving problem on unordered chromosomes (section 13.3.5.2).

16.3 Tightness of Bounds

When neither a formula for computing a given rearrangement distance nor an algorithm for exactly solving the rearrangement problem is known, we nevertheless have lower and upper bounds on the number of operations that we will need to use. An immediate question is that of characterizing which instances are tight with respect to our bounds and which are not. We will call this the "tightness problem":

Problem 16.2 (tightness) Given a distance d and a bound b on a set S, characterize the elements s of S such that $d(s) = b(s)$.

The tightness problem is open for the following:

1. All bounds in the case of sorting a permutation by transpositions; Christie [115] contributed a new lower bound on the transposition distance by characterizing a few permutations that are not tight with respect to the lower bound of theorem 3.4; Hartman and Verbin [204] characterized 3-permutations that are tight with respect to the lower bound of theorem 3.1;

2. All bounds in the case of sorting an unsigned permutation by prefix transpositions, except for the lower bound of lemma 3.5 (page 38), for which one can check in polynomial time whether a given permutation is tight (see section 3.2.4);

3. All bounds in the case of sorting an unsigned permutation by reversals, except for the following:

- The lower bound of theorem 3.17 (page 40), for which one can check in polynomial time whether a given permutation is tight (see section 3.3.3.2);
- The lower bound of theorem 3.18 (page 42), since computing that lower bound's value is **NP**-hard;

4. All bounds in the case of sorting an unsigned permutation by prefix, weighted, fixed-length, or bounded reversals, as well as for strip moves and combined reversals and transpositions.

APPENDICES

A
Graph Theory

Not surprisingly, in this field of combinatorics, we use a lot of notions from graph theory. In order to be unambiguous about all notions on graphs, directed or undirected, we briefly present here everything needed in this book. For more developments on graph theory, the reader may refer to Berge [43] and Diestel [143].

A.1 Undirected Graphs

A.1.1 Basic Definitions

An **undirected graph** is a pair $G = (V, E)$, where V (sometimes also denoted by $V(G)$) is a set whose elements are called **vertices** or **nodes** and E (sometimes also denoted by $E(G)$) is a collection of two-element subsets of V called **edges** (see figure A.1 for an illustration). Undirected graphs are often simply referred to as **graphs**. Two vertices connected by an edge are the **extremities** of that edge. A graph is **simple** if it contains no **loop** (i.e., an edge whose extremities coincide) and has no more than one edge connecting any two vertices.

The number of vertices (resp. edges) of G is its **order** (resp. its **size**). A vertex u is **incident** with an edge e if $u \in e$. Two edges that share a common extremity are said to be **incident**. Two distinct vertices are **adjacent** if there is an edge connecting them. Pairwise nonadjacent vertices (resp. nonincident edges) are called **independent**. A set of pairwise adjacent vertices is called a **clique**. The **complete graph** K_n is a clique of order n, that is, a simple graph in which every pair of distinct vertices is adjacent, and has size $\binom{n}{2} = n(n-1)/2$. The **complement** of $G = (V, E)$ is the graph $\bar{G}$ with vertex set V and edge set $\{\{u, v\} \in V \times V : \{u, v\} \notin E \text{ and } u \neq v\}$.

The **degree** $\deg(u)$ of a vertex $u \in V$ is the number of edges incident to u, where loops count twice. A vertex of degree 0 is said to be **isolated**. The **neighborhood** of $u \in V$, denoted $N(u)$, is the set of all vertices adjacent to u. More generally, for $V' \subseteq V$, $N(V') = \{v : \{u, v\} \in E \text{ and } u \in V'\}$.

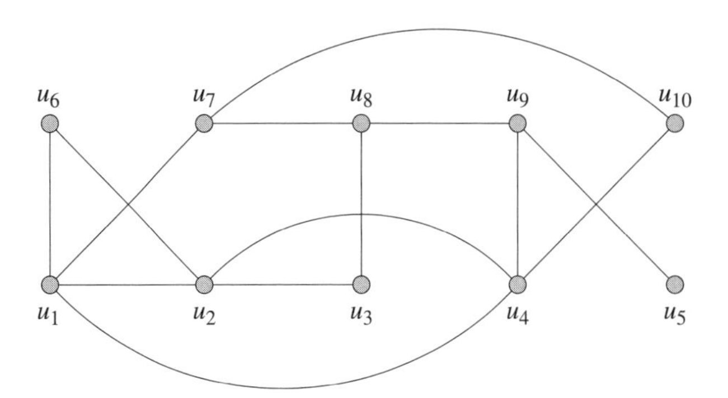

Figure A.1
A simple graph of order 10 and size 14

Property A.1 In a graph, the number of odd-degree vertices is even.

The **minimum degree** $\delta(G)$ and **maximum degree** $\Delta(G)$ of G are defined by $\delta(G) = \min_{u \in V} \deg(u)$ and $\Delta(G) = \max_{u \in V} \deg(u)$, respectively. If all vertices of a simple graph G have the same degree k, that is, $\delta(G) = \Delta(G) = k$, then G is k-**regular**. The complete graph K_n is thus an $(n-1)$-regular graph.

Property A.2 A k-regular graph G of order n has $nk/2$ edges.

Let $G = (V, E)$ and $G' = (V', E')$. If $V' \subseteq V$ and $E' \subseteq E$, then G' is a **subgraph** of G, written as $G' \subseteq G$. If $G' \subseteq G$ and E' contains all edges $\{u, v\} \in E$ with $u, v \in V'$, then G' is an **induced subgraph** of G, and more specifically V' **induces** G' in G, written $G' = G[V']$. Therefore, if $X \subseteq V$ is any set of vertices, then $G[X]$ denotes the graph on X whose edges are precisely the edges of G with both ends in X. By extension, if $Y \subseteq E$ is any set of edges, then $G[Y]$ denotes the graph on the extremities of Y whose edges are precisely Y. Finally, $G' \subseteq G$ is a **spanning subgraph** of G if $V' = V$.

Let $G = (V, E)$ be a graph. For any $e \in E$, the graph obtained from G by deleting the edge e is written $G - e$. Similarly, for any two nonadjacent vertices $u, v \in V$, $G + \{u, v\}$ denotes the graph obtained from G by adding the edge $\{u, v\}$.

A.1.2 Paths and Cycles

A **path** $P = (V, E)$ is a graph of the form

$$V = \{u_1, u_2, \ldots, u_n\},$$

$$E = \{\{u_1, u_2\}, \{u_2, u_3\}, \ldots, \{u_{n-1}, u_n\}\}.$$

The vertices that begin and end the path are termed the **extremities** of the path. A **cycle** is a path whose extremities coincide. The **length** of a path or a cycle is the num-

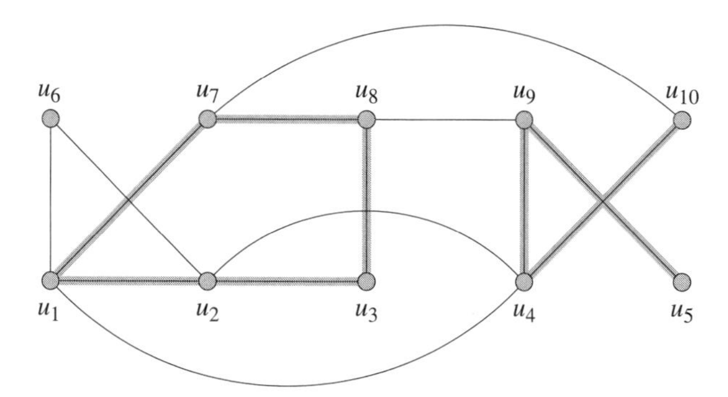

Figure A.2
A vertex-induced elementary path of length 3 and a vertex-induced elementary cycle of length 5

ber of edges traversed by the cycle or path. A path or a cycle is **elementary** if all the vertices are distinct. (See figure A.2 for an illustration). An **odd cycle** is a cycle of odd length; otherwise it is an **even cycle**. A graph is **acyclic** if it does not contain any induced cycle.

Property A.3 An acyclic graph of order n has at most $n - 1$ edges.

The problem of finding a minimum-length path between two vertices is known as the SHORTEST PATH problem and is solvable in polynomial time. It is, however, **NP**-complete to determine, for a graph G and a parameter k, whether the graph has an induced path of length at least k (this problem is known as the LONGEST PATH) problem. A **Hamiltonian path** (resp. a **Hamiltonian cycle**) is a path (resp. a cycle) in a graph that visits each vertex exactly once. Determining whether such paths and cycles exist in graphs is the HAMILTONIAN PATH problem, which is **NP**-complete. An **Eulerian path** (resp. an **Eulerian cycle**) is a path (resp. a cycle) in a graph that visits each edge exactly once.

Theorem A.1 (Euler's theorem) [167] A graph has a Eulerian cycle if and only if all vertices have even degree. A graph has a Eulerian path if and only if all but two (i.e., the two end-point) vertices have an even degree.

The **distance** $d_G(u, v)$ in a graph G between two vertices u and v is the length of a shortest path from u to v in G; if no such path exists, then $d_G(u, v) = \infty$. The greatest distance between any two vertices in G is called the **diameter** of G.

A.1.3 Connectivity

A graph is called **connected** if any two of its vertices are linked by a path, and **disconnected** otherwise. A **connected component** is a maximal connected subgraph of G. Each vertex belongs to exactly one connected component, as does each edge.

Property A.4 A connected graph of order n has at least $n - 1$ edges.

It can be tested in polynomial time whether a graph is connected (a simple depth-first search suffices to identify all components in linear time).

A.1.4 Bipartite Graphs

A graph $G = (V, E)$ is **bipartite** if its vertices can be partitioned into two disjoint sets or **classes** V' and V'', and each edge connects a vertex of V' to a vertex of V''.

A bipartite graph G with bipartition $V = V' \cup V''$ is often written $G = (V', V'', E)$. The **complete bipartite graph** $K_{p,q} = (V', V'', E)$ is defined by $|V'| = p$, $|V''| = q$, and $E = \{\{u, v\} : u \in V' \text{ and } v \in V''\}$. If $|V'| = |V''|$, G is called a **balanced bipartite graph**.

A.1.5 Trees and Forests

An undirected acyclic graph is also called a **forest**, and a **tree** is a connected acyclic graph (i.e., a connected forest). Vertices of degree 1 in a tree are called **leaves**. Every forest is bipartite and every nontrivial tree has at least two leaves. Trees are acyclic connected graphs, so properties A.3 and A.4 yield the following.

Property A.5 A connected graph of order n is a tree if and only if it has $n - 1$ edges.

A tree is called a **rooted tree** if one vertex has been designated as the **root**, in which case all edges have a natural orientation toward or away from the root. For each vertex of a rooted tree, its neighbor that is closer to the root is called its **parent**, and its neighbors that are farther from the root are called its **children**. An **ordered tree** is a tree for which an ordering is specified for the children of each node.

A.1.6 Matching

Given a graph $G = (V, E)$, a **matching** $\mathcal{M}$ is a set of nonincident edges in G (i.e., $\mathcal{M}$ is a set of independent edges). Every vertex incident with an edge in $\mathcal{M}$ is said to be a **matched vertex**. By abuse of language, a graph whose edge set is a matching is said to be a **matching graph**. A matching is **maximal** if it is maximal for inclusion. A matching is **maximum** if it contains the largest possible number of edges. Every maximum matching must be maximal, but not every maximal matching must be maximum. A **perfect matching** is a matching that covers all vertices of the graph. That is, every vertex of the graph is incident to exactly one edge of the matching (see figure A.3). Every perfect matching is maximum, and hence is maximal.

A maximum matching can be found in a graph $G = (V, E)$ in $O(|E| \sqrt{|V|})$ time (see Micali and Vazirani [268]). Note that the dual problem of finding a maximum cardinality subset of independent vertices, the so-called MAXIMUM INDEPENDENT SET problem, is **NP**-complete.

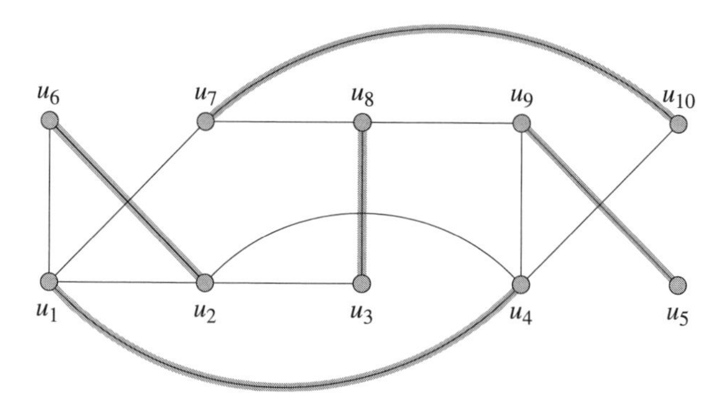

Figure A.3
A perfect matching in a graph

Matching problems are often concerned with bipartite graphs. Maximum matching in bipartite graphs is characterized by a duality condition.

Theorem A.2 (König) In any bipartite graph G, the maximum cardinality of a matching in G is equal to the minimum cardinality of a vertex cover.

A.1.7 Adjacency Matrix

The **adjacency matrix** $A = [a_{i,j}]$ of a graph G is defined by

$$a_{i,j} = \begin{cases} 1 & \text{if } \{i, j\} \in E, \\ 0 & \text{otherwise.} \end{cases}$$

The adjacency matrix of a simple undirected graph has 0's on the diagonal (no self-loops). The adjacency matrix $A(G)$ of the graph depicted in figure A.1 is

$$\begin{pmatrix} 0 & 1 & 0 & 0 & 0 & 1 & 1 & 0 & 0 & 0 \\ 1 & 0 & 1 & 1 & 0 & 1 & 0 & 0 & 0 & 0 \\ 0 & 1 & 0 & 0 & 0 & 0 & 0 & 1 & 0 & 0 \\ 0 & 1 & 0 & 0 & 0 & 0 & 0 & 0 & 1 & 1 \\ 0 & 0 & 0 & 0 & 0 & 0 & 0 & 0 & 1 & 0 \\ 1 & 1 & 0 & 0 & 0 & 0 & 0 & 0 & 0 & 0 \\ 1 & 0 & 0 & 0 & 0 & 0 & 0 & 0 & 0 & 1 \\ 0 & 0 & 1 & 0 & 0 & 0 & 0 & 0 & 1 & 0 \\ 0 & 0 & 0 & 1 & 1 & 0 & 0 & 1 & 0 & 0 \\ 0 & 0 & 0 & 1 & 0 & 0 & 1 & 0 & 0 & 0 \end{pmatrix}.$$

The adjacency matrix of a complete graph is all 1's except for 0's on the diagonal. The adjacency matrix of a complete bipartite graph $K_{p,q}$ has the form

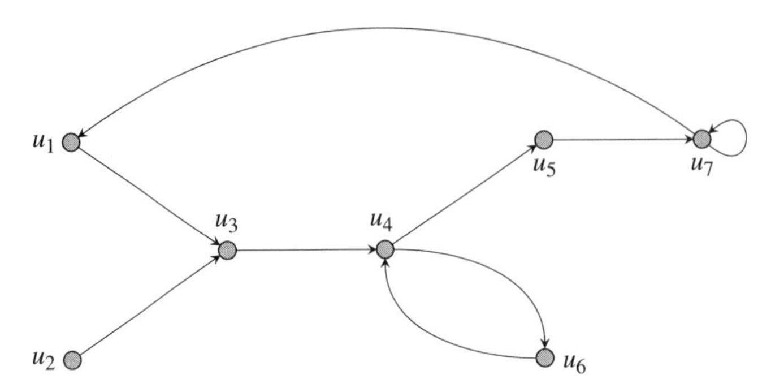

Figure A.4
A directed graph of order n and size m

$$\begin{pmatrix} O & J \\ J^T & O \end{pmatrix},$$

where J is a $p \times q$ matrix of all 1's and O denotes an all-zero matrix.

A.2 Directed Graphs

A.2.1 Basic Definitions

A **directed graph** (or **digraph**) is a pair $D = (V, A)$, where V is a set whose elements are called **vertices** and $A \subset V \times V$ is a multiset of ordered pairs of vertices whose elements are called **arcs**. (See figure A.4 for an illustration.) The **order** (**size**) of D is the number of vertices (arcs) in D. For an arc $(u, v) \in A$, the first vertex is its **tail** and the second vertex is its **head**. The head and the tail of an arc are its **extremities**. A directed graph may have arcs with the same extremities. A **source** is a vertex with no incoming edges, and a **sink** is a vertex with no outgoing edges.

The **out-degree** (resp. **in-degree**) $\deg^+(u)$ (resp. $\deg^-(u)$) of a vertex $u \in V$ is the number of arcs with tail (head) u. For a vertex $u \in V$, $N^+(u) = \{v \in V \backslash \{u\} : (u, v) \in A\}$ and $N^-(u) = \{v \in V \backslash \{u\} : (v, u) \in A\}$ stand for the **out-neighborhood** and the **in-neighborhood** of u, respectively. The **neighborhood** of $u \in V$ is defined to be $N(u) = N^+(u) \cup N^-(u)$. By extension, for a set $X \in V$, $N^+(X) = (\bigcup_{v \in X} N^+(v)) \backslash X$ and $N^-(X) = (\bigcup_{v \in X} N^-(v)) \backslash X$ stand for the out-neighborhood and the in-neighborhood of X, respectively. The **minimum out-degree** and **minimum in-degree** of $D = (V, A)$ are defined by

$$\delta^+(D) = \min_{u \in V} \deg^+(u) \quad \text{and}$$

$$\delta^-(D) = \min_{u \in V} \deg^-(u),$$

respectively. Similarly, the **maximum out-degree** and **maximum in-degree** of $D = (V, A)$ are defined by

$$\Delta^+(D) = \max_{u \in V} \deg^+(u) \quad \text{and}$$

$$\Delta^-(D) = \max_{u \in V} \deg^-(u),$$

respectively.

A.2.2 Paths and Cycles

A **directed path** is an oriented simple path such that all arcs go in the same direction, meaning all internal vertices have in- and out-degrees 1. A vertex u is **reachable** from another vertex v if there is a directed path that starts from v and ends at u. In contrast to undirected graphs, the condition that u is reachable from v does not imply that v is also reachable from u.

A **directed cycle** (or **circuit**) is an oriented simple cycle such that all arcs go in the same direction, meaning all vertices have in- and out-degrees 1. A digraph is **acyclic** if it does not contain any directed cycle.

Hamiltonian and Eulerian paths and directed cycles are defined as undirected graphs are, by replacing paths with directed paths and cycles with directed cycles.

A.2.3 Connectivity

A digraph is **strongly connected** if every vertex is reachable from every other, following the directions of the arcs. A digraph is **weakly connected** if its underlying undirected graph is connected.

A **strongly connected component** of a directed graph is a subgraph whose vertices are reachable from all other vertices in the subgraph. Reachability between vertices is established by the existence of a path between those vertices. The strongly connected components (SCC) of a directed graph are its maximal strongly connected subgraphs. These form a partition of the graph. If each strongly connected component is contracted to a single vertex, the resulting graph is a directed acyclic graph. The strongly connected components of a digraph $D = (V, A)$ can be computed in $\Theta(|V| + |A|)$ time (provided the directed graph is represented as an adjacency list).

A.2.4 Directed Acyclic Graphs

A **directed acyclic graph** (**DAG**) is a directed graph with no directed cycles. A finite DAG has at least one source and at least one sink. A **topological sort** of a DAG is a linear ordering of its vertices in which each vertex comes before all vertices to which it has outbound arcs. In general, this ordering is not unique. Any two graphs representing the same partial order have the same set of topological sort orders.

The **depth** of a vertex in a finite DAG is the length of the longest path from a source to that vertex, and its height is the length of the longest path from that vertex to a sink. The **length** of a finite DAG is the length (number of edges) of a longest directed path. It is equal to the maximum height of all sources and to the maximum depth of all sinks.

The **transitive closure** of a DAG G is the DAG obtained from G by adding an arc from u to v whenever v is reachable from u.

B
Complexity Theory

This survey on combinatrics of genome rearrangements is strongly oriented toward algorithmics and combinatorial optimization. We make an intensive use of the theory of complexity of algorithms. In this appendix we recall the basics of this theory, and introduce all the notions we use throughout the book. For more developments on complexity theory, the reader may refer to Garey and Johnson [181] and Papadimitriou [290].

B.1 The Class NP

A **decision problem** is a problem whose answer is either YES or NO. For any decision problem $\mathcal{P}$, we write $I(\mathcal{P})$ for the set of all instances of problem $\mathcal{P}$. The class **P** (polynomial time) contains all decision problems that can be solved by a deterministic polynomial-time algorithm. The class **NP** (nondeterministic polynomial time) contains all decision problems that can be solved by a nondeterministic polynomial-time algorithm. In an equivalent definition, the class **NP** contains those decision problems for which the YES answers have simple proofs of the fact that the answer is indeed YES (the theory refers to them as problems having a **polynomial-time verifier**). **NP** contains many important problems of practical interest, called **NP**-complete problems, for which no polynomial-time algorithms are known.

Example B.1 An independent set in a graph $G = (V, E)$ is a subgraph wherein every two vertices are not connected by an edge. A k-independent set is an independent set of size k. The INDEPENDENT SET problem is thus to determine whether a graph contains an independent set of a specified size k. This problem is in **NP**.

• Proof 1 (nondeterministic algorithm). Nondeterministically select a subset $V' \subseteq V$ of k vertices and test whether no edge connecting vertices in V' exists (the latter step is clearly a polynomial-time procedure). Return YES if the test passes, and NO otherwise.

• Proof 2 (polynomial-time verifier). Given a set V' of size k, (1) test whether V' is indeed a subset of V and (2) test whether no edge connecting vertices in V' exist. Return YES if the test passes, and NO otherwise.

Arguably, the greatest open question in theoretical computer science concerns the relationship between those two classes: Is **P** equal to **NP**? In essence, the **P** versus **NP** question is to determine whether any polynomial-time *verifiable* problem is polynomial-time *decidable*. Most computer scientists believe that $\mathbf{P} \neq \mathbf{NP}$.

Reducing one problem to another is a key concept in complexity theory, and **polynomial-time reductions** play a crucial role in the context of time complexity.

Definition B.1 A decision problem $\mathcal{P}$ is **polynomial-time reducible** to a problem $\mathcal{P}'$—in symbols, $\mathcal{P} \leq_P \mathcal{P}'$—if there is a deterministic algorithm that transforms instances $x \in I(\mathcal{P})$ into instances $x' \in I(\mathcal{P}')$, in such a way that the answer to x is YES if and only if the answer to x' is YES.

Polynomial-time reductions take into account the efficiency of computation. As a consequence, if one problem $\mathcal{P}$ is polynomial-time reducible to a problem $\mathcal{P}'$ known to have a polynomial-time solution, we obtain a polynomial-time solution for $\mathcal{P}$. More formally, if $\mathcal{P} \leq_P \mathcal{P}'$ and $\mathcal{P}' \in \mathbf{P}$, then $\mathcal{P} \in \mathbf{P}$.

In computational complexity theory, a computational problem is **complete** for a complexity class when it is, in a formal sense, one of the hardest in the complexity class. Hardness and completeness for the class **NP** are defined as follows.

Definition B.2 A problem $\mathcal{P}$ is **NP-hard** if every problem $\mathcal{P}' \in \mathbf{NP}$ is polynomial-time reducible to $\mathcal{P}$.

Definition B.3 A decision problem $\mathcal{P}$ is **NP-complete** if $\mathcal{P}$ is **NP**-hard and $\mathcal{P} \in \mathbf{NP}$.

NP-complete problems are thus the most difficult problems in **NP**. Indeed, a deterministic polynomial-time solution to any **NP**-complete problem would provide a solution to every other problem in **NP**.

Theorem B.1 If $\mathcal{P}$ is **NP**-complete and $\mathcal{P} \in \mathbf{P}$, then $\mathbf{P} = \mathbf{NP}$.

Theorem B.2 If $\mathcal{P}$ is **NP**-complete and $\mathcal{P} \leq_P \mathcal{P}'$ for some problem $\mathcal{P}' \in \mathbf{NP}$, then $\mathcal{P}'$ is **NP**-complete.

Based on the definition alone, however, it is not obvious that **NP**-complete problems do exist. The first problem proved to be **NP**-complete was the Boolean satisfiability (SAT) problem. This result came to be known as the Cook-Levin theorem [120]. A Boolean variable is one that can take on the value TRUE or FALSE. Usually, TRUE is represented by 1 and FALSE by 0. The Boolean operations AND,

OR, and NOT, represented by the symbols $\wedge$, $\vee$ and $\neg$, respectively, are described as follows:

$$0 \wedge 0 = 0 \quad 0 \vee 0 = 0 \quad \neg 0 = 1$$

$$0 \wedge 1 = 0 \quad 0 \vee 1 = 1 \quad \neg 1 = 0$$

$$1 \wedge 0 = 0 \quad 1 \vee 0 = 1$$

$$1 \wedge 1 = 1 \quad 1 \vee 1 = 1.$$

A Boolean formula is an expression written using only AND, OR, NOT, variables, and parentheses. It is satisfiable if some assignment of 0's and 1's to the variables makes the formula evaluate to TRUE. The SAT problem is to test whether a given Boolean formula is satisfiable.

Theorem B.3 (Cook-Levin Theorem [120]) The SAT problem is **NP**-complete.

Karp [230] showed that the SAT problem remains **NP**-complete even if the given Boolean formula is in conjunctive normal form (a conjunction of clauses, where a **clause** is a disjunction of literals) and each clause has three literals. This last problem is called 3-CNF-SAT.

According to definition B.3, proving a problem $\mathcal{P}$ is **NP**-complete requires (1) proving that $\mathcal{P}$ is in **NP** and (2) proving that every problem in **NP** reduces to $\mathcal{P}$. Proving membership in **NP**-complete is usually straightforward, but the latter step involves technical details about Turing machines. However, combining theorem B.2 with the Cook-Levin theorem is now enough to prove that a known **NP**-complete problem $\mathcal{P}'$ (see section B.2 for a partial list) polynomial-time reduces to $\mathcal{P}$. In 1972, Karp [230] showed in a breakthrough paper that twenty-one miscellaneous combinatorial and graph-theoretical problems are **NP**-complete.

Example B.2 As an illustration, we demonstrate that the INDEPENDENT SET problem is **NP**-complete. According to example B.1, we need to show its hardness. The polynomial-time reduction is from the 3-CNF-SAT problem. Let $\phi = C_1 \wedge C_2 \wedge \cdots \wedge C_m$ be a Boolean formula in conjunctive normal form where each clause C_i has three literals. Write $X = \{x_1, x_2, \ldots, x_n\}$, the set of Boolean variables involved in ϕ. The reduction generates (G, k), where $G = (V, E)$ is a graph and k is a positive integer.

The vertices of G are organized into m groups of three vertices each. Each group corresponds to one of the clauses in ϕ and each vertex in a group corresponds to a literal in the associated clause. The edges of G connect all vertices that belong to the same group and vertices from different groups that correspond to opposite literals. (Figure B.1 illustrates the construction.)

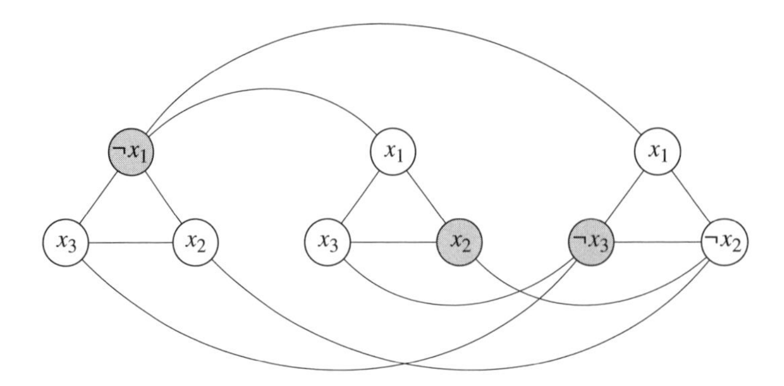

Figure B.1
The graph of the reduction produced from $\phi = (\neg x_1 \vee x_2 \vee x_3) \wedge (x_1 \vee x_2 \vee x_3) \wedge (x_1 \vee \neg x_2 \vee \neg x_3)$. A feasible solution is given by the assignment $x_1 = 0$, $x_2 = 1$, and $x_3 = 1$

We now want to show that ϕ is satisfiable if and only if G contains an independent set of size $k = m$. First, assume that ϕ has a satisfying assignment. In that satisfying assignment, at least one literal is true in every clause. Selecting in each group a vertex corresponding to a true literal in the satisfying assignment results in an independent set $V' \subseteq V$ of size $k = m$.

Conversely, assume G has an independent set $V' \subseteq V$ of size $k = m$. No vertex of V' occurs in the same group more than once, since vertices in a group are fully connected by edges. Therefore, V' contains exactly one vertex in each group. Assign truth values to the variables of ϕ in such a way that each literal labeling a vertex in V' is made true (note that this is always possible because opposite literals are connected in G). This assignment satisfies ϕ since each group contains a vertex in V'.

B.1.1 NP-Optimization Problems: From PTAS to APX

In **optimization problems**, we seek the *best* solution among a collection of possible solutions. Optimization problems are either minimization or maximization problems.

Definition B.4 An **NP-optimization problem** $\mathcal{P}$ is a quadruple $(I, \mathsf{sol}, m, \mathsf{goal})$ where

- I is a set of instances that is recognizable in polynomial time.
- Given an instance $x \in I$, $\mathsf{sol}(x)$ is the set of feasible solutions of x. These solutions are short, that is, a polynomial p exists such that, for any $y \in \mathsf{sol}(x)$, $|y| \leq p(|x|)$. Moreover, it is decidable in polynomial time whether, for any x and for any y such that $|y| \leq p(|x|)$, $y \in \mathsf{sol}(x)$.
- Given an instance x and a feasible solution y to x, $m(x, y)$ denotes the measure of y. The function m is computable in polynomial time and is also called the **objective function**.

- Goal is the goal function, and is either min or max.

The class **NPO** is the set of all **NP**-optimization problems.

Approximation algorithms are very often associated with **NP**-hard problems. However, since it is unlikely that there can ever be efficient polynomial-time exact algorithms solving **NP**-hard problems, one settles for polynomial-time nonoptimal solutions. Approximation algorithms are thus concerned with provable solution quality and provable run-time bounds. More formally, the goal of an **NPO** problem with respect to an instance x is thus to find an optimum solution, that is, a feasible solution y such that

$$m(x, y) = \mathsf{goal}\{m(x, y') : y' \in \mathsf{sol}(x)\}.$$

In the following, **opt** will denote the function mapping an instance x to the measure of an optimum solution.

Definition B.5 Let $\mathcal{P} = (I, \mathsf{sol}, m, \mathsf{goal})$ be an **NPO** problem. Given an instance $x \in I(\mathcal{P})$ and a feasible solution $y \in \mathsf{sol}(x)$, the *performance ratio* of y with respect to x is defined as

$$R(x, y) = \max\left\{\frac{m(x, y)}{\mathbf{opt}(x)}, \frac{\mathbf{opt}(x)}{m(x, y)}\right\}.$$

The performance ratio is always a number greater than or equal to 1; the closer it is to 1, the closer y is to an optimum solution.

Definition B.6 Let $\mathcal{P} = (I, \mathsf{sol}, m, \mathsf{goal})$ be an **NPO** problem and let A be an algorithm that, for any $x \in I$, returns a feasible solution $A(x)$ of x. Given an arbitrary function $f : \mathbb{N}^* \to \mathbb{R}^*$, A is called an $f(n)$-**approximate algorithm** (or an $f(n)$-**approximation algorithm**) for $\mathcal{P}$ if, for any $x \in I$, the performance ratio of the feasible solution $f(x)$ with respect to x verifies $R(x, A(x)) \leq f(|x|)$.

If an **NPO** problem admits an $f(n)$-approximate polynomial-time algorithm, it is said to be approximable within (or to have approximation ratio bounded by) $f(n)$. Based on the above definitions, several complexity classes can be defined.

Definition B.7 Let $\mathcal{P} = (I, \mathsf{sol}, m, \mathsf{goal})$ be an **NPO** problem. An algorithm A is said to be a **polynomial-time approximation scheme** (PTAS) for $\mathcal{P}$ if, for any $x \in I$ and for any rational $\varepsilon > 1$, $A(x, \varepsilon)$ returns a feasible solution of x whose performance ratio is at most ε.

Definition B.8 The class **PTAS** is the set of **NPO** problems that admit a polynomial-time approximation scheme.

The running time of a PTAS is required to be polynomial in n for every fixed ε but can be different for different ε. Therefore, algorithms running in $O(n^{1/\varepsilon})$ or even $O(n^{\exp(1/\varepsilon)})$ time are PTAS. Even more restrictive is the **fully polynomial-time approximation scheme** (**FPTAS**), which requires the algorithm to be polynomial in both the problem size n and $1/\varepsilon$.

Definition B.9 The class **FPTAS** is the set of **NPO** problems that admit a fully polynomial-time approximation scheme.

Relaxing the constraint that there is a polynomial-time algorithm to solve a problem within every fixed percentage (one algorithm for each percentage) results first in the class **APX**.

Definition B.10 The class **APX** is the set of **NPO** problems that allow polynomial-time approximation algorithms with approximation ratio bounded by a constant (or constant-factor approximation algorithms for short).

Example B.3 Given a graph $G = (V, E)$, the VERTEX COVER problem seeks to find a minimum cardinality subset $V' \subseteq V$ such that every edge of G has at least one end point in V'. The VERTEX COVER problem is 2-approximable, as shown by the following simple algorithm: "*Until all edges in G touch a marked edge, select an edge in G untouched by any marked edge and mark that edge. The algorithm returns all vertices that are end points of marked edges.*" An obvious implementation of the above algorithm runs in polynomial time. Let E' be the set of marked edges and let V' be the set of vertices that the algorithm outputs. The set V' is a vertex cover because E' contains or touches every edge in G, and hence V' touches all edges in G. What is left is to show that V' is at most twice as large as a minimum cardinality vertex cover, say $V_{\mathbf{opt}}$. This is indeed the case, since on the one hand $|V'| = 2|E'|$ because the edges in E' do not touch each other and, on the other hand, $V_{\mathbf{opt}}$ is a vertex cover, and hence every edge in E' is touched by some vertex in $V_{\mathbf{opt}}$. Furthermore, no such vertex in $V_{\mathbf{opt}}$ touches two edges in E' (again, this follows from the fact that the edges in E' do not touch each other), and hence $|V_{\mathbf{opt}}| \geq |E'|$. It follows that $|V'|/|V_{\mathbf{opt}}| \leq 2$, and V' is no more than twice as large as a minimum cardinality vertex cover of G.

By definition, **FPTAS** $\subseteq$ **PTAS** $\subseteq$ **APX**. It can be shown, however, that unless **P** $=$ **NP**, there are problems that are in **PTAS** but not in **FPTAS**, and that there are problems that are in **APX** but not in **PTAS**.

Theorem B.4 Unless **P** $=$ **NP**, **FPTAS** $\subsetneq$ **PTAS** $\subsetneq$ **APX**.

In the context of approximation problems, an approximation preserving reduction is needed. **PTAS reduction** is a reduction that is often used to perform reductions between solutions to optimization problems (see, e.g., Ausiello et al. [22]). It preserves the property that a problem has a polynomial-time approximation scheme (PTAS). A PTAS reduction from a problem $\mathcal{P}_1$ to a problem $\mathcal{P}_2$ is denoted by $\mathcal{P}_1 \leq_{\text{PTAS}} \mathcal{P}_2$. Hardness and completeness for the **APX** are defined as follows.

Definition B.11 An **NPO** problem is **APX**-hard if there is a PTAS reduction from every problem in **APX** to the **NPO** problem.

Definition B.12 An **NPO** problem $\mathcal{P}$ is **APX**-complete if $\mathcal{P}$ is **APX**-hard and $\mathcal{P} \in$ **APX**.

Theorem B.5 If $\mathcal{P}_1 \leq_{\text{PTAS}} \mathcal{P}_2$ and $\mathcal{P}_2 \in$ **APX** (respectively, $\mathcal{P}_2 \in$ **PTAS**), then $\mathcal{P}_1 \in$ **APX** (respectively, $\mathcal{P}_1 \in$ **PTAS**).

From theorem B.5, it immediately follows that if an **NPO** problem $\mathcal{P}$ is **APX**-hard, then it does not belong to **PTAS**. In other words, to say a problem is **APX**-hard is generally bad news, because it denies the existence of a PTAS, which is the most useful sort of approximation algorithm.

To show that a problem is **APX**-hard, it usually suffices to show a special type of polynomial-time reduction from some problem already known to be **APX**-hard. The most frequently used type of reduction is the L-reduction, introduced by Papadimitriou and Yannakakis [291].

Definition B.13 Let $\mathcal{P}_1 = (I_1, \mathsf{sol}_1, m_1, \mathsf{goal}_1)$ and $\mathcal{P}_2 = (I_2, \mathsf{sol}_2, m_2, \mathsf{goal}_2)$ be two **NPO** optimization problems. A pair f and g is an **L-reduction** if all of the following conditions are met:

- Both f and g are computable in a logarithmic amount of space.
- If $x \in I_1$, then $f(x) \in I_2$.
- If $y \in \mathsf{sol}_2(f(x))$, then $g(y) \in \mathsf{sol}_1(x)$.
- There exists a positive constant α such that

$$OPT_2(f(x)) \leq \alpha\, \mathbf{opt}_1(x).$$

- There exists a positive constant β such that

$$|OPT_1(x) - m_1(x, g(y))| \leq \beta |OPT_2(f(x)) - m_2(f(x), y)|.$$

The rationale of this concept stems from the following theorem.

Theorem B.6 [291] If (f, g) is an L-reduction from problem $\mathcal{P}_1$ to problem $\mathcal{P}_2$ with constants α and β, and there exists a polynomial-time ρ-approximation algorithm for $\mathcal{P}_2$, then there also exists a polynomial-time $(\rho\alpha\beta)$-approximation algorithm for $\mathcal{P}_1$.

It follows from theorem B.6 that if $\mathcal{P}_2$ has a polynomial-time approximation scheme, then so does $\mathcal{P}_1$. Papadimitriou and Yannakakis [291] and Ausiello et al. [22] give examples of L-reductions.

B.1.2 NP-Optimization Problems: Beyond APX

Unfortunately, not all **NPO** problems are in **APX** (i.e., have a fixed approximation ratio). The MINIMUM SET COVER and MAXIMUM INDEPENDENT SET problems are two well-known examples.

Given a collection C of subsets of a finite set X, the MINIMUM SET COVER seeks to find a minimum cardinality subset $C' \subseteq C$ such that every element in X belongs to at least one member of C'. This problem is approximable within ratio $1 + \log|X|$ [223], but not within ratio $c \log|X|$ for some $c > 0$ [311]. (Feige [169] shows a stronger lower bound under a plausible complexity hypothesis.) The MINIMUM SET COVER problem is, however, approximable within ratio $c \log|X|$ if every element of X belongs to at least $\varepsilon|C|$ sets from C for any $c > 0$ and $\varepsilon > 0$ (see Karpinski and Zelikovsky [231]).

Given a graph $G = (V, E)$, the MAXIMUM INDEPENDENT SET problem seeks to find a maximum cardinality independent set in G. This problem is approximable within $O(|V|/(\log|V|)^2)$ [79], but is not approximable within ratio $|V|^{1-\varepsilon}$ for any $\varepsilon > 0$ under a plausible complexity hypothesis [205].

In the light of such bad news, researchers have turned to considering restricted instances for better approximation ratios:

- Duh and Fürer [149] show that the MINIMUM SET COVER problem is **APX**-complete and is approximable within $\sum_{i=1}^{k} \frac{1}{i} - 1/2$ when the cardinality of all sets in C is bounded from above by a constant k. Papadimitriou and Yannakakis [291] show that the variant in which the number of occurrences in C of any element is bounded by a constant $B \geq 2$ is **APX**-complete and approximable within ratio B.
- The MAXIMUM INDEPENDENT SET problem is **APX**-complete for fixed $\Delta(G) \geq 3$ [291], and is approximable within $(\Delta(G) + 3)/5$ for small $\Delta(G)$ [54] and within ratio $O(\Delta \log \log \Delta/\log \Delta)$ for larger $\Delta = \Delta(G)$ [229]. Of particular importance, the MAXIMUM INDEPENDENT SET problem has a PTAS for planar graphs [32].

B.1.3 Parameterized Complexity

Parameterized complexity [148] is a measure of complexity of problems with multiple input parameters. The basic idea is that the combinatorial explosion that occurs in exact algorithms for many intractable problems can systematically be addressed by seeking parameters to fix to contain this explosion. Many hard computational problems have the following general form: Given an object x and a nonnegative integer k, does x have some property that depends on k? For example, given a graph

$G = (V, E)$ and a nonnegative integer k, the **NP**-complete VERTEX COVER problem seeks to find a vertex cover of size k. In parameterized complexity theory, k is called the **parameter**. In many practical applications, the parameter k can be considered "*small*" in comparison with the size $|x|$ of the input object x. Hence, it is of great practical interest to know whether these problems have deterministic algorithms that are exponential only with respect to k and polynomial with respect to $|x|$.

Definition B.14 A *parameterized problem* is a pair $(x, k) \in \Sigma^* \times \mathbb{N}^+$, where Σ is a finite alphabet. The first component, x, is the description of the instance, and the second component, k, is the parameter of the problem.

The first key notion in this context is the concept of fixed-parameter tractability.

Definition B.15 A parameterized problem $\mathcal{P}$ is **fixed-parameter tractable** if the question "$(x, k) \in \mathcal{P}$?" can be decided in running time $f(k)\,|x|^{O(1)}$, where f is an arbitrary function. The class **FPT** contains all problems that are fixed-parameter tractable.

FPT is thus the class of those parameterized problems for which the seemingly inherent "combinatorial explosion" really can be restricted to a (hopefully) "small part" of the input, the parameter. Extensive research in fixed-parameter algorithms has led to the development of a rich toolbox of parameterized techniques (see Niedermeier [282] for a recent account), a few of which are the following:

- *Depth-bounded search tree*: explore exhaustively all paths in a tree whose height depends solely on the parameter.
- *Data reduction and problem kernel*: reduce the parameterized problem to a kernel whose size depends solely on the parameter.
- *Dynamic programming*: combine properties of overlapping subproblems and optimality of substructures.
- *Tree decomposition*: algorithmic feasibility for many problems on graphs that are "almost" trees.
- *Iterative compression*: iteratively compress a feasible solution.

As a theoretic counterpart to fixed-parameter tractability, a theory of **parameterized intractability** has been developed. Unfortunately, the landscape of fixed-parameter intractable problems is not as simple as the landscape of classical intractable problems provided by the theory of **NP**-completeness. Central in parameterized intractability is the notion of **parameterized reduction** [148].

Definition B.16 Let $\mathcal{P}_1$ and $\mathcal{P}_2$ be two parameterized problems. A **standard parameterized m-reduction** transforms an input (x, k) of $\mathcal{P}_1$ into an input $(f(x, k), g(k))$ of $\mathcal{P}_2$ such that

- $(x, k) \in \mathcal{P}_1$ if and only if $(f(x, k), g(k)) \in \mathcal{P}_2$, and
- f runs in $(p(|x|)\, h(k))$ time for some polynomial p and some function h.

Note that polynomial-time reductions are rarely parameterized reductions. For example, the classical polynomial-time reduction from the INDEPENDENT SET problem to the CLIQUE problem (transformation of a graph into its complement) is *not* a parameterized reduction.

The most important parameterized complexity classes of parameterized intractable problems are the classes in the **W**-hierarchy [148].

$$\mathbf{FPT} \subseteq \mathbf{W[1]} \subseteq \mathbf{W[2]} \subseteq \cdots$$

For example, the class $\mathbf{W}[t]$, the t-th level of the **W**-hierarchy, is defined to be the class of all problems that are parameterized reducible to a parameterized version of the satisfiability problem for a certain class of Boolean circuits. Many natural parameterized problem that are not known to be fixed-parameter tractable have been shown to be complete for one of these classes (most of them for the first level $\mathbf{W[1]}$ or the second level $\mathbf{W[2]}$ of the hierarchy).

B.2 Some NP-Complete Problems

We list here in alphabetical order the definitions of some **NP**-complete problems that are mentioned in the book. Bracketed references before the problems' names are the numbers assigned to those problems by Garey and Johnson [181].

[SP1] 3-DIMENSIONAL-MATCHING

INSTANCE: Set $M \subseteq X \times Y \times Z$, where X, Y, and Z are disjoint sets having the same number q of elements.

QUESTION: Does M contain a matching (i.e., is there a subset of M of size q and such that no two elements of that subset agree in any coordinate)?

Reference: Garey and Johnson [181]. Transformation from 3-SATISFIABILITY.

[SP15] 3-PARTITION

INSTANCE: Set A of $3m$ elements, a natural B, and a size $s(a)$ for each object a in A such that $B/4 < s(a) < B/2$ and that $\sum_{a \in A} s(a) = mB$.

QUESTION: Can A be partitioned into m disjoint sets $A_1, A_2, \ldots, A_m$, each of cardinality 3, such that $\sum_{a \in A} s(a) = B$?

Reference: Garey and Johnson [181]. Transformation from 3-DIMENSIONAL MATCHING.

[LO2] 3-SATISFIABILITY

INSTANCE: Set U of variables, collection C of clauses over U such that each clause involves three variables.

QUESTION: Is there a satisfying truth assignment for C?

Reference: Garey and Johnson [181]. Transformation from SATISFIABILITY.

ALTERNATING CYCLE DECOMPOSITION

INSTANCE: The breakpoint graph $G = (V, E)$ of an unsigned permutation, natural K.

QUESTION: Can E be partitioned into at least K alternating cycles?

Reference: Caprara [93]. Transformation from EULERIAN CYCLE DECOMPOSITION. The problem was shown to be **APX**-hard by Berman and Karpinski [57].

[SR1] BIN-PACKING

INSTANCE: An integer capacity B, a finite set of integers $X = \{x_1, x_2, \ldots, x_n\}$ where $x_i \leq B$, a positive integer $k \leq n$.

QUESTION: Is there a partition of X into disjoint sets $X_1, X_2, \ldots, X_k$ such that the sum of elements in each X_i is at most B?

Reference: Garey and Johnson [181].

BREAKPOINT GRAPH DECOMPOSITION

See ALTERNATING CYCLE DECOMPOSITION.

[GT19] CLIQUE

INSTANCE: Graph $G = (V, E)$, positive integer $K \leq |V|$.

QUESTION: Does G contain a clique of size at least K?

Reference: Garey and Johnson [181]. Transformation from VERTEX COVER.

EULERIAN CYCLE DECOMPOSITION

INSTANCE: Eulerian graph $G = (V, E)$, natural K.

QUESTION: Can E be partitioned into at least K cycles?

Reference: Holyer [213].

[SP2] EXACT COVER BY 3-SETS

INSTANCE: Set X of size $3q$ and a collection C of three-element subsets of X.

QUESTION: Does C contain an exact cover for X (i.e., a subcollection $C' \subseteq C$ such that every element of X occurs in exactly one member of C')?

Reference: Garey and Johnson [181]. Transformation from 3-DIMENSIONAL MATCHING.

[GT38] HAMILTONIAN CYCLE

INSTANCE: A graph $G = (V, E)$.

QUESTION: Does G contain a Hamiltonian cycle (i.e., a cycle that visits each vertex exactly once)?

Reference: Garey and Johnson [181]. Transformation from VERTEX COVER.

[GT39] HAMILTONIAN PATH

INSTANCE: A graph $G = (V, E)$.

QUESTION: Does G contain a Hamiltonian path (i.e., a path that visits each vertex exactly once)?

Reference: Garey and Johnson [181]. Transformation from VERTEX COVER.

[GT20] INDEPENDENT SET

INSTANCE: Graph $G = (V, E)$, natural $K \leq |V|$.

QUESTION: Does G contain an independent set of size K or more (i.e., a subset $V' \subseteq V$ such that $|V'| \geq K$ and such that no two vertices of V' are adjacent to an edge in E)?

Reference: Garey and Johnson [181]. Transformation from VERTEX COVER.

LARGEST BALANCED INDEPENDENT SET

INSTANCE: Connected balanced bipartite graph $G = (V', V'', E)$, natural $1 \leq K \leq |V'|$.

QUESTION: Do there exist subsets $U' \subseteq V'$, $U'' \subseteq V''$ with $|U'| = |U''| = K$ such that no edge between vertices of U' and of U'' appears in E?

Reference: As noted by DasGupta et al. [131], the LARGEST BALANCED INDEPENDENT SET is equivalent to the LARGEST BALANCED COMPLETE BIPARTITE SUBGRAPH problem, which is shown to be **NP**-complete (see Garey and Johnson [181], problem [GT24]) by a transformation from CLIQUE.

[LO1] SATISFIABILITY

INSTANCE: Set U of variables, collection C of clauses over U.

QUESTION: Is there a satisfying truth assignment for C?

Reference: Garey and Johnson [181]. Generic transformation.

[SP3] SET PACKING

INSTANCE: Collection C of finite sets, natural $K \leq |C|$.

QUESTION: Does C contain at least K mutually disjoint sets?

Reference: Garey and Johnson [181]. Transformation from EXACT COVER BY 3-SETS.

[ND22] TRAVELING SALESMAN

INSTANCE: A set C of m cities, a distance d that takes natural values for each pair of cities, and a natural B.

QUESTION: Is there a tour of C of length B or less?

Reference: Garey and Johnson [181]. Transformation from HAMILTONIAN CYCLE.

[GT1] VERTEX COVER

INSTANCE: A graph $G = (V, E)$, a natural $K \leq |V|$.

QUESTION: Is there a vertex cover of size K or less for G (i.e., a subset of V of size K or less that contains an end point of each edge in E)?

Reference: Garey and Johnson [181]. Transformation from 3-SATISFIABILITY.

Glossary

$A(G)$	Adjacency matrix of graph G	239
$B(S)$	Number of uniform blocks in string S	141
$C^S D^S$	Block covering constraints	124
$C_{d,\pi^\circ}(\mathcal{T})$	Circular lower bound of tree topology $\mathcal{T}$ with parameters d and π°	214
$E(G)$	Edge set of the graph G	235
G/v	Local complementation of the neighborhood of vertex v in graph G	85
$G = (V, E)$	Graph with vertex set V and edge set E	235
$G[X]$	Induced subgraph of G defined by X	236
$IL(\pi)$	Interleaving graph of permutation π	83
K_n	Clique of order n	235
LCS	Length of the longest common substring of strings S and T	93
$MAD(\pi, \sigma)$	MAD number of permutations π and σ	118
$N(u)$	Neighborhood of the vertex u	235
$N^+(u)$	Out-neighborhood of vertex u	240
$N^-(u)$	In-neighborhood of vertex u	240
$N_{\mathcal{P}}$	Number of conserved intervals of set $\mathcal{P}$ of permutations	115
$OV(\pi)$	Overlap graph of permutation π	84
$SAD(\pi, \sigma)$	SAD number of permutations π and σ	118
$S_n^{\pm}$	Hyperoctahedral group on $\{1, 2, \ldots, n\}$	15
S_n	Symmetric group on $\{1, 2, \ldots, n\}$	14
T	Telomeric marker	162
$V(G)$	Vertex set of the graph G	235
$W_w(\mathcal{T})$	Weight of tree $\mathcal{T}$ with edge weights w	207

$BG(\Pi, \Gamma)$	Breakpoint graph of genomes Π and Γ	167
$BG(\pi)$	Breakpoint graph of permutation π	41
$BG(\pi)$	Breakpoint graph of signed permutation π	65
$count(g)$	Number of sets of $\mathcal{S}(h,k)$ in which g appears in an instance of synteny	184
$\Delta(G)$	Maximum degree of graph G	236
$\Delta^+(D)$	Maximum out-degree of digraph D	241
$\Delta^-(D)$	Maximum in-degree of digraph D	241
Id_Π	Identity genome of Π	166
$occ(S)$	Maximum number of occurrences of a gene in string S	92
$occ(a, S)$	Number of occurrences of a in string S	92
$\Omega(\Pi)$	Synteny graph of genome Π	203
$\Pi \oplus \Pi$	Doubled genome Π	200
$SG(\mathcal{S}(h,k))$	Synteny graph of $\mathcal{S}(h,k)$	184
$bd(S, T)$	Number of breakpoints between strings S and T	136
$bid(S, T)$	Block-interchange distance between strings S and T	153
$bidD_k(n)$	Block-interchange diameter for n length balanced strings	154
$bisd(S)$	Block-interchange sorting distance of string S	154
$bp(\pi, \sigma)$	Breakpoint distance between permutations π and σ	21
$cid(\mathcal{P}, \mathcal{Q})$	Conserved interval distance between sets of permutations $\mathcal{P}$ and $\mathcal{Q}$	115
$\circ$	Composition of two permutations	13
$dcj(\Pi, \Gamma)$	DCJ distance between genomes Π and Γ	170
$\deg^+(u)$	Out-degree of vertex u	240
$\deg^-(u)$	In-degree of vertex u	240
$\delta(G)$	Minimum degree of the graph G	236
$\delta^+(D)$	Minimum out-degree of digraph D	240
$\delta^-(D)$	Minimum in-degree of digraph D	240
$ebd(S, T)$	Exemplar breakpoint distance between strings S and T	106
$\equiv^\circ$	Relation constructing circular strings	93
$\equiv^\circ$	Relation defining circular permutations	19
$\equiv^\circ$	Relation defining genomic circular permutations	19
$erd(S, T)$	Exemplar reversal distance between strings S and T	106
$fbd(S, T)$	Full breakpoint distance between strings S and T	103

$hvd(\Pi)$	Halving distance of genome Π	203
$hvdD(N)$	Halving diameter for genomes with N chromosomes	205
ι	Identity permutation	14
κ_k	Permutation κ with parameter k	34
$kbd(\Pi, \Gamma)$	k-break rearrangement distance between Π and Γ	172
$lbed_E(S, T)$, $lbed_S(S, T)$	Large block edit distances between strings S and T	130
$lsyd(\mathcal{S}(h, k))$	Linear syntenic distance of $\mathcal{S}(h, k)$	187
$\mathcal{A}$	Alphabet	92
$\mathcal{A}^*$	The set of all strings on alphabet $\mathcal{A}$	92
$\mathcal{A}^n$	The set of all strings of length n on alphabet $\mathcal{A}$	92
$\mathcal{S}(h, k)$	Genome with k chromosomes on alphabet $\{1, 2, \ldots, h\}$	183
$\overline{G}$	Complement of the graph G	235
$\bar{\pi}$	$(0, \pi_n, \pi_{n-1}, \ldots, \pi_1) \circ (0, 1, 2, \ldots, n)$	177
π^l	Linear extension of permutation π	20
π^{-1}	Inverse of permutation π	14
π°	Circular permutation	19
$prD_k(n)$	Prefix reversal diameter	145
$prd(S, T)$	Prefix reversal distance between strings S and T	145
$prgd(S)$	Prefix reversal grouping distance of string S	146
$prsd(S)$	Prefix reversal sorting distance of string S	145
$ptd(S, T)$	Prefix transposition distance between strings S and T	152
$ptdD_k(n)$	Prefix transposition diameter for n-length balanced strings	152
$rd(\pi, \sigma)$	Reversal distance between signed permutations π and σ	106
$rdD_k(n)$	Reversal diameter for n-length strings	139
$rsd(S)$	Reversal sorting distance of string S	141
$rtd(\Pi, \Gamma)$	Reversal and translocation distance between genomes Π and Γ	172
$sbed(S, T)$	Symmetric block edit distance between strings S and T	128
$syd(\mathcal{S}(h, k))$	Syntenic distance of $\mathcal{S}(h, k)$	184
$\tau(i, j, k)$	Transposition with parameters i, j, k	25
$td(S, T)$	Transposition distance between strings S and T	148
$tdD_k(n)$	Transposition diameter for n-length balanced strings	148
$tld(\Pi)$	Translocation distance between genomes Π and Id_Π	166

$tld(\Pi, \Gamma)$	Translocation distance between genomes Π and Γ	165
$tsd(S)$	Transposition sorting distance of string S	149
$xbed(S, T)$	Exclusive breakpoint distance between strings S and T	103
$as(\pi, \sigma)$	Adjacency similarity between permutations π and σ	110
$b(S, T)$	Number of breakpoints between binary strings S and T	139
$bcd(S, T)$	Block covering distance between strings S and T	124
$bd(P, Q)$	Breakpoint distance between posets P and Q	80
$bp(\pi)$	Number of breakpoints of permutation π	21
$bp(\pi, \sigma)$	Breakpoint distance of permutations π and σ	21
$cis(\pi, \sigma)$	Conserved intervals similarity between permutations π and σ	117
$dcj(\pi)$	DCJ distance between permutations π and Id	73
$dd(\Pi, \Gamma)$	Double distance between genomes Π and Γ	201
$f(S)$	Number of gene families with at least two members in string S	92
$f_{ab}(S)$	Number of times ab occurs in string S	139
$hd(S, T)$	Hamming distance between strings S and T	128
$inv_2(\pi)$	"Bubble sort distance" of permutation π	51
$lcp(S, T)$	Length of the longest common prefix of strings S and T	93
$lcs(S, T)$	Length of the longest common suffix of strings S and T	93
$p(\pi)$	Number of points of permutation π	21
$rd(P, Q)$	Reversal distance between posets P and Q	79
$sb(\pi)$	Number of strong breakpoints of permutation π	21
$ulam(\pi)$	Ulam distance of permutation π	31
$z(S)$	Number of uniform blocks filled with zeros in string S	141
$\pi^{e\circ}$	Circular extention of permutation π	19
(S, T)	Pruning of strings S and T	99
$(P, \leq)$	A partially ordered set	76
$(\pi_1\ \pi_2\ \cdots\ \pi_n)$	A linear permutation	13
A_n	The alternating group on $\{1, 2, \ldots, n\}$	15
$G(\pi)$	Cycle graph of permutation π	28
$I(\mathcal{P})$	The set of all instances of problem $\mathcal{P}$	243
$\mathcal{L}(a_1, \ldots, a_k)$	The set of all balanced strings containing a_i members of the gene family i, for $1 \leq i \leq k$	94
$M_{*,i}$	The ith column of matrix M	85

Symbol	Description	Page
M_I	The largest element of interval I	63
$M_{i,*}$	The ith row of matrix M	85
$R(x, y)$	Performance ratio of solution y with respect to an instance x of an **NPO** problem	247
$[\pi_1\ \pi_2\ \cdots\ \pi_n]$	A circular permutation	19
$\downarrow Q$	The lower set of Q	77
$\Gamma(\pi)$	Γ-graph of permutation π	30
$\left[{n \atop k}\right]$	Stirling number of the first kind	51
$\Upsilon(\pi)$	Set of all clans of permutation π of length at least 3	38
$cid(\pi, \sigma)$	The conserved interval distance between π and σ	64
$fs(\pi)$	The number of conserved intervals of permutation π	64
γ_n	Gollan permutation of n elements	40
$\leq_P$	Polynomial-time reducibility between decision problems	244
$\leq_{\mathrm{PTAS}}$	PTAS reduction	248
$\lessdot$	Covering relation in posets	76
$\lvert LIS(\pi)\rvert$	Length of a longest increasing subsequence of permutation π	31
$\bar{x}^m$	Shorthand notation for $(x+m) \pmod{n+1}$	32
$\vec{\pi}$	A spin of permutation π	43
$\vec{\pi}$	Set of all spins of a permutation π	43
$perfd_S(\pi)$	Perfect reversal distance of permutation π, with respect to a subset S of its common intervals	69
$rd(\pi)$	Reversal distance of permutation π	40
$\rho(i, j)$	A reversal	40
$sk(\Pi, \Gamma)$	The maximum number of disjoint breakable subsets in $BG(\Pi, \Gamma)$	172
$srd(\pi)$	The reversal distance of signed permutation π	67
$syd(\Pi_1, \Pi_2)$	Syntenic distance between genomes Π_1 and Π_2	182
$syd_{ff}(\mathcal{S}(h, k))$	Syntenic distance of $\mathcal{S}(h, k)$ using only fusions and fissions	188
$tdrld_\alpha(\pi)$	Tandem-duplication random-loss distance of permutation π, with parameter α	59
a_P	The number of antichains in poset P	81
$bid(\pi)$	Block-interchange distance of permutation π	49
$c(\Gamma(\pi))$	Number of cycles in $\Gamma(\pi)$	30
$c^*(BG(\pi))$	Number of cycles in a maximal alternating cycle decomposition of $BG(\pi)$	42

$c_1(G(\pi))$	Number of 1-cycles in $G(\pi)$	38
$c_{even}(G(\pi))$	Number of even cycles in $G(\pi)$	29
$c_{even}(\Gamma(\pi))$	Number of even cycles in $\Gamma(\pi)$	30
$c_{odd}(G(\pi))$	Number of odd cycles in $G(\pi)$	29
$c_{old}(\Gamma(\pi))$	Number of odd cycles in $\Gamma(\pi)$	30
$des(\pi)$	Number of descents in a permutation π	27
$exc(\pi)$	Cayley distance of permutation π	50
$lc(\pi)$	Left code of permutation π	31
$lc(\pi_i)$	Left code of element π_i	31
$m(x, y)$	Measure of solution y to instance x of a problem	247
m_I	The smallest element of interval I	63
$pexc(\pi)$	Prefix exchange distance of permutation π	52
$plat(S)$	Number of plateaus in sequence S	31
$prd(\pi)$	Prefix reversal distance of permutation π	47
$ptb(\pi)$	Number of prefix transposition breakpoints of permutation π	37
$ptd(\pi)$	Prefix transposition distance of permutation π	37
$rc(\pi)$	Right code of permutation π	31
$rc(\pi_i)$	Right code of element π_i	31
$sol(x)$	Set of feasible solutions of instance x of a given problem	246
$t(\pi)$	Number of leaves of the tree of unoriented components of permutation π plus a correcting term	67
$td(\pi)$	Transposition distance of permutation π	26
$\mathcal{P}$	A (decision or optimization) problem	243
$\pi_\circ^\circ$	Toric permutation	32
S	$\mathcal{M}$-pruned string obtained from string S using matching $\mathcal{M}$	95
$c(G(\pi))$	Number of alternating cycles in $G(\pi)$	28
FPT	The class of all problems that are fixed-parameter tractable.	251
NP	The class of all decision problems that can be solved by a non-deterministic polynomial-time algorithm.	243
P	The class of all decision problems that can be solved by a deterministic polynomial-time algorithm.	243

Bibliography

[1] Z. Adam and D. Sankoff, *The ABCs of MGR with DCJ*, Evolutionary Bioinformatics, 4 (2008), pp. 69–74.

[2] S. Ahn and S. Tanksley, *Comparative linkage maps of the rice and maize genomes*, Proceedings of the National Academy of Sciences (USA), 90 (1993), pp. 7980–7984.

[3] M. Aigner and D. B. West, *Sorting by insertion of leading element*, Journal of Combinatorial Theory, ser. A, 45 (1987), pp. 306–309.

[4] Y. Ajana, J.-F. Lefebvre, E. R. M. Tillier, and N. El-Mabrouk, *Exploring the set of all minimal sequences of reversals—an application to test the replication-directed reversal hypothesis*, in Proceedings of the Second International Workshop on Algorithms in Bioinformatics (WABI), R. Guigo and D. Gusfield, eds., vol. 2452 of Lecture Notes in Computer Science, Springer-Verlag, 2002, pp. 300–315.

[5] S. B. Akers and B. Krishnamurthy, *A group-theoretic model for symmetric interconnection networks*, IEEE Transactions on Computers, 38 (Apr. 1989), pp. 555–566.

[6] S. B. Akers, D. Harel, and B. Krishnamurthy, *The star graph: An attractive alternative to the n-cube*, in Proceedings of the Fourth International Conference on Parallel Processing (ICPP), Pennsylvania State University Press, 1987, pp. 393–400.

[7] M. A. Alekseyev, *Multi-break rearrangements: From circular to linear genomes*, in Proceedings of the Fifth RECOMB Comparative Genomics Satellite Workshop (RECOMB-CG), G. Tesler and D. Durand, eds., vol. 4751 of Lecture Notes in Computer Science, Springer-Verlag, 2007, pp. 1–15.

[8] M. A. Alekseyev and P. A. Pevzner, *Genome halving problem revisited*, in Proceedings of the Twenty-fourth Conference on Foundations of Software Technology and Theoretical Computer Science (FSTTCS), K. Lodaya and M. Mahajan, eds., vol. 3328 of Lecture Notes in Computer Science, Springer-Verlag, 2004, pp. 1–15.

[9] M. A. Alekseyev and P. A. Pevzner, *Colored de Bruijn graphs and the genome halving problem*, IEEE/ACM Transactions on Computational Biology and Bioinformatics, 4, no. 1 (2007), pp. 98–107.

[10] M. A. Alekseyev and P. A. Pevzner, *Whole genome duplications, multi-break rearrangements, and genome halving problem*, in Proceedings of the Eighteenth Annual ACM-SIAM Symposium on Discrete Algorithms (SODA), N. Bansal, K. Pruhs, and C. Stein, eds., Society for Industrial and Applied Mathematics, 2007, pp. 665–679.

[11] M. A. Alekseyev and P. A. Pevzner, *Whole genome duplications and contracted breakpoint graphs*, SIAM Journal of Computing, 36 (2007), pp. 1748–1763.

[12] A. Amir, Y. Aumann, G. Benson, A. Levy, O. Lipsky, E. Porat, S. Skiena, and U. Vishne, *Pattern matching with address errors: Rearrangement distances*, in Proceedings of the Seventeenth Annual ACM-SIAM Symposium on Discrete Algorithms (SODA), ACM, 2006, pp. 1221–1229.

[13] A. Amir, Y. Aumann, P. Indyk, A. Levy, and E. Porat, *Efficient computations of l_1 and l_∞ rearrangement distances*, in Proceedings of the Fourteenth International Symposium on String Processing and Information Retrieval (SPIRE), N. Ziviani and R. Baeza-Yates, eds., vol. 4726 of Lecture Notes in Computer Science, Springer-Verlag, 2007, pp. 39–49.

[14] A. Amir, T. Hartman, O. Kapah, A. Levy, and E. Porat, *On the cost of interchange rearrangement in strings*, in Proceedings of the Fifteenth Annual European Symposium on Algorithms (ESA), L. Arge, M. Hoffmann, and E. Welzl, eds., vol. 4698 of Lecture Notes in Computer Science, Springer-Verlag, 2007, pp. 99–110.

[15] S. Angibaud, G. Fertin, I. Rusu, A. Thévenin, and S. Vialette, *Efficient tools for computing the number of breakpoints and the number of adjacencies between two genomes with duplicate genes*, Journal of Computational Biology, 15, no. 8 (2008), pp. 1093–1115.

[16] S. Angibaud, G. Fertin, I. Rusu, and S. Vialette, *How pseudo-Boolean programming can help genome rearrangement distance computation*, in Proceedings of the Fourth RECOMB Comparative Genomics Satellite Workshop (RECOMB-CG), G. Bourque and N. El-Mabrouk, eds., vol. 4205 of Lecture Notes in Computer Science, Springer-Verlag, 2006, pp. 75–86.

[17] S. Angibaud, G. Fertin, I. Rusu, A. Thévenin, and S. Vialette, *A pseudo-Boolean programming approach for computing the breakpoint distance between two genomes with duplicate genes*, in Proceedings of the Fifth RECOMB Comparative Genomics Satellite Workshop (RECOMB-CG), G. Tesler and D. Durand, eds., vol. 4751 of Lecture Notes in Computer Science, Springer-Verlag, 2007, pp. 16–29.

[18] S. Angibaud, G. Fertin, I. Rusu, and S. Vialette, *A pseudo-Boolean framework for computing rearrangement distances between genomes with duplicates*, Journal of Computational Biology, 14 (2007), pp. 379–393.

[19] S. Angibaud, G. Fertin, I. Rusu, A. Thévenin, and S. Vialette, *On the approximability of comparing genomes with duplicates*, Journal of Graph Algorithms and Applications (2008). In press.

[20] D. L. Applegate, R. E. Bixby, V. Chvátal, and W. J. Cook, The Traveling Salesman Problem: A Computational Study, Princeton Series in Applied Mathematics, Princeton University Press, 2007.

[21] W. Arndt and J. Tang, *Improving inversion median computation using commuting reversals and cycle information*, in Proceedings of the Fifth RECOMB Comparative Genomics Satellite Workshop (RECOMB-CG), G. Tesler and D. Durand, eds., vol. 4751 of Lecture Notes in Computer Science, Springer-Verlag, 2007, pp. 30–44.

[22] G. Ausiello, P. Crescenzi, G. Gambosi, V. Kann, A. Marchetti-Spaccamela, and M. Protasi, Complexity and Approximation: Combinatorial Optimization Problems and Their Approximability Properties, Springer-Verlag, 1999.

[23] A. Auyeung and A. Abraham, *Estimating genome reversal distance by genetic algorithm*, in IEEE Congress on Evolutionary Computation, vol. 2, IEEE Computer Society Press, 2003, pp. 1157–1161.

[24] D. Avis and M. Newborn, *On pop-stacks in series*, Utilitas Mathematica, 19 (1981), pp. 129–140.

[25] A. Bachrach, K. Chen, C. Harrelson, R. Mihaescu, S. Rao, and A. Shah, *Lower bounds for maximum parsimony with gene order data*, in Proceedings of the Third RECOMB Comparative Genomics Satellite Workshop (RECOMB-CG), A. McLysaght and D. H. Huson, eds., vol. 3678 of Lecture Notes in Computer Science, Springer-Verlag, 2005, pp. 1–10.

[26] D. A. Bader, B. M. E. Moret, and M. Yan, *A linear-time algorithm for computing inversion distance between signed permutations with an experimental study*, Journal of Computational Biology, 8, no. 5 (2001), pp. 483–491.

[27] M. Bader and E. Ohlebusch, *Sorting by weighted reversals, transpositions, and inverted transpositions*, in Proceedings of the Tenth Annual International Conference on Research in Computational Molecular Biology (RECOMB), A. Apostolico, C. Guerra, S. Istrail, P. A. Pevzner, and M. S. Waterman, eds., vol. 3909 of Lecture Notes in Computer Science, Springer-Verlag, 2006, pp. 563–577.

[28] M. Bader and E. Ohlebusch, *Sorting by weighted reversals, transpositions, and inverted transpositions*, Journal of Computational Biology, 14, no. 5 (2007), pp. 615–636.

[29] V. Bafna and P. A. Pevzner, *Genome rearrangements and sorting by reversals*, in Proceedings of the Thirty-fourth Annual Symposium on Foundations of Computer Science (FOCS), IEEE Computer Society Press, 1993, pp. 148–157.

[30] V. Bafna and P. A. Pevzner, *Sorting by transpositions*, SIAM Journal on Discrete Mathematics, 11 (May 1998), pp. 224–240 (electronic).

[31] V. Bafna, D. Beaver, M. Fürer, and P. A. Pevzner, *Circular permutations and genome shuffling*, in Comparative Genomics: Empirical and Analytical Approaches to Gene Order Dynamics, Map Alignment and the Evolution of Gene Families, vol. 1 of Sankoff and Nadeau [322], 2000, pp. 199–206.

[32] B. Baker, *Approximation algorithms for NP-complete problems on planar graphs*, Journal of the ACM, 41 (Jan. 1994), pp. 153–180.

[33] D. W. Bass and I. H. Sudborough, *On the shuffle-exchange permutation network*, in Proceedings of the Third International Symposium on Parallel Architectures, Algorithms, and Networks (ISPAN), IEEE Computer Society Press, 1997, pp. 165–171.

[34] D. W. Bass and I. H. Sudborough, *Pancake problems with restricted prefix reversals and some corresponding Cayley networks*, Journal of Parallel and Distributed Computing, 63, no. 3 (2003), pp. 327–336.

[35] T. Batu and S. C. Sahinalp, *Locally consistent parsing with applications to approximate string comparisons*, in Developments in Language Theory, C. de Felice and A. Restivo, eds., vol. 3572 of Lecture Notes in Computer Science, Springer-Verlag, 2005, pp. 22–35.

[36] W. W. Bein, L. L. Larmore, S. Latifi, and I. H. Sudborough, *Block sorting is hard*, International Journal of Foundations of Computer Science, 14, no. 3 (2003), pp. 425–437.

[37] W. W. Bein, L. L. Larmore, L. Morales, and I. H. Sudborough, *A faster and simpler 2-approximation algorithm for block sorting*, in Proceedings of the Fifteenth International Symposium on Fundamentals of Computation Theory (FCT), M. Liskiewicz and R. Reischuk, eds., vol. 3623 of Lecture Notes in Computer Science, Springer-Verlag, 2005, pp. 115–124.

[38] M. A. Bender, D. Ge, S. He, H. Hu, R. Pinter, S. Skiena, and F. Swidan, *Improved bounds on sorting with length-weighted reversals*, in Proceedings of the Fifteenth Annual ACM-SIAM Symposium on Discrete Algorithms (SODA), J. I. Munro, ed., Society for Industrial and Applied Mathematics, 2004, pp. 919–928.

[39] M. A. Bender, D. Ge, S. He, H. Hu, R. Pinter, S. Skiena, and F. Swidan, *Improved bounds on sorting by length-weighted reversals*, Journal of Computer and System Sciences, 74, no. 5 (2008), pp. 744–774.

[40] M. Benoît-Gagné and S. Hamel, *A new and faster method of sorting by transpositions*, in Proceedings of the Eighteenth Annual Symposium on Combinatorial Pattern Matching (CPM), B. Ma and K. Zhang, eds., vol. 4580 of Lecture Notes in Computer Science, Springer-Verlag, 2007, pp. 131–141.

[41] S. Bérard, A. Bergeron, C. Chauve, and C. Paul, *Perfect sorting by reversals is not always difficult*, IEEE/ACM Transactions on Computational Biology and Bioinformatics, 4 (Jan.–Mar. 2007), pp. 4–16.

[42] S. Bérard, C. Chauve, and C. Paul, *A more efficient algorithm for perfect sorting by reversals*, Information Processing Letters, 106 (2008), pp. 90–95.

[43] C. Berge, *Graphes et hypergraphes*, Monographies Universitaires de Mathématiques, no. 37, Dunod, 1970.

[44] A. Bergeron, *A very elementary presentation of the Hannenhalli-Pevzner theory*, Discrete Applied Mathematics, 146, no. 2 (2005), pp. 134–145.

[45] A. Bergeron and J. Stoye, *On the similarity of sets of permutations and its applications to genome comparison*, in Proceedings of the Ninth International Computing and Combinatorics Conference (COCOON), T. Warnow and B. Zhu, eds., vol. 2697 of Lecture Notes in Computer Science, Springer-Verlag, 2003, pp. 68–79.

[46] A. Bergeron and F. Strasbourg, *Experiments in computing sequences of reversals*, in Proceedings of the First Workshop on Algorithms in Bioinformatics (WABI), O. Gascuel and B. M. E. Moret, eds., vol. 2149 of Lecture Notes in Computer Science, Springer-Verlag, 2001, pp. 164–174.

[47] A. Bergeron, J. Mixtacki, and J. Stoye, *Reversal distance without hurdles and fortresses*, in Proceedings of the Fifth Annual Symposium on Combinatorial Pattern Matching (CPM), M. Crochemore and D. Gusfield, eds., vol. 807 of Lecture Notes in Computer Science, Springer-Verlag, 1994, pp. 388–399.

[48] A. Bergeron, C. Chauve, T. Hartman, and K. St-Onge, *On the properties of sequences of reversals that sort a signed permutation*, JOBIM, 2002, pp. 99–108.

[49] A. Bergeron, S. Heber, and J. Stoye, *Common intervals and sorting by reversals: A marriage of necessity*, Bioinformatics, 18, supp. 2 (2002), pp. S54–S63.

[50] A. Bergeron, J. Mixtacki, and J. Stoye, *The reversal distance problem*, in Gascuel [182].

[51] A. Bergeron, J. Mixtacki, and J. Stoye, *On sorting by translocations*, in Proceedings of the Ninth Annual International Conference on Research in Computational Molecular Biology (RECOMB), S. Miyano, J. Mesirov, S. Kasif, S. Istrail, P. A. Pevzner, and M. Waterman, eds., vol. 3500 of Lecture Notes in Bioinformatics, Springer-Verlag, 2005, pp. 615–629.

[52] A. Bergeron, J. Mixtacki, and J. Stoye, *A unifying view of genome rearrangements*, in Proceedings of the Sixth International Workshop on Algorithms in Bioinformatics (WABI), R. L. Malmberg, L. Cai, P. Berman, S. Rahmann, L. Zhang, and R. Sharan, eds., vol. 4175 of Lecture Notes in Computer Science, Springer-Verlag, 2006, pp. 163–173.

[53] A. Bergeron, J. Mixtacki, and J. Stoye, *HP distance via double cut and join distance*, in Proceedings of the Nineteenth Annual Symposium on Combinatorial Pattern Matching (CPM), P. Ferragina and G. M. Landau, eds., vol. 5029 of Lecture Notes in Computer Science, Springer-Verlag, 2008, pp. 56–68.

[54] P. Berman and T. Fujito, *On approximation properties of the independent set problem for degree 3 graphs*, in Proceedings of the Fourth International Workshop on Algorithms and Data Structures (WADS), S. G. Akl, F. K. H. A. Dehne, J.-R. Sack, and N. Santoro, eds., vol. 955 of Lecture Notes in Computer Science, Springer-Verlag, 1995, pp. 449–460.

[55] P. Berman and M. Fürer, *Approximating maximum independent set in bounded degree graphs*, in Proceedings of the Fifth Annual ACM-SIAM Symposium on Discrete Algorithms (SODA), Society for Industrial and Applied Mathematics, 1994, pp. 365–371.

[56] P. Berman and S. Hannenhalli, *Fast sorting by reversal*, in Proceedings of the Seventh Annual Symposium on Combinatorial Pattern Matching (CPM), D. S. Hirschberg and E. W. Myers, eds., vol. 1075 of Lecture Notes in Computer Science, Springer-Verlag, 1996, pp. 168–185.

[57] P. Berman and M. Karpinski, *On some tighter inapproximability results (extended abstract)*, in Proceedings of the Twenty-sixth International Colloquium on Automata, Languages and Programming (ICALP), J. Wiedermann, P. van Emde Boas, and M. Nielsen, eds., vol. 1644 of Lecture Notes in Computer Science, Springer-Verlag, 1999, pp. 200–209.

[58] P. Berman, S. Hannenhalli, and M. Karpinski, *1.375-approximation algorithm for sorting by reversals*, in Proceedings of the Tenth Annual European Symposium on Algorithms (ESA), R. H. Möhring and R. Raman, eds., vol. 2461 of Lecture Notes in Computer Science, Springer-Verlag, 2002, pp. 200–210.

[59] M. Bern and P. Plassmann, *The Steiner problem with edge lengths 1 and 2*, Information Processing Letters, 32 (1989), pp. 171–176.

[60] M. Bernt, D. Merkle, and M. Middendorf, *Genome rearrangement based on reversals that preserve conserved intervals*, IEEE/ACM Transactions on Computational Biology and Bioinformatics, 3, no. 3 (2006), pp. 275–288.

[61] M. Bernt, D. Merkle, and M. Middendorf, *A parallel algorithm for solving the reversal median problem*, in Proceedings of the Sixth International Conference on Parallel Processing and Applied Mathematics (PPAM), R. Wyrzykowski, J. Dongarra, N. Meyer, and J. Wasniewski, eds., vol. 3911 of Lecture Notes in Computer Science, Springer-Verlag, 2006, pp. 1089–1096.

[62] M. Bernt, D. Merkle, and M. Middendorf, *The reversal median problem, common intervals, and mitochondrial gene orders*, in Proceedings of the Second International Symposium on Computational Life Sciences (COMPLIFE), M. Berthold, R. Glen, and I. Fischer, eds., vol. 4216 of Lecture Notes in Bioinformatics, 2006, pp. 52–63.

[63] M. Bernt, D. Merkle, and M. Middendorf, *A fast and exact algorithm for the perfect reversal median problem*, in Proceedings of the Third International Symposium on Bioinformatics Research and Applications (ISBRA), I. Măndoiu and A. Zelikovsky, eds., vol. 4463 of Lecture Notes in Bioinformatics, Springer-Verlag, 2007, pp. 305–316.

[64] M. Bernt, D. Merkle, and M. Middendorf, *Using median sets for inferring phylogenetic trees*, Bioinformatics, 23 (2007), pp. e129–e135. Special issue of ECCB 2006.

[65] M. Bernt, D. Merkle, K. Ramsch, G. Fritzsch, M. Perseke, D. Bernhard, M. Schlegel, P. F. Stadler, and M. Middendorf, *CREx: Inferring genomic rearrangements based on common intervals*, Bioinformatics, 23, no. 21 (2007), pp. 2957–2958.

[66] M. Blanchette, T. Kunisawa, and D. Sankoff, *Parametric genome rearrangement*, Gene, 172 (1996), pp. GC11–GC17.

[67] M. BLANCHETTE, G. BOURQUE, AND D. SANKOFF, *Breakpoint phylogenies*, in vol. 8 of S. Miyano and Tagaki, eds., Genome Informatics, Universal Academy Press, 1997, pp. 25–34.

[68] G. BLIN AND R. RIZZI, *Conserved interval distance computation between non-trivial genomes*, in Proceedings of the Eleventh International Computing and Combinatorics Conference (COCOON), L. Wang, ed., vol. 3595 of Lecture Notes in Computer Science, Springer-Verlag, 2005, pp. 22–31.

[69] G. BLIN, C. CHAUVE, AND G. FERTIN, *The breakpoint distance for signed sequences*, in Proceedings of the First International Conference on Algorithms and Computational Methods for Biochemical and Evolutionary Networks, vol. 3 of Texts in Algorithms, King's College (London) and KCL Publications, 2004, pp. 3–16.

[70] G. BLIN, C. CHAUVE, AND G. FERTIN, *Genes order and phylogenetic reconstruction: Application to gamma-proteobacteria*, in Proceedings of the Third RECOMB Comparative Genomics Satellite Workshop (RECOMB-CG), A. McLysaght and D. H. Huson, eds., vol. 3678 of Lecture Notes in Computer Science, Springer-Verlag, 2005, pp. 11–20.

[71] G. BLIN, A. CHATEAU, C. CHAUVE, AND Y. GINGRAS, *Inferring positional homologs with common intervals of sequences*, in Proceedings of the Fourth RECOMB Comparative Genomics Satellite Workshop (RECOMB-CG), G. Bourque and N. El-Mabrouk, eds., vol. 4205 of Lecture Notes in Computer Science, Springer-Verlag, 2006, pp. 24–38.

[72] G. BLIN, E. BLAIS, D. HERMELIN, P. GUILLON, M. BLANCHETTE, AND N. EL-MABROUK, *Gene maps linearization using genomic rearrangement distances*, Journal of Computational Biology, 14, no. 4 (2007), pp. 394–407.

[73] G. BLIN, C. CHAUVE, G. FERTIN, R. RIZZI, AND S. VIALETTE, *Comparing genomes with duplications: A computational complexity point of view*, IEEE/ACM Transactions on Computational Biology and Bioinformatics, 4, no. 4 (2007), pp. 523–534.

[74] H.-J. BÖCKENHAUER AND D. BONGARTZ, eds., Algorithmic Aspects of Bioinformatics, Natural Computing Series, Springer-Verlag, 2007. Original German edition published by Teubner.

[75] M. BÓNA, *A survey of stack-sorting disciplines*, Electronic Journal of Combinatorics, 9 (2003).

[76] M. BÓNA, *Combinatorics of permutations*, in Discrete Mathematics and Its Applications, Chapman & Hall/CRC, 2004.

[77] D. BONGARTZ, *Algorithmic Aspects of Some Combinatorial Problems in Bioinformatics*, Ph.D. thesis, University of Viersen, Germany, 2006.

[78] J. BONIN AND A. DE MIER, *Lattice path matroids: Structural properties*, European Journal of Combinatorics, 27 (2006), pp. 701–738.

[79] R. BOPPANA AND M. HALLDÓRSSON, *Approximating maximum independent sets by excluding subgraphs*, BIT, 32, no. 2 (1992), pp. 180–196.

[80] P. BOSE, J. F. BUSS, AND A. LUBIW, *Pattern matching for permutations*, Information Processing Letters, 65 (1998), pp. 277–283.

[81] G. BOURQUE AND P. A. PEVZNER, *Genome-scale evolution: Reconstructing gene orders in the ancestral species*, Genome Research, 12 (2002), pp. 26–36.

[82] G. BOURQUE, P. A. PEVZNER, AND G. TESLER, *Reconstructing the genomic architecture of ancestral mammals: Lessons from human, mouse, and rat genomes*, Genome Research, 14 (2004), pp. 507–516.

[83] G. BOURQUE, Y. YACEF, AND N. EL-MABROUK, *Maximizing synteny blocks to identify ancestral homologs*, in Proceedings of the Third RECOMB Comparative Genomics Satellite Workshop (RECOMB-CG), A. McLysaght and D. H. Huson, eds., vol. 3678 of Lecture Notes in Computer Science, Springer-Verlag, 2005, pp. 21–34.

[84] G. BOURQUE, E. M. ZDOBNOV, P. BORK, P. A. PEVZNER, AND G. TESLER, *Comparative architectures of mammalian and chicken genomes reveal highly variable rates of genomic rearrangements across different lineages*, Genome Research, 15 (Jan. 2005), pp. 98–110.

[85] G. BOURQUE, G. TESLER, AND P. A. PEVZNER, *The convergence of cytogenetics and rearrangement-based models for ancestral genome reconstruction*, Genome Research, 16 (2006), pp. 311–313.

[86] M. BOUVEL AND D. ROSSIN, *A variant of the tandem duplication-random loss model of genome rearrangement*. Available at http://arxiv.org/abs/0801.2524, 2008.

[87] M. Braga, M.-F. Sagot, C. Scornavacca, and E. Tannier, *The solution space of sorting by reversals*, in Proceedings of the Third International Symposium on Bioinformatics Research and Applications (ISBRA), I. Măndoiu and A. Zelikovsky, eds., vol. 4463 of Lecture Notes in Computer Science, Springer-Verlag, 2007, pp. 293–304.

[88] D. Bryant, The Complexity of the Breakpoint Median Problem, tech. rep. CRM-2579, Centre de Recherches Mathématiques, Université de Montréal, 1998.

[89] D. Bryant, *The complexity of calculating exemplar distances*, in vol. 1 of Sankoff and Nadeau [322], pp. 207–212.

[90] D. Bryant, *A lower bound for the breakpoint phylogeny problem*, Journal of Discrete Algorithms, 2 (2004), pp. 229–255.

[91] B.-M. Bui-Xuan, M. Habib, and C. Paul, *Revisiting T. Uno and M. Yagiura's algorithm*, in Proceedings of the Sixteenth Annual International Symposium on Algorithms and Computation (ISAAC), X. Deng and D. Du, eds., Springer-Verlag, 2005, pp. 146–155.

[92] A. Caprara, *Formulations and hardness of multiple sorting by reversals*, in Proceedings of the Third Annual International Conference on Computational Molecular Biology (RECOMB), S. Istrail, P. Pevzner, and M. Waterman, eds., ACM, 1999, pp. 84–93.

[93] A. Caprara, *Sorting permutations by reversals and Eulerian cycle decompositions*, SIAM Journal on Discrete Mathematics, 12 (Feb. 1999), pp. 91–110 (electronic).

[94] A. Caprara, *On the tightness of the alternating-cycle lower bound for sorting by reversals*, Journal of Combinatorial Optimization, 3 (1999), pp. 149–182.

[95] A. Caprara, *On the practical solution of the reversal median problem*, in Proceedings of the First Workshop on Algorithms in Bioinformatics (WABI), O. Gascuel and B. M. E. Moret, eds., vol. 2149 of Lecture Notes in Computer Science, Springer-Verlag, 2001, pp. 238–251.

[96] A. Caprara, *Additive bounding, worst-case analysis and the breakpoint median problem*, SIAM Journal on Optimization, 13 (2002), pp. 508–519.

[97] A. Caprara, *The reversal median problem*, INFORMS Journal on Computing, 15, no. 1 (2003), pp. 93–113.

[98] A. Caprara and R. Rizzi, *Improved approximation for breakpoint graph decomposition and sorting by reversals*, Journal of Combinatorial Optimization, 6 (2002), pp. 157–182.

[99] A. Caprara, G. Lancia, and S.-K. Ng, *Sorting permutations by reversals through branch-and-price*, INFORMS Journal on Computing, 13 (2001), pp. 224–244.

[100] A. Cayley, *Note on the theory of permutations*, Philosophical Magazine, 34 (1849), pp. 527–529.

[101] M. Charikar, K. Makarychev, and Y. Makarychev, *Near-optimal algorithms for maximum constraint satisfaction problems*, in Proceedings of the Eighteenth Annual ACM-SIAM Symposium on Discrete Algorithms (SODA), N. Bansal, K. Pruhs, and C. Stein, eds., Society for Industrial and Applied Mathematics, 2007, pp. 62–68.

[102] K. Chaudhuri, K. Chen, R. Mihaescu, and S. Rao, *On the tandem duplication-random loss model of genome rearrangement*, in Proceedings of the Seventeenth Annual ACM-SIAM Symposium on Discrete Algorithms (SODA), ACM, 2006, pp. 564–570.

[103] C. Chauve and G. Fertin, *On maximal instances for the original syntenic distance*, Theoretical Computer Science, 326, no. 1–3 (2004), pp. 29–43.

[104] C. Chauve, G. Fertin, R. Rizzi, and S. Vialette, *Genomes containing duplicates are hard to compare*, in Proceedings of the Second International Workshop on Bioinformatics Research and Applications (IWBRA), V. N. Alexandrov, G. D. van Albada, P. M. A. Sloot, and J. Dongarra, eds., vol. 3992 of Lecture Notes in Computer Science, Springer-Verlag, 2006, pp. 783–790.

[105] T. Chen and S. Skiena, *Sorting with fixed-length reversals*, Discrete Applied Mathematics, 71 (1996), pp. 269–295.

[106] X. Chen, J. Zheng, Z. Fu, P. Nan, Y. Zhong, S. Lonardi, and T. Jiang, *Assignment of orthologous genes via genome rearrangement*, IEEE/ACM Transactions on Computational Biology and Bioinformatics, 2, no. 4 (2005), pp. 302–315.

[107] Z. Chen, R. H. Fowler, B. Fu, and B. Zhu, *Lower bounds on the approximation of the exemplar conserved distance problem of genomes*, in Proceedings of the Twelfth International Computing and Combinatorics Conference (COCOON), D. Z. Chen and D. T. Lee, eds., vol. 4112 of Lecture Notes in Computer Science, Springer-Verlag, 2006, pp. 245–254.

[108] Z. Chen, B. Fu, and B. Zhu, *The approximability of the exemplar breakpoint distance problem*, in Proceedings of the Second International Conference on Algorithmic Aspects in Information and Management (AAIM), S.-W. Cheng and C. K. Poon, eds., vol. 4041 of Lecture Notes in Computer Science, Springer-Verlag, 2006, pp. 291–302.

[109] Z. Chen, B. Fu, J. Xu, B. Yang, Z. Zhao, and B. Zhu, *Non-breaking similarity of genomes with gene repetitions*, in Proceedings of the Eighteenth Annual Symposium on Combinatorial Pattern Matching (CPM), B. Ma and K. Zhang, eds., vol. 4580 of Lecture Notes in Computer Science, Springer-Verlag, 2007, pp. 119–130.

[110] B. Chitturi and I. H. Sudborough, *Bounding prefix transposition distance for strings and permutations*, in Proceedings of the Forty-first Annual Hawaii International Conference on System Sciences (HICSS), IEEE Computer Society Press, 2008, p. 468.

[111] B. Chitturi, W. Fahle, Z. Meng, L. Morales, C. Shields, I. Sudborough, and W. Voit, *An (18/11)n upper bound for sorting by prefix reversals*, Theoretical Computer Science, (2008).

[112] B. Chitturi, H. Sudborough, W. Voit, and X. Feng, *Adjacent swaps on strings*, in Proceedings of the Fourteenth International Computing and Combinatorics Conference (COCOON), X. Hu and J. Wang, eds., vol. 5092 of Lecture Notes in Computer Science, Springer-Verlag, 2008, pp. 299–308.

[113] D. A. Christie, *Sorting permutations by block-interchanges*, Information Processing Letters, 60 (1996), pp. 165–169.

[114] D. A. Christie, *A 3/2-approximation algorithm for sorting by reversals*, in Proceedings of the Ninth Annual ACM-SIAM Symposium on Discrete Algorithms (SODA), Society for Industrial and Applied Mathematics, 1998, pp. 244–252.

[115] D. A. Christie, *Genome Rearrangement Problems*, Ph.D. thesis, University of Glasgow, Scotland, 1998.

[116] D. A. Christie and R. W. Irving, *Sorting strings by reversals and by transpositions*, SIAM Journal on Discrete Mathematics, 14, no. 2 (2001), pp. 193–206.

[117] M. Chrobak, P. Kolman, and J. Sgall, *The greedy algorithm for the minimum common string partition problem*, in Proceedings of the Seventh International Workshop on Approximation Algorithms for Combinatorial Optimization Problems (APPROX), vol. 3122 of Lecture Notes in Computer Science, Springer-Verlag, 2004, pp. 84–95.

[118] D. S. Cohen and M. Blum, *On the problem of sorting burnt pancakes*, Discrete Applied Mathematics, 61 (1995), pp. 105–120.

[119] E. Coissac, E. Maillier, and P. Netter, *A comparative study of duplication in bacteria and eukaryotes: The importance of telomeres*, Molecular Biology and Evolution, 14 (1997), pp. 1062–1074.

[120] S. Cook, *The complexity of theorem proving procedures*, in Proceedings of the Third Annual ACM Symposium on Theory of Computing (STOC), ACM, 1971, pp. 151–158.

[121] T. H. Cormen, C. E. Leiserson, R. L. Rivest, and C. Stein, *Introduction to Algorithms*, 2nd ed., MIT Press, 2001.

[122] G. Cormode and S. Muthukrishnan, *The string edit distance matching problem with moves*, in Proceedings of the Thirteenth Annual ACM-SIAM Symposium on Discrete Mathematics (SODA), Society for Industrial and Applied Mathematics, 2002, pp. 667–676.

[123] G. Cormode, S. Muthukrishnan, M. Paterson, S. C. Sahinalp, and U. Vishkin, *Techniques and applications for approximating string distances—rough draft*, 2000. http://citeseer.ist.psu.edu/320221.html.

[124] M. E. Cosner, *Phylogenetic and Molecular Evolutionary Studies of Chloroplast DNA Variations in the Campanulaceae*, Ph.D. thesis, Ohio State University, 1993.

[125] M. E. Cosner, R. K. Jansen, B. M. E. Moret, L. A. Raubeson, L.-S. Wang, T. Warnow, and S. Wyman, *An empirical comparison of phylogenetic methods on chloroplast gene order data in Campanulaceae*, in vol. 1 of Sankoff and Nadeau [322], pp. 99–121.

[126] D. Cranston, I. H. Sudborough, and D. B. West, *Short proofs for cut-and-paste sorting of permutations*, Discrete Mathematics, 307, no. 22 (2007), pp. 2866–2870.

[127] M. Crochemore, D. Hermelin, G. Landau, and S. Vialette, *Approximating the 2-interval pattern problem*, in Proceedings of the Thirteenth Annual European Symposium on Algorithms (ESA), G. S. Brodal and S. Leonardi, eds., vol. 3669 of Lecture Notes in Computer Science, Springer-Verlag, 2005, pp. 426–437.

[128] Y. Cui, L. Wang, and D. Zhu, *A 1.75-approximation algorithm for unsigned translocation distance*, Journal of Computer and System Sciences, 73 (2007), pp. 1045–1059.

[129] Y. Cui, L. Wang, D. Zhu, and X. Liu, *A $(1.5+\varepsilon)$-approximation algorithm for unsigned translocation distance*, IEEE/ACM Transactions on Computational Biology and Bioinformatics, 5, no. 1 (2008), pp. 56–66.

[130] A. C. E. Darling, B. Mau, F. R. Blattner, and N. T. Perna, *GRIL: Genome rearrangement and inversion locator*, Bioinformatics, 20 (Jan. 2004), pp. 122–124.

[131] B. DasGupta, T. Jiang, S. Kannan, M. Li, and E. Sweedyk, *On the complexity and approximation of syntenic distance*, Discrete Applied Mathematics, 88 (1998), pp. 59–82.

[132] B. Davey and H. Priestley, Introduction to Lattices and Order, 2nd ed., Cambridge University Press, 2002.

[133] W. H. E. Day, *Computationally difficult parsimony problems in phylogenetic systematics*, Journal of Theoretical Biology, 103 (1983), pp. 429–438.

[134] C. Demetrescu and I. Finocchi, *Combinatorial algorithms for feedback problems in directed graphs*, Information Processing Letters, 86 (2003), pp. 129–136.

[135] P. Diaconis, Group Representations in Probability and Statistics, vol. 11 of Lecture Notes—Monograph, Institute of Mathematical Sciences, 1988.

[136] Z. Dias and C. Carvalho de Souza, *Polynomial-sized ILP models for rearrangement distance problems*, in Proceedings of the Second Brazilian Symposium on Bioinformatics (BSB), M.-F. Sagot and M. E. M. T. Walter, eds., vol. 4643 in Lecture Notes in Computer Science, Springer-Verlag, 2007, pp. 74–85.

[137] Z. Dias and J. Meidanis, *Genome rearrangements distance by fusion, fission, and transposition is easy*, in Proceedings of the Eighth International Symposium on String Processing and Information Retrieval (SPIRE), IEEE Computer Society Press, 2001, pp. 250–253.

[138] Z. Dias and J. Meidanis, *Sorting by prefix transpositions*, in Proceedings of the Ninth International Symposium on String Processing and Information Retrieval (SPIRE), A. H. F. Laender and A. L. Oliveira, eds., vol. 2476 of Lecture Notes in Computer Science, Springer-Verlag, 2002, pp. 65–76.

[139] Z. Dias and J. Meidanis, *The syntenic distance problem using only fusions and fissions*, in Proceedings of the Second Brazilian Workshop on Bioinformatics (WOB), S. Lifschitz, N. F. Almeida Jr., G. J. Pappas Jr., and R. Linden, eds., 2003, pp. 72–79.

[140] Z. Dias, J. Meidanis, and M. E. M. T. Walter, *A new approach for approximating the transposition distance*, in Proceedings of the Seventh International Symposium on String Processing and Information Retrieval (SPIRE), IEEE Computer Society Press, 2000, pp. 199–208.

[141] G. Didier, *Common intervals of two sequences*, in Proceedings of the Third Workshop on Algorithms in Bioinformatics (WABI), G. Benson and R. Page, eds., vol. 2812 of Lecture Notes in Computer Science, Springer-Verlag, 2003, pp. 17–24.

[142] Y. Diekmann, M.-F. Sagot, and E. Tannier, *Evolution under reversals: Parsimony and conservation of common intervals*, IEEE/ACM Transactions on Computational Biology and Bioinformatics, 4, no. 2 (2007), pp. 301–309.

[143] R. Diestel, Graph Theory, vol. 173 of Graduate Texts in Mathematics, 3rd ed., Springer-Verlag, 2005.

[144] R. Dilworth, *A decomposition theorem for partially ordered sets*, Annals of Mathematics, 51 (1950), pp. 161–166.

[145] T. Dobzhansky and A. H. Sturtevant, *Inversions in the third chromosome of wild races of Drosophila pseudoobscura, and their use in the study of the history of the species*, Proceedings of the National Academy of Sciences (USA), 22, no. 7 (1936), 448–450.

[146] T. Dobzhansky and A. H. Sturtevant, *Inversions in the chromosomes of Drosophila pseudoobscura*, Genetics, 23 (1938), pp. 28–64.

[147] J.-P. Doignon and A. Labarre, *On Hultman numbers*, Journal of Integer Sequences, 10 (2007), art. 07.6.2.

[148] R. Downey and M. R. Fellows, Parameterized Complexity, Springer-Verlag, 1999.

[149] T. Duh and M. Fürer, *Approximation of k-set cover by semi-local optimization*, in Proceedings of the Twenty-ninth Annual ACM Symposium on the Theory of Computing (STOC), ACM, 1997, pp. 256–265.

[150] R. Durrett, *Genome rearrangement: Recent progress and open problems*, 2003. http://www.math.cornell.edu/~durrett/FGR.

[151] H. Dweighter, *Elementary problems and solutions, problem E2569*, American Mathematical Monthly, 82 (1975), 1010.

[152] J. V. Earnest-DeYoung, E. Lerat, and B. M. E. Moret, *Reversing gene erosion: Reconstructing ancestral bacterial genomes from gene-content and order data*, in Proceedings of the Fourth Workshop on Algorithms in Bioinformatics (WABI), I. Jonassen and J. Kim, eds., vol. 3240 of Lecture Notes in Computer Science, Springer-Verlag, 2004, pp. 1–13.

[153] J. Edmonds, *Paths, trees and flowers*, Canadian Journal of Mathematics, 17 (1965), 449–467.

[154] N. El-Mabrouk, *Sorting signed permutations by reversals and insertions/deletions of contiguous segments*, Journal of Discrete Algorithms, 1 (2000), pp. 105–121.

[155] N. El-Mabrouk, *Reconstructing an ancestral genome using minimum segments duplications and reversals*, Journal of Computer and System Sciences, 65, no. 3 (2002), pp. 442–464.

[156] N. El-Mabrouk, *Genome rearrangements with gene families*, in Gascuel [182], pp. 291–320.

[157] N. El-Mabrouk and D. Sankoff, *Hybridization and genome rearrangement*, in Proceedings of the Tenth Annual Symposium on Combinatorial Pattern Matching (CPM), M. Crochemore and M. Paterson, eds., vol. 1645 of Lecture Notes in Computer Science, Springer-Verlag, 1999, pp. 78–87.

[158] N. El-Mabrouk and D. Sankoff, *The reconstruction of doubled genomes*, SIAM Journal of Computing, 32, no. 3 (2003), pp. 754–792.

[159] N. El-Mabrouk, J. Nadeau, and D. Sankoff, *Genome halving*, in Proceedings of the Ninth Annual Symposium on Combinatorial Pattern Matching (CPM), G. Goos, J. Hartmanis, and J. van Leeuwen, eds., vol. 1448 of Lecture Notes in Computer Science, Springer-Verlag, 1998, pp. 235–250.

[160] I. Elias and T. Hartman, *A 1.375-approximation algorithm for sorting by transpositions*, IEEE/ACM Transactions on Computational Biology and Bioinformatics, 3, no. 4 (2006), pp. 369–379.

[161] F. Ergun, S. Muthukrishnan, and S. C. Sahinalp, *Comparing sequences with segment rearrangements*, in Proceedings of the Twenty-third Conference on Foundations of Software Technology and Theoretical Computer Science (FSTTCS), P. K. Pandya and J. Radhakrishnan, eds., vol. 2914 of Lecture Notes in Computer Science, Springer-Verlag, 2003, pp. 183–194.

[162] N. Eriksen, *$(1+\varepsilon)$-approximation of sorting by reversals and transpositions*, Theoretical Computer Science, 289, no. 1 (2002), pp. 517–529.

[163] N. Eriksen, *Reversal and transposition medians*, Theoretical Computer Science, 374, no. 1–3 (2007), pp. 111–126.

[164] H. Eriksson, K. Eriksson, J. Karlander, L. Svensson, and J. Wästlund, *Sorting a bridge hand*, Discrete Mathematics, 241 (2001), pp. 289–300. Selected papers in honor of Helge Tverberg.

[165] K. Eriksson, *Statistical and combinatorial aspects of comparative genomics*, Scandinavian Journal of Statistics, 31, no. 2 (2004), pp. 203–216.

[166] V. Estivill-Castro and D. Wood, *A survey of adaptive sorting algorithms*, ACM Computing Surveys, 24, no. 4 (1992), pp. 441–476.

[167] L. Euler, *Solutio problematis ad geometriam situs pertinentis*, Commentarii Academiae Scientiarum Iperialis Petropolitanae, 8 (1736), pp. 128–140.

[168] S. Even and O. Goldreich, *The minimum-length generator sequence problem is NP-hard*, Journal of Algorithms, 2 (1981), pp. 311–313.

[169] U. FEIGE, *A threshold of* log *n for approximating set cover*, Journal of the ACM, 45, no. 4 (1998), pp. 634–652.

[170] J. FELSENSTEIN, Inferring Phylogenies, 2nd ed., Sinauer Associates, 2003.

[171] J. FENG AND D. ZHU, *Faster algorithms for sorting by transpositions and sorting by block interchanges*, ACM Transactions on Algorithms, 3, no. 3 (2007), pp. 1–14.

[172] W. FENG, L. WANG, AND D. ZHU, *CTRD: A fast applet for computing signed translocation distance between genomes*, Bioinformatics, 20, no. 17 (2004), pp. 3256–3257.

[173] X. FENG, I. H. SUDBOROUGH, AND E. LU, *A fast algorithm for sorting by short swap*, in Proceedings of the IASTED International Conference on Computational and Systems Biology (CASB), D.-Z. Du, ed., ACTA Press, 2006, pp. 62–67.

[174] V. FERRETTI, J. H. NADEAU, AND D. SANKOFF, *Original synteny*, in Proceedings of the Seventh Annual Symposium on Combinatorial Pattern Matching (CPM), D. S. Hirschberg and E. W. Myers, eds., vol. 1075 of Lecture Notes in Computer Science, Springer-Verlag, 1996, pp. 159–167.

[175] M. FIGEAC AND J.-S. VARRÉ, *Sorting by reversals with common intervals*, in Proceedings of the Fourth Workshop on Algorithms in Bioinformatics (WABI), I. Jonassen and J. Kim, eds., vol. 3240 of Lecture Notes in Computer Science, Springer-Verlag, 2004, pp. 26–37.

[176] J. FISCHER AND S. W. GINZINGER, *A 2-approximation algorithm for sorting by prefix reversals*, in Proceedings of the Thirteenth Annual European Symposium on Algorithms (ESA), G. S. Brodal and S. Leonardi, eds., vol. 3669 of Lecture Notes in Computer Science, Springer-Verlag, 2005, pp. 415–425.

[177] V. J. FORTUNA, *Distâncias de Transposição entre Genomas*, master's thesis, Universidade Estadual de Campinas, Brazil, 2005.

[178] L. R. FOULDS AND R. L. GRAHAM, *The Steiner tree problem in phylogeny is NP-complete*, Advances in Applied Mathematics, 3 (1982), pp. 43–49.

[179] L. FROENICKE, M. G. CALDÉS, A. GRAPHODATSKY, S. MÜLLER, L. A. LYONS, T. J. ROBINSON, M. VOLLETH, F. YANG, AND J. WIENBERG, *Are molecular cytogenetics and bioinformatics suggesting diverging models of ancestral mammalian genomes?*, Genome Research, 16, no. 3 (2006), pp. 306–310.

[180] Z. FU AND T. JIANG, *Computing the breakpoint distance between partially ordered genomes*, Journal of Bioinformatics and Computational Biology, 5, no. 5 (2007), pp. 1087–1101.

[181] M. R. GAREY AND D. S. JOHNSON, Computers and Intractability: A Guide to the Theory of NP-Completeness, W. H. Freeman, 1979.

[182] O. GASCUEL, ed., Mathematics of Evolution and Phylogeny, Oxford University Press, 2005.

[183] W. H. GATES AND C. H. PAPADIMITRIOU, *Bounds for sorting by prefix reversal*, Discrete Mathematics, 27 (1979), pp. 47–57.

[184] B. GAUT AND J. DOEBLEY, *DNA sequence evidence for the segmental allotetraploid origin of maize*, Proceedings of the National Academy of Sciences (USA), 94 (1997), pp. 6809–6814.

[185] T. GIBSON AND J. SPRING, *Evidence in favor of ancient octaploidy in the vertebrate genome*, Biochemical Society Transactions, 28 (2000), pp. 259–264.

[186] S. GOG AND M. BADER, *How to achieve an equivalent simple permutation in linear time*, in Proceedings of the Fifth RECOMB Comparative Genomics Satellite Workshop (RECOMB-CG), G. Tesler and D. Durand, eds., vol. 4751 of Lecture Notes in Computer Science, Springer-Verlag, 2007, pp. 58–68.

[187] S. GOG, M. BADER, AND E. OHLEBUSCH, *GENESIS: Genome evolution scenarios*, Bioinformatics, 24, no. 5 (2008), pp. 711–712.

[188] A. GOLDSTEIN, P. KOLMAN, AND J. ZHENG, *Minimum common string partition problem: Hardness and approximations*, Electronic Journal of Combinatorics, 12 (2005), no. R50.

[189] J. GRAMM AND R. NIEDERMEIER, *Breakpoint medians and breakpoint phylogenies: A fixed-parameter approach*, Bioinformatics, 18, supp. 1 (2002), pp. S128–S139.

[190] Q.-P. GU, S. PENG, AND Q. M. CHEN, *Sorting permutations and its applications in genome analysis*, Lectures on Mathematics in the Life Sciences, 26 (1999), pp. 191–201.

[191] Q.-P. GU, S. PENG, AND I. H. SUDBOROUGH, *A 2-approximation algorithm for genome rearrangements by reversals and transpositions*, Theoretical Computer Science, 210 (1999), pp. 327–339.

[192] S. A. GUYER, L. S. HEATH, AND J. P. VERGARA, *Subsequence and run heuristics for sorting by transpositions*, Technical Report, Virginia State University, 1997.

[193] E. GYÖRI AND G. TURÁN, *Stack of pancakes*, Studia Scientiarum Mathematicarum Hungarica, 13 (1978), pp. 133–137.

[194] Y. HAN, *Improving the efficiency of sorting by reversals*, in Proceedings of the 2006 International Conference on Bioinformatics and Computational Biology, CSREA Press, 2006, pp. 406–409.

[195] S. HANNENHALLI, *Polynomial-time algorithm for computing translocation distance between genomes*, Discrete Applied Mathematics, 71 (1996), pp. 137–151.

[196] S. HANNENHALLI AND P. A. PEVZNER, *Transforming men into mice: Polynomial algorithm for genomic distance problem*, in Proceedings of the Thirty-sixth Annual Symposium on Foundations of Computer Science (FOCS), IEEE Computer Society Press, 1995, pp. 581–592.

[197] S. HANNENHALLI AND P. A. PEVZNER, *Towards a computational theory of genome rearrangements*, in Computer Science Today: Recent Trends and Developments, Jan van Leeuwen, ed., vol. 1000 of Lecture Notes in Computer Science, Springer-Verlag, 1995, pp. 184–202.

[198] S. HANNENHALLI AND P. A. PEVZNER, *To cut... or not to cut (applications of comparative physical maps in molecular evolution)*, in Proceedings of the Seventh Annual ACM-SIAM Symposium on Discrete Algorithms (SODA), Society for Industrial and Applied Mathematics, 1996, pp. 304–313.

[199] S. HANNENHALLI AND P. A. PEVZNER, *Transforming cabbage into turnip: Polynomial algorithm for sorting signed permutations by reversals*, Journal of the ACM, 46, no. 1 (1999), pp. 1–27.

[200] S. HANNENHALLI, C. CHAPPEY, E. V. KOONIN, AND P. A. PEVZNER, *Genome sequence comparison and scenarios for gene rearrangements: A test case*, Genomics, 30, no. 2 (1995), pp. 299–311.

[201] T. HARTMAN, *A simpler 1.5-approximation algorithm for sorting by transpositions*, in Proceedings of the Fourteenth Annual Symposium on Combinatorial Pattern Matching (CPM), R. A. Baeza-Yates, E. Chávez, and M. Crochemore, eds., vol. 2676 of Lecture Notes in Computer Science, Springer-Verlag, 2003, pp. 156–169.

[202] T. HARTMAN AND R. SHAMIR, *A simpler and faster 1.5-approximation algorithm for sorting by transpositions*, Information and Computation, 204, no. 2 (2006), pp. 275–290.

[203] T. HARTMAN AND R. SHARAN, *A 1.5-approximation algorithm for sorting by transpositions and transreversals*, Journal of Computer and System Sciences, 70 (2005), pp. 300–320.

[204] T. HARTMAN AND E. VERBIN, *Matrix tightness: A linear-algebraic framework for sorting by transpositions*, in Proceedings of the Thirteenth International Conference on String Processing and Information Retrieval (SPIRE), F. Crestani, P. Ferragina, and M. Sanderson, eds., vol. 4209 of Lecture Notes in Computer Science, Springer-Verlag, 2006, pp. 279–290.

[205] J. HÅSTAD, *Clique is hard to approximate within* $n^{1-\varepsilon}$, Acta Mathematica, 182 (1999), pp. 105–142.

[206] D. HE, *A novel greedy algorithm for the minimum common string partition problem*, in Proceedings of the Third International Workshop on Bioinformatics Research and Applications (IWBRA), I. Măndoiu and A. Zelikovsky, eds., vol. 4463 of Lecture Notes in Computer Science, Springer-Verlag, 2007, pp. 441–452.

[207] L. S. HEATH AND J. P. C. VERGARA, *Sorting by bounded block-moves*, Discrete Applied Mathematics, 88 (1998), pp. 181–206.

[208] L. S. HEATH AND J. P. C. VERGARA, *Sorting by short block-moves*, Algorithmica, 28 (2000), pp. 323–352.

[209] L. S. HEATH AND J. P. C. VERGARA, *Sorting by short swaps*, Journal of Computational Biology, 10, no. 2 (2003), pp. 775–789.

[210] S. HEBER AND J. STOYE, *Finding all common intervals of k permutations*, in Proceedings of the Twelfth Annual Symposium on Combinatorial Pattern Matching (CPM), A. Amir and G. M. Landau, eds., vol. 2089 of Lecture Notes in Computer Science, Springer-Verlag, 2001, pp. 207–218.

[211] S. HEDETNIEMI, S. HEDETNIEMI, AND A. LIESTMAN, *A survey of gossiping and broadcasting in communication networks*, Networks, 18 (1988), pp. 319–349.

[212] M. H. HEYDARI AND I. H. SUDBOROUGH, *On the diameter of the pancake network*, Journal of Algorithms, 25, no. 1 (1997), pp. 67–94.

[213] I. Holyer, *The NP-completeness of some edge-partition problems*, SIAM Journal of Computing, 10 (1981), pp. 713–717.

[214] A. Hultman, *Toric Permutations*, master's thesis, Department of Mathematics, Kungliga Tekniska Högskolan, Stockholm, 1999.

[215] C. Hurkens, L. van Iersel, J. Keijsper, S. Kelk, L. Stougie, and J. Tromp, *Prefix reversals on binary and ternary strings*, SIAM Journal on Discrete Mathematics, 21, no. 3 (2007), pp. 592–611.

[216] Y. Interian and R. Durrett, *Genomic midpoints: Computation and evolutionary implications*. Submitted.

[217] Y. Interian and R. Durrett, *Computing genomic midpoints*. http://www.math.cornell.edu/~durrett/mdpt/mdpt.html.

[218] B. N. Jackson, P. S. Schnable, and S. Aluru, *Consensus genetic maps as median orders from inconsistent sources*, IEEE/ACM Transactions on Computational Biology and Bioinformatics, 5, no. 2 (2008), pp. 161–171.

[219] O. Jaillon, J.-M. Aury, F. Brunet, J.-L. Petit, N. Stange-Thomann, E. Mauceli, L. Bouneau, C. Fischer, C. Ozouf-Costaz, A. Bernot, S. Nicaud, D. Jaffe, S. Fisher, G. Lutfalla, C. Dossat, B. Segurens, C. Dasilva, M. Salanoubat, M. Levy, N. Boudet, S. Castellano, V. Anthouard, C. Jubin, V. Castelli, M. Katinka, B. Vacherie, C. Biémont, Z. Skalli, L. Cattolico, J. Poulain, V. de Berardinis, C. Cruaud, S. Duprat, P. Brottier, J.-P. Coutanceau, J. Gouzy, G. Parra, G. Lardier, C. Chapple, K. J. McKernan, P. McEwan, S. Bosak, M. Kellis, J.-N. Volff, R. Guigó, M. C. Zody, J. Mesirov, K. Lindblad-Toh, B. Birren, C. Nusbaum, D. Kahn, M. Robinson-Rechavi, V. Laudet, V. Schachter, F. Quétier, W. Saurin, C. Scarpelli, P. Wincker, E. S. Lander, J. Weissenbach, and H. R. Crollius, *Genome duplication in the teleost fish Tetraodon nigroviridis reveals the early vertebrate proto-karyotype*, Nature, 431, no. 7011 (2004), pp. 946–957.

[220] G. Jean and M. Nikolski, *Genome rearrangements: A correct algorithm for optimal capping*, Information Processing Letters, 104, no. 1 (2007), pp. 14–20.

[221] M. R. Jerrum, *The complexity of finding minimum-length generator sequences*, Theoretical Computer Science, 36 (1985), pp. 265–289.

[222] T. Jiang, Y. Xu, and M. Q. Zhang, eds., Current Topics in Computational Molecular Biology, MIT Press, 2002.

[223] D. Johnson, *Approximation algorithms for combinatorial problems*, Journal of Computer and System Sciences, 9 (1974), pp. 256–278.

[224] N. C. Jones and P. A. Pevzner, An Introduction to Bioinformatics Algorithms, MIT Press, 2004.

[225] V. Kann, *On the Approximability of NP-Complete Optimization Problems*, Ph.D. thesis, Department of Numerical Analysis and Computing Science, Royal Institute of Technology, Stockholm, 1992.

[226] H. Kaplan and N. Shafrir, *The greedy algorithm for edit distance with moves*, Information Processing Letters, 97, no. 1 (2006), pp. 23–27.

[227] H. Kaplan and E. Verbin, *Sorting signed permutations by reversals, revisited*, Journal of Computer and System Sciences, 70, no. 3 (2005), pp. 321–341.

[228] H. Kaplan, R. Shamir, and R. E. Tarjan, *A faster and simpler algorithm for sorting signed permutations by reversals*, SIAM Journal of Computing, 29, no. 3 (2000), pp. 880–892 (electronic).

[229] D. Karger, R. Motwani, and M. Sudan, *Approximate graph coloring by semidefinite programming*, Journal of the ACM, 45, no. 2 (1998), pp. 246–265.

[230] R. M. Karp, *Reducibility among combinatorial problems*, in Complexity of Computer Computations, R. E. Miller and J. W. Thatcher, eds., Plenum, 1972, pp. 85–103.

[231] M. Karpinski and A. Zelikovsky, *Approximating dense cases of covering problems*, Electronic Colloquium on Computational Complexity, tech. rep. TR97-004 (1997), p. 4.

[232] J. Kececioglu and D. Sankoff, *Exact and approximation algorithms for sorting by reversals, with application to genome rearrangement*, Algorithmica, 13 (1995), pp. 180–210.

[233] J. D. Kececioglu and R. Ravi, *Of mice and men: Algorithms for evolutionary distances between genomes with translocation*, in Proceedings of the Sixth Annual ACM-SIAM Symposium on Discrete Algorithms (SODA), Society for Industrial and Applied Mathematics, 1995, pp. 604–613.

[234] M. KELLIS, B. BIRREN, AND E. LANDER, *Proof and evolutionary analysis of ancient genome duplication in the yeast* Saccharomyces cerevisiae, Nature, 428 (2004), pp. 617–624.

[235] J. D. KLEIN, *Problem Kernels for Genome Rearrangement Problems*, diplomarbeit, Friedrich-Schiller-Universit ät Jena, 2007.

[236] J. KLEINBERG AND D. LIBEN-NOWELL, *The syntenic diameter of the space of N-chromosome genomes*, in vol. 1 of Sankoff and Nadeau [322], pp. 185–197.

[237] D. E. KNUTH, Sorting and Searching, vol. 3 of his The Art of Computer Programming, 2nd ed., Addison-Wesley, 1995.

[238] P. KOLMAN, *Approximating reversal distance for strings with bounded number of duplicates*, in Proceedings of the Thirtieth International Symposium on Mathematical Foundations of Computer Science (MFCS), J. Jedrzejowicz and A. Szepietowski, eds., vol. 3618 of Lecture Notes in Computer Science, Springer-Verlag, 2005, pp. 580–590.

[239] P. KOLMAN AND T. WALEŃ, *Reversal distance for strings with duplicates: Linear time approximation using hitting set*, in Proceedings of the Fourth International Workshop on Approximation and Online Algorithms (WAOA), T. Erlebach and C. Kaklamanis, eds., vol. 4368 of Lecture Notes in Computer Science, Springer-Verlag, 2006, pp. 279–289.

[240] A. LABARRE, *New bounds and tractable instances for the transposition distance*, IEEE/ACM Transactions on Computational Biology and Bioinformatics, 3, no. 4 (2006), pp. 380–394.

[241] A. LABARRE, *Edit distances and factorisations of even permutations*, in Proceedings of the Sixteenth Annual European Symposium on Algorithms (ESA), Lecture Notes in Computer Science, Springer-Verlag, 5193 (2008), pp. 635–646.

[242] S. LAKSHMIVARAHAN, J.-S. JWO, AND S. K. DHALL, *Symmetry in interconnection networks based on Cayley graphs of permutation groups: A survey*, Parallel Computing, 19 (1993), pp. 361–407.

[243] R. LANGKJÆR, P. CLIFTEN, M. JOHNSTON, AND J. PISKUR, *Yeast genome duplication was followed by asynchronous differentiation of duplicated genes*, Nature, 421 (2003), pp. 848–852.

[244] A. LEMPEL, *Matrix factorization over GF(2) and trace-orthogonal bases of* $GF(2^n)$, SIAM Journal of Computing, 4, no. 2 (1975), pp. 175–186.

[245] R. LENNE, C. SOLNON, T. STÜTZLE, E. TANNIER, AND M. BIRATTARI, *Reactive stochastic local search algorithms for the genomic median problem*, in Proceedings of the Eighth European Conference on Evolutionary Computation in Combinatorial Optimization (EVOCOP), J. I. van Hemert and C. Cotta, eds., vol. 4972 of Lecture Notes in Computer Science, Springer-Verlag, 2008, pp. 266–276.

[246] G. LI, X. QI, X. WANG, AND B. ZHU, *A linear-time algorithm for computing translocation distance between signed genomes*, in Proceedings of the Fifteenth Annual Symposium on Combinatorial Pattern Matching (CPM), S. C. Sahinalp, S. Muthukrishnan, and U. Dogrusoz, eds., vol. 3109 of Lecture Notes in Computer Science, Springer-Verlag, 2004, pp. 323–332.

[247] Z. LI, L. WANG, AND K. ZHANG, *Algorithmic approaches for genome rearrangement: A review*, IEEE Transactions on Systems, Man and Cybernetics, part C, 36, no. 5 (2006), pp. 636–648.

[248] D. LIBEN-NOWELL, *On the structure of syntenic distance*, Journal of Computational Biology, 8, no. 1 (2001), pp. 53–67.

[249] D. LIBEN-NOWELL, *Gossip is synteny: Incomplete gossip and the syntenic distance between genomes*, Journal of Algorithms, 43, no. 2 (2002), pp. 264–283.

[250] D. LIBEN-NOWELL AND J. KLEINBERG, *Structural properties and tractability results for linear synteny*, Journal of Discrete Algorithms, 2, no. 2 (2004), pp. 207–228.

[251] G.-H. LIN AND T. JIANG, *A further improved approximation algorithm for breakpoint graph decomposition*, Journal of Combinatorial Optimization, 8, no. 4 (2004), pp. 183–194.

[252] G.-H. LIN AND G. XUE, *Signed genome rearrangement by reversals and transpositions: Models and approximations*, Theoretical Computer Science, 259, no. 1–2 (2001), pp. 513–531.

[253] G.-H. LIN AND G. XUE, *On the terminal Steiner tree problem*, Information Processing Letters, 84, no. 2 (2002), pp. 103–107.

[254] Y. C. LIN, C. L. LU, H.-Y. CHANG, AND C. Y. TANG, *An efficient algorithm for sorting by block-interchanges and its application to the evolution of Vibrio species*, Journal of Computational Biology, 12, no. 1 (2005), pp. 102–112.

[255] Y. C. Lin, C. L. Lu, Y.-C. Liu, and C. Y. Tang, *SPRING: A tool for the analysis of genome rearrangement using reversals and block-interchanges*, Nucleic Acids Research, 34 (July 1, 2006), pp. W696–699.

[256] Y. C. Lin, C. L. Lu, and C. Y. Tang, *Sorting permutation by reversals with fewest block-interchanges*, manuscript, 2006.

[257] D. Lopresti and A. Tomkins, *Block edit models for approximate string matching*, Theoretical Computer Science, 181 (1997), pp. 159–179.

[258] C. L. Lu, T. C. Wang, Y. C. Lin, and C. Y. Tang, *ROBIN: A tool for genome rearrangement of block-interchanges*, Bioinformatics, 21 (2005), pp. 2780–2782.

[259] C. L. Lu, Y. L. Huang, T. C. Wang, and H.-T. Chiu, *Analysis of circular genome rearrangement by fusions, fissions and block-interchanges*, BMC Bioinformatics, 7, no. 1 (2006).

[260] L. Lundin, *Evolution of vertebrate genome reflected in paralogous chromosomal regions in man and in the house mouse*, Genomics, 16 (1993), pp. 1–19.

[261] M. Mahajan, R. Rama, V. Raman, and S. Vijayakumar, *Approximate block sorting*, International Journal of Foundations of Computer Science, 17, no. 2 (2006), pp. 337–355.

[262] M. Mahajan, R. Rama, and S. Vijayakumar, *On sorting by 3-bounded transpositions*, Discrete Mathematics, 306, no. 14 (2006), pp. 1569–1585.

[263] M. Marron, K. M. Swenson, and B. M. E. Moret, *Genomic distances under deletions and insertions*, Theoretical Computer Science, 325, no. 3 (2004), pp. 347–360.

[264] J. Meidanis and Z. Dias, *An alternative algebraic formalism for genome rearrangements*, in vol. 1 of Sankoff and Nadeau [322], pp. 213–223.

[265] J. Meidanis, M. Walter, and Z. Dias, *Reversal distance of signed circular chromosomes*, tech. rep. IC-00-23, University of Campinas, Brazil, 2000.

[266] J. Meidanis, M. E. M. T. Walter, and Z. Dias, *A lower bound on the reversal and transposition diameter*, Journal of Computational Biology, 9, no. 5 (2002), pp. 743–745.

[267] H. Mewes, K. Albermann, M. Bahr, D. Frishman, A. Gleissner, J. Hani, K. Heumann, K. Kleine, A. Maierl, S. Oliver, F. Pfeiffer, and A. Zollner, *Overview of the yeast genome*, Nature, 387, 6632, supp. (1997), pp. S7–S65.

[268] S. Micali and V. V. Vazirani, *An* $O(\sqrt{|V|}\,|E|)$ *algorithm for finding maximum matching in general graphs*, in Proceedings of the Twenty-first Annual Symposium on Foundations of Computer Science (FOCS), IEEE Computer Society Press, 1980, pp. 17–27.

[269] C. Mira and J. Meidanis, *Sorting by block-interchanges and signed reversals*, in Proceedings of the Fourth International Conference on Information Technology: New Generations (ITNG), IEEE Computer Society Press, 2007, pp. 670–676.

[270] C. Mira, Q. Peng, J. Meidanis, and P. A. Pevzner, *A shortest-cycle based heuristics for the multiple genome rearrangement problem*, tech. rep. IC-06-23, Instituto de Computação, Universidade Estadual de Campinas, Brazil, 2006.

[271] J. Mixtacki, *Genome halving under DCJ revisited*, in Proceedings of the Fourteenth International Computing and Combinatorics Conference (COCOON), X. Hu and J. Wang, eds., vol. 5092 of Lecture Notes in Computer Science, Springer-Verlag, 2008, pp. 276–286.

[272] Z. Mo and T. Zeng, *An improved genetic algorithm for problem of genome rearrangement*, Wuhan University Journal of Natural Sciences, 11, no. 3 (2006), pp. 498–502.

[273] B. M. E. Moret, L.-S. Wang, T. Warnow, and S. K. Wyman, *New approaches for reconstructing phylogenies from gene order data*, Bioinformatics, 17 (2001), pp. S165–S173.

[274] B. M. E. Moret, S. K. Wyman, D. A. Bader, T. Warnow, and M. Yan, *A new implementation and detailed study of breakpoint analysis*, in Proceedings of the Sixth Pacific Symposium on Biocomputing (PSB), World Scientific Press, 2001, pp. 583–594.

[275] B. M. E. Moret, A. C. Siepel, J. Tang, and T. Liu, *Inversion medians outperform breakpoint medians in phylogeny reconstruction from gene-order data*, in Proceedings of the Second International Workshop on Algorithms in Bioinformatics (WABI), R. Guigo and D. Gusfield, eds., vol. 2452 of Lecture Notes in Computer Science, Springer-Verlag, 2002, pp. 521–536.

[276] B. M. E. Moret, J. Tang, L.-S. Wang, and T. Warnow, *Steps toward accurate reconstructions of phylogenies from gene-order data*, Journal of Computer and System Sciences, 65, no. 3 (2002), pp. 508–525.

[277] W. J. Murphy, D. M. Larkin, A. Everts-van der Wind, G. Bourque, G. Tesler, L. Auvil, J. E. Beever, B. P. Chowdhary, F. Galibert, L. Gatzke, C. Hitte, S. N. Meyers, D. Milan, E. A. Ostrander, G. Pape, H. G. Parker, T. Raudsepp, M. B. Rogatcheva, L. B. Schook, L. C. Skow, M. Welge, J. E. Womack, S. J. O'Brien, P. A. Pevzner, and H. A. Lewin, *Dynamics of mammalian chromosome evolution inferred from multi-species comparative maps*, Science, 309 (2005), pp. 613–617.

[278] S. Muthukrishnan and S. C. Sahinalp, *An efficient algorithm for sequence comparison with block reversals*, Theoretical Computer Science, 321, no. 1 (2004), pp. 95–101.

[279] C. T. Nguyen, *Algorithms for Calculating Exemplar Distances*, honours year project report, National University of Singapore, 2005.

[280] C. T. Nguyen, Y. C. Tay, and L. Zhang, *Divide-and-conquer approach for the exemplar breakpoint distance*, Bioinformatics, 21, no. 10 (2005), pp. 2171–2176.

[281] T. C. Nguyen, H. T. Ngo, and N. B. Nguyen, *Sorting by restricted-length-weighted reversals*, Genomics Proteomics Bioinformatics, 3, no. 2 (2005), pp. 120–127.

[282] R. Niedermeier, Invitation to Fixed-Parameter Algorithms, Oxford Lecture Series in Mathematics and Its Applications, Oxford University Press, 2006.

[283] E. Ohlebusch, M. I. Abouelhoda, K. Hockel, and J. Stallkamp, *The median problem for the reversal distance in circular bacterial genomes*, in Proceedings of the Sixteenth Annual Symposium on Combinatorial Pattern Matching (CPM), A. Apostolico, M. Crochemore and K. Park, eds., vol. 3537 of Lecture Notes in Computer Science, Springer-Verlag, 2005, pp. 116–127.

[284] S. Ohno, Evolution by Gene Duplication, Springer-Verlag, 1970.

[285] M. Ozery-Flato and R. Shamir, *Two notes on genome rearrangement*, Journal of Bioinformatics and Computational Biology, 1, no. 1 (2003), pp. 71–94.

[286] M. Ozery-Flato and R. Shamir, *An $O(n^{3/2}\sqrt{\log(n)})$ algorithm for sorting by reciprocal translocations*, in Proceedings of the Seventeenth Annual Symposium on Combinatorial Pattern Matching (CPM), M. Lewenstein and G. Valiente, eds., vol. 4009 of Lecture Notes in Computer Science, Springer-Verlag, 2006, pp. 259–269.

[287] M. Ozery-Flato and R. Shamir, *Rearrangements in genomes with centromeres part I: Translocations*, in Proceedings of the Eleventh Annual International RECOMB Conference (RECOMB), T. Speed and H. Huang, eds., vol. 4453 of Lecture Notes in Computer Science, Springer-Verlag, 2007, pp. 339–353.

[288] M. Ozery-Flato and R. Shamir, *Sorting by reciprocal translocations via reversals theory*, Journal of Computational Biology, 14, no. 4 (2007), pp. 408–422.

[289] J. D. Palmer and L. A. Herbon, *Plant mitochondrial DNA evolves rapidly in structure, but slowly in sequence*, Journal of Molecular Evolution, 28, no. 1–2 (1988), pp. 87–97.

[290] C. H. Papadimitriou, Computational Complexity, Addison-Wesley, 1994.

[291] C. H. Papadimitriou and M. Yannakakis, *Optimization, approximation and complexity classes*, Journal of Computer and System Sciences, 43 (1991), pp. 425–440.

[292] I. Pe'er and R. Shamir, *The median problems for breakpoints are NP-complete*, Electronic Colloquium on Computational Complexity, tech. rep. 71, 1998.

[293] I. Pe'er and R. Shamir, *Approximation algorithms for the permutations median problem in the Breakpoint model*, in Sankoff and Nadeau [322], pp. 225–241.

[294] Q. Peng, P. A. Pevzner, and G. Tesler, *The fragile breakage versus random breakage models of chromosome evolution*, PLoS Computational Biology, 2, no. 2 (2006), p. e14.

[295] P. Pevzner and G. Tesler, *Transforming men into mice: The Nadeau-Taylor chromosomal breakage model revisited*, in Proceedings of the Seventh Annual International Conference on Research in Computational Molecular Biology (RECOMB), ACM Press, 2003, pp. 247–256.

[296] P. A. Pevzner, *Computational Molecular Biology: An Algorithmic Approach*, MIT Press, 2000.

[297] P. A. Pevzner and G. Tesler, *Human and mouse genomic sequences reveal extensive breakpoint reuse in mammalian evolution*, Proceedings of the National Academy of Sciences (USA), 100 (2003), pp. 7672–7677.

[298] P. A. Pevzner and M. S. Waterman, *Open combinatorial problems in computational molecular biology*, in Third Israel Symposium on the Theory of Computing and Systems, IEEE Computer Society Press, 1995, pp. 158–173.

[299] R. Y. Pinter and S. Skiena, *Genomic sorting with length-weighted reversals*, in Proceedings of the Thirteenth International Conference on Genome Informatics (GIW), R. Lathrop, K. Nakai, S. Miyano, T. Takagi, and M. Kanehisa, eds., vol. 13, Universal Academy Press, 2002, pp. 103–111.

[300] N. Pisanti and M.-F. Sagot, *Further thoughts on the syntenic distance between genomes*, Algorithmica, 34, no. 2 (2002), pp. 157–180.

[301] P. Popescu, H. Hayes, and B. Dutrillaux, Techniques in Animal Cytogenetics, Springer-Verlag, 2000.

[302] V. Y. Popov, *Multiple genome rearrangement by swaps and by element duplications*, Theoretical Computer Science, 385, no. 1–3 (2007), pp. 115–126.

[303] F. J. Portier and T. P. Vaughan, *Whitney numbers of the second kind for the star poset*, European Journal of Combinatorics, 11 (1990), pp. 277–288.

[304] X. Qi, G. Li, J. Wu, and B. Liu, *Sorting signed permutations by fixed-length reversals*, International Journal of Foundations of Computer Science, 17, no. 4 (2006), pp. 933–948.

[305] X. Qi, G. Li, S. Li, and Y. Xu, *Sorting genomes by translocations and deletions*, in Proceedings of the Fifth International IEEE Computer Society Computational Systems Bioinformatics Conference (CSB), P. Markstein and Y. Xu, eds., IEEE Computer Society Press, 2006, pp. 157–166.

[306] X.-Q. Qi, *Combinatorial Algorithms of Genome Rearrangements in Bioinformatics*, Ph.D. thesis, University of Shandong, China, 2006.

[307] X.-Q. Qi, J. Cao, and C. Zhang, *Sorting circular binary strings with length weighted transpositions*, Journal of Shandong Normal University (Natural Science), 41 (2006), pp. 82–85. In Chinese.

[308] K. Qiu, H. Meijer, and S. Akl, *Parallel routing and sorting of the pancake network*, in Proceedings of the Third International Conference on Computing and Information (ICCI), F. K. H. A. Dehne, F. Fiala, and W. W. Koczkodaj, eds., vol. 497 of Lecture Notes in Computer Science, Springer-Verlag, 1991, pp. 360–371.

[309] A. J. Radcliffe, A. D. Scott, and E. L. Wilmer, *Reversals and transpositions over finite alphabets*, SIAM Journal on Discrete Mathematics, 19, no. 1 (2005), pp. 224–244 (electronic).

[310] A. Rahman, S. Shatabda, and M. Hasan, *Approximation algorithm for sorting by reversals and transpositions*, in Proceedings of the First Workshop on Algorithms and Computation (WALCOM), M. Kaykobad and M. S. Rahman, eds., Bangladesh Academy of Sciences, 2007, pp. 97–108.

[311] R. Raz and S. Safra, *A sub-constant error-probability low-degree test, and sub-constant error-probability PCP characterization of NP*, in Proceedings of the Twenty-ninth Annual ACM Symposium on the Theory of Computing (STOC), ACM, 1997, pp. 475–484.

[312] K. Rice and T. Warnow, *Parsimony is hard to beat*, in Proceedings of the Third International Computing and Combinatorics Conference (COCOON), T. Jiang and D. T. Lee, eds., vol. 1276 of Lecture Notes in Computer Science, Springer-Verlag, 1997, pp. 124–133.

[313] G. Robins and A. Zelikovsky, *Improved Steiner tree approximation in graphs*, in Proceedings of the Eleventh Annual ACM-SIAM Symposium on Discrete Algorithms (SODA), San Francisco, PA, USA, Society for Industrial and Applied Mathematics, 2000, pp. 770–779.

[314] M.-F. Sagot and E. Tannier, *Perfect sorting by reversals*, in Proceedings of the Eleventh International Computing and Combinatorics Conference (COCOON), L. Wang, ed., vol. 3595 of Lecture Notes in Computer Science, Springer-Verlag, 2005, pp. 42–51.

[315] D. Sankoff, *Edit distance for genome comparison based on non-local operations*, in Proceedings of the Third Annual Symposium on Combinatorial Pattern Matching (CPM), A. Apostolico, M. Crochemore, Z. Galil, and U. Manber, eds., vol. 644 of Lecture Notes in Computer Science, Springer-Verlag, 1992, pp. 121–135.

[316] D. Sankoff, *Genome rearrangement with gene families*, Bioinformatics, 15, no. 11 (1999), pp. 909–917.

[317] D. Sankoff and M. Blanchette, *The median problem for breakpoints in comparative genomics*, in Proceedings of the Third International Computing and Combinatorics Conference (COCOON), T. Jiang and D. T. Lee, eds., vol. 1276 of Lecture Notes in Computer Science, Springer-Verlag, 1997, pp. 251–263.

[318] D. Sankoff and M. Blanchette, *Multiple genome rearrangement and breakpoint phylogeny*, Journal of Computational Biology, 5, no. 3 (1998), pp. 555–570.

[319] D. Sankoff and M. Blanchette, *Probability models for genome rearrangement and linear invariants for phylogenetic inference*, in Proceedings of the Third Annual International Conference on Computational Molecular Biology (RECOMB), S. Istrail, P. Pevzner, and M. Waterman, eds., ACM, 1999, pp. 302–309.

[320] D. Sankoff and N. El-Mabrouk, *Genome rearrangement*, in Jiang et al. [222].

[321] D. Sankoff and L. Haque, *Power boosts for cluster tests*, in Proceedings of the Third RECOMB Comparative Genomics Satellite Workshop (RECOMB-CG), A. McLysaght and D. H. Huson, eds., vol. 3678 of Lecture Notes in Bioinformatics, Springer-Verlag, 2005, pp. 121–130.

[322] D. Sankoff and J. H. Nadeau, eds., Comparative Genomics: Empirical and Analytical Approaches to Gene Order Dynamics, Map Alignment and the Evolution of Gene Families, vol. 1 of Computational Biology, Kluwer Academic Press, 2000.

[323] D. Sankoff and P. Trinh, *Chromosomal breakpoint re-use in genome sequence rearrangement*, Journal of Computational Biology, 12 (2005), pp. 812–821.

[324] D. Sankoff, C. Morel, and R. J. Cedergren, *Evolution of 5S RNA and the non-randomness of base replacement*, Nature New Biology, 245 (October 1973), pp. 232–234.

[325] D. Sankoff, R. J. Cedergren, and G. Lapalme, *Frequency of insertion-deletion, transversion, and transition in the evolution of 5s ribosomal RNA*, Journal of Molecular Evolution, 7 (1976), pp. 133–149.

[326] D. Sankoff, G. Leduc, N. Antoine, B. Paquin, F. Lang, and R. Cedergren, *Gene order comparisons for phylogenetic inference: Evolution of the mitochondrial genome*, Proceedings of the National Academy of Sciences (USA), 89 (1992), pp. 6575–6579.

[327] D. Sankoff, G. Sundaram, and J. D. Kececioglu, *Steiner points in the space of genome rearrangements*, International Journal of the Foundations of Computer Science, 7 (1996), pp. 1–9.

[328] D. Sankoff, D. Bryant, M. Deneault, B. F. Lang, and G. Burger, *Early eukaryote evolution based on mitochondrial gene order breakpoints*, in Proceedings of the Fourth Annual International Conference on Computational Molecular Biology (RECOMB), R. Shamir, S. Miyano, S. Istrail, P. Pevzner, and M. Waterman, eds., ACM, 2000, pp. 254–262.

[329] D. Sankoff, J.-F. Lefebvre, E. Tillier, A. Maler, and N. El-Mabrouk, *The distribution of inversion lengths in bacteria*, in Proceedings of the Second RECOMB Comparative Genomics Satellite Workshop (RECOMB-CG), J. Lagergren, ed., vol. 3388 of Lecture Notes in Computer Science, Springer-Verlag, 2004, pp. 97–108.

[330] T. Schmidt and J. Stoye, *Quadratic time algorithms for finding common intervals in two and more sequences*, in Proceedings of the Fifteenth Annual Symposium on Combinatorial Pattern Matching (CPM), S. C. Sahinalp, S. Muthukrishnan, and U. Dogrusoz, eds., vol. 3109 of Lecture Notes in Computer Science, Springer-Verlag, 2004, pp. 347–358.

[331] C. Seoighe and K. Wolfe, *Extent of genomic rearrangement after genome duplication in yeast*, Proceedings of the National Academy of Sciences (USA), 95 (1998), pp. 4447–4452.

[332] G. Seroussi and A. Lempel, *Factorization of symmetric matrices and trace-orthogonal bases in finite fields*, SIAM Journal of Computing, 9 (1980), pp. 758–767.

[333] J. Meidanis and J. Setubal, Introduction to Computational Molecular Biology, PWS Publishing, 1997.

[334] P. Seymour, *Packing directed circuits fractionally*, Combinatorica, 15, no. 2 (1995), pp. 281–288.

[335] D. Shapira and J. A. Storer, *Edit distance with move operations*, in Proceedings of the Thirteenth Annual Symposium on Combinatorial Pattern Matching (CPM), vol. 2373 of Lecture Notes in Computer Science, Springer-Verlag, 2002, pp. 85–98.

[336] D. Shapira and J. A. Storer, *Large edit distance with multiple block operations*, in Proceedings of the Tenth International Symposium on String Processing and Information Retrieval (SPIRE), M. A.

Nascimento, E. S. de Moura, and A. L. Oliveira, eds., vol. 2857 of Lecture Notes in Computer Science, Springer-Verlag, 2003, pp. 369–377.

[337] D. Shapira and J. A. Storer, *Edit distance with move operations*, Journal of Discrete Algorithms, 5, no. 2 (2007), pp. 380–392.

[338] Y.-F. She and G.-L. Chen, *Parallel algorithm for computing reversal distance*, Proceedings of the Sixth International Conference on Parallel and Distributed Computing Applications and Technologies, 2005, pp. 950–953.

[339] A. C. Siepel, *Exact Algorithms for the Reversal Median Problem*, master's thesis, University of New Mexico, 2001.

[340] A. C. Siepel, *An algorithm to enumerate sorting reversals for signed permutations*, Journal of Computational Biology, 10 (2003), pp. 575–597.

[341] A. C. Siepel and B. M. E. Moret, *Finding an optimal inversion median: Experimental results*, in Proceedings of the First International Workshop on Algorithms in Bioinformatics (WABI), O. Gascuel and B. M. E. Moret, eds., vol. 2149 of Lecture Notes in Computer Science, Springer-Verlag, 2001, pp. 189–203.

[342] A. Solomon, P. Sutcliffe, and R. Lister, *Sorting circular permutations by reversal*, in Proceedings of the Eighth International Workshop on Algorithms and Data Structures (WADS), F. K. H. A. Dehne, J.-R. Sack, and M. H. M. Smid, eds., vol. 2748 of Lecture Notes in Computer Science, Springer-Verlag, 2003, pp. 319–328.

[343] A. H. Sturtevant and E. Novitski, *The homologies of the chromosome elements in the genus Drosophila*, Genetics, 26, no. 5 (1941), pp. 517–541.

[344] J. Suksawatchon, C. Lursinsap, and M. Bodén, *Computing the reversal distance between genomes in the presence of multi-gene families via binary integer programming*, Journal of Bioinformatics and Computational Biology, 5, no. 1 (2007), pp. 117–133.

[345] K. Swenson, M. Marron, J. Earnest-DeYoung, and B. M. E. Moret, *Approximating the true evolutionary distance between two genomes*, in Proceedings of the Seventh Workshop on Algorithm Engineering and Experiments and the Second Workshop on Analytic Algorithmics and Combinatorics (ALENEX/ANALCO), C. Demetrescu, R. Sedgewick, and R. Tamassia, eds., SIAM, 2005, pp. 121–129.

[346] F. Swidan, M. A. Bender, D. Ge, S. He, H. Hu, and R. Pinter, *Sorting by length-weighted reversals: Dealing with signs and circularity*, in Proceedings of the Fifteenth Annual Symposium on Combinatorial Pattern Matching (CPM), S. C. Sahinalp, S. Muthukrishnan, and U. Dogrusoz, eds., vol. 3109 of Lecture Notes in Computer Science, Springer-Verlag, 2004, pp. 32–46.

[347] J. Tang and B. M. E. Moret, *Phylogenetic reconstruction from gene-rearrangement data with unequal gene content*, in Proceedings of the Eighth International Workshop on Algorithms and Data Structures (WADS), F. K. H. A. Dehne, J.-R. Sack, and M. H. M. Smid, eds., vol. 2748 of Lecture Notes in Computer Science, Springer-Verlag, 2003, pp. 37–46.

[348] J. Tang and B. M. E. Moret, *Linear programming for phylogenetic reconstruction based on gene rearrangements*, in Proceedings of the Sixteenth Annual Symposium on Combinatorial Pattern Matching (CPM), A. Apostolico, M. Crochemore and K. Park, eds., vol. 3537 of Lecture Notes in Computer Science, Springer-Verlag, 2005, pp. 406–416.

[349] J. Tang and L.-S. Wang, *Improving genome rearrangement phylogeny using sequence-style parsimony*, in Proceedings of the Fifth IEEE International Symposium on Bioinformatics and Bioengineering (BIBE), IEEE Computer Society Press, 2005, pp. 137–144.

[350] J. Tang, B. M. E. Moret, L. Cui, and C. W. dePamphilis, *Phylogenetic reconstruction from arbitrary gene-order data*, in Proceedings of the Fourth IEEE International Symposium on Bioinformatics and Bioengineering (BIBE), IEEE Computer Society Press, 2004, pp. 592–599.

[351] E. Tannier and M.-F. Sagot, *Sorting by reversals in subquadratic time*, in Proceedings of the Fifteenth Annual Symposium on Combinatorial Pattern Matching (CPM), S. C. Sahinalp, S. Muthukrishnan, and U. Dogrusoz, eds., vol. 3109 of Lecture Notes in Computer Science, Springer-Verlag, 2004, pp. 1–13.

[352] E. Tannier, A. Bergeron, and M.-F. Sagot, *Advances on sorting by reversals*, Discrete Applied Mathematics, 155, no. 6–7 (2007), pp. 881–888.

[353] E. Tannier, C. Zheng, and D. Sankoff, *Multichromosomal median and halving problems*, in Proceedings of the Eighth Workshop on Algorithms in Bioinformatics (WABI), vol. 5251 of Lecture Notes in Bioinformatics, Springer-Verlag, 2008.

[354] B. Tenner, *Database of permutation pattern avoidance*, published electronically at http://math.depaul.edu/~bridget/patterns.html.

[355] G. Tesler, *Efficient algorithms for multichromosomal genome rearrangements*, Journal of Computer and System Sciences, 65, no. 3 (2002), pp. 587–609.

[356] G. Tesler, *GRIMM: Genome rearrangements web server*, Bioinformatics, 18, no. 3 (2002), pp. 492–493.

[357] W. F. Tichy, *The string-to-string correction problem with block moves*, ACM Transactions on Computer Systems, 2, no. 4 (1984), pp. 309–321.

[358] C. Ting and H. E. Yong, *Optimal algorithms for uncovering synteny problem*, Journal of Combinatorial Optimization, 12 (2006), pp. 421–432.

[359] N. Tran, *An easy case of sorting by reversals*, in Proceedings of the Eighth Annual Symposium on Combinatorial Pattern Matching (CPM), A. Apostolico and J. Hein, eds., vol. 1264 of Lecture Notes in Computer Science, Springer-Verlag, 1997, pp. 83–89.

[360] C.-W. Tseng and M. V. Zelkowitz, eds., Computational Biology and Bioinformatics, vol. 68 of Advances in Computers, Elsevier Academic Press, 2006.

[361] T. Uno and M. Yagiura, *Fast algorithms to enumerate all common intervals of two permutations*, Algorithmica, 26, no. 2 (2000), pp. 290–309.

[362] F. Viduani Martinez, J. Coelho de Pina, and J. Soares, *Algorithms for terminal Steiner trees*, in Proceedings of the Eleventh International Computing and Combinatorics Conference (COCOON), L. Wang, ed., vol. 3595 of Lecture Notes in Computer Science, Springer-Verlag, 2005, pp. 369–379.

[363] M. E. M. T. Walter, Z. Dias, and J. Meidanis, *Reversal and transposition distance of linear chromosomes*, in Proceedings of the Fifth International Symposium on String Processing and Information Retrieval (SPIRE), IEEE Computer Society Press, 1998, pp. 96–102.

[364] L. Wang, *Translocation distance: Algorithms and complexity*, in Tseng and Zelkowitz [360], pp. 106–127.

[365] L. Wang, D. Zhu, X. Liu, and S. Ma, *An $O(n^2)$ algorithm for signed translocation*, Journal of Computer and System Sciences, 70, no. 3 (2005), pp. 284–299.

[366] L.-S. Wang, R. K. Jansen, B. M. E. Moret, L. A. Raubeson, and T. Warnow, *Fast phylogenetic methods for the analysis of genome rearrangement data: An empirical study*, in Proceedings of the Seventh Pacific Symposium on Biocomputing (PSB), R. B. Altman, A. K. Dunker, L. Hunter, K. Lauderdale, and T. E. Klein, eds., World Scientific Press, 2002, pp. 524–535.

[367] L.-S. Wang, T. Warnow, B. M. E. Moret, R. K. Jansen, and L. A. Raubeson, *Distance-based genome rearrangement phylogeny*, Journal of Molecular Evolution, 63, no. 4 (2006), pp. 473–483.

[368] R. Warren and D. Sankoff, *Genome halving with general operations*, in Proceedings of the Sixth Asia-Pacific Bioinformatics Conference (APBC), A. Brazma, S. Miyano, and T. Akutsu, eds., vol. 6 of Advances in Bioinformatics and Computational Biology, Imperial College Press, 2008, pp. 231–240.

[369] G. A. Watterson, W. J. Ewens, T. E. Hall, and A. Morgan, *The chromosome inversion problem*, Journal of Theoretical Biology, 99 (1982), pp. 1–7.

[370] H. Wielandt, Finite Permutation Groups, translated from German by R. Bercov, Academic Press, 1964.

[371] J. Wienberg, *The evolution of eutherian chromosomes*, Current Opinion in Genetics & Development, 14 (2004), pp. 657–666.

[372] K. Wolfe and D. Shields, *Molecular evidence for an ancient duplication of the entire yeast genome*, Nature, 387 (1997), pp. 708–713.

[373] S. Wu and X. Gu, *Multiple genome rearrangement by reversals*, in Proceedings of the Seventh Pacific Symposium on Biocomputing (PSB), R. B. Altman, A. K. Dunker, L. Hunter, and T. E. Klein, eds., World Scientific Press, 2002, pp. 259–270.

[374] S. Wu and X. Gu, *Algorithms for multiple genome rearrangement by signed reversals*, in Proceedings of the Eighth Pacific Symposium on Biocomputing (PSB), World Scientific Press, 2003, pp. 363–374.

[375] S. Yancopoulos, O. Attie, and R. Friedberg, *Efficient sorting of genomic permutations by translocation, inversion and block interchange*, Bioinformatics, 21, no. 16 (2005), pp. 3340–3346.

[376] I. V. Yap, D. Schneider, J. Kleinberg, D. Matthews, S. Cartinhour, and S. R. McCouch, *A graph-theoretic approach to comparing and integrating genetic, physical and sequence-based maps*, Genetics, 165 (2003), pp. 2235–2247.

[377] P. Yin and A. J. Hartemink, *Theoretical and practical advances in genome halving*, Bioinformatics, 21 (2005), pp. 869–879.

[378] F. Yue, M. Zhang, and J. Tang, *A heuristic for phylogenetic reconstruction using transposition*, in Proceedings of the Seventh IEEE International Conference on Bioinformatics and Bioengineering (BIBE), J. Y. Yang, M. Qu Yang, M. M. Zhu, Y. Zhang, H. R. Arabnia, Y. Deng, and N. G. Bourbakis, eds., IEEE Computer Society Press, 2007, pp. 802–808.

[379] C. Zheng and D. Sankoff, *Genome rearrangements with partially ordered chromosomes*, Journal of Combinatorial Optimization, 11, no. 2 (2006), pp. 133–144.

[380] C. Zheng, A. Lenert, and D. Sankoff, *Reversal distance for partially ordered genomes*, ISMB (supplement of Bioinformatics), 21 (2005), pp. i502–i508.

[381] C. Zheng, Q. Zhu, and D. Sankoff, *Genome halving with an outgroup*, Evolutionary Bioinformatics, 2 (2006), pp. 295–302.

[382] C. Zheng, Q. Zhu, Z. Adam, and D. Sankoff, *Guided genome halving: Hardness, heuristics and the history of the hemiascomycetes*, ISMB (supplement to Bioinformatics), 24, no. 13 (2008), pp. i96–i104.

[383] D. Zhu and S. Ma, *Improved polynomial-time algorithm for computing translocation distance between genomes*, J. Comput., 25, no. 2 (2002), pp. 189–196. In Chinese.

[384] D. Zhu and L. Wang, *On the complexity of unsigned translocation distance*, Theoretical Computer Science, 352, no. 1–3 (2006), pp. 322–328.

Index

2-break rearrangement, 73
2-gene, 203
k-break rearrangement, 172
3-DIMENSIONAL MATCHING, 252
3-PARTITION, 252
3-SATISFIABILITY, 253

ALTERNATING CYCLE DECOMPOSITION, 253
BIN-PACKING, 253
BREAKPOINT GRAPH DECOMPOSITION, 253
BREAKPOINT PHYLOGENY PROBLEM, 210
CLIQUE, 253
EULERIAN CYCLE DECOMPOSITION, 253
EXACT COVER BY 3-SETS, 253
FULL STEINER TREE PROBLEM, 208
HAMILTONIAN CIRCUIT, 254
HAMILTONIAN PATH, 254
INDEPENDENT SET, 254
LARGE PARSIMONY PROBLEM, 208
LARGEST BALANCED INDEPENDENT SET, 254
LONGEST PATH PROBLEM, 237
MAXIMUM PARSIMONY ON BINARY ENCODINGS, 218
MINIMUM COMMON STRING PARTITION, 133
REVERSAL PHYLOGENY PROBLEM, 211
REVERSED MINIMUM COMMON STRING PARTITION, 137
SATISFIABILITY, 254
SEQUENCE-BASED MAXIMUM PARSIMONY PROBLEM, 218
SET PACKING, 254
SHORTEST PATH PROBLEM, 237
SIGNED MINIMUM COMMON STRING PARTITION, 136
SMALL PARSIMONY PROBLEM, 212
STEINER TREE PROBLEM, 208
TRANSPOSITION PHYLOGENY PROBLEM, 211
TRAVELING SALESMAN, 255
VERTEX COVER, 255

Adjacency, 20, 162
  belongs, 173
  in a genome, 162
  matrix, 239
  reverse, 20
  telomeric, 162
  vertices, 235
Alphabet, 92
Alternating group, 15
Arc, 240
  black, 28
  desire, 28
  gray, 28
  head, 240
  reality, 28
  tail, 240
Ascent, 27

Block, 93
  edit, 93
  family, 124
  move, 128
  uncopy, 128
  uniform, 141
Block interchange
  permutations, 49
  strings, 153
Breakable, 172
Breakpoint, 20
  binary string, 139
  distance, 21
  external, 162, 163
  internal, 162, 163
  phylogeny, 210
  prefix transposition, 37
  strong, 20
Breakpoint diagram, 35
Breakpoint graph
  contracted, 201
  of a genome, 167
  of permutation, 41
  signed permutation, 65

Cayley graph, 17
Center, 206
Centromere, 168

Character encodings
  MPBE, 218
  MPME, 218
  SB encoding, 218
Character moving, 151
Character swaps, 155
Chromosome, 1, 162
  circular, 162
  legal, 169
  linear, 162
Circular lower bound, 214
Clan, 38
Clique, 235
Collection, block edit, 123
Complement of a graph, 235
Component
  genomes, 164
    chain, 173
    tree, 173
  minimal, 164
  oriented, 173
  of an overlap graph, 86
  permutation, 66
    chain, 66
    tree, 66
Conjugacy class, 14
Connected
  component, 237
  graph, 237
Cover
  cost, 174
  of a poset, 76
  of a PQ-tree, 173
$\mathcal{CS}$-Factor, 113
Cut-and-paste, 55
Cycle
  even, 29
  graph, elementary, 237
  in a graph, 236
    even, 237
    length, 236
    odd, 237
  Hamiltonian, 237
  interleaving, 83
  length, 30
  odd, 29
  of a permutation, 14

DCJ, 170
  cut, 170
  join, 170
Degree
  of a graph
    maximum, 236
    minimum, 236
  of a vertex, 236
Deletion
  block, 94
    exclusive, 102
  of a nucleotide, 2
  of a segment, 3
Dense instance, 140
Descent, 27
Diameter, 6
  of a graph, 237
  of a metric space, 6
  syntenic, 185
Digraph
  connected
    strongly, 241
    weakly, 241
Directed acyclic graph, 241
  topological sort, 241
  transitive closure, 242
Directed graph, 240
  acyclic, 241
  cycle, 241
Disjoint cycle decomposition, 14
Distance
  k-break, 172
  between two genomes, 6
  between vertices, 237
  block covering, 124
  block edit, 123
    exclusive, with reversals, 102, 103
  breakpoint, 163
    poset, 80
  Cayley, 50
  conserved interval, 64, 115
  DCJ, 170
    permutation, 73
  double, 201
  Hamming, 123
  left-invariant, 18
  Levenshtein, 123
  monotonic, 104
  prefix exchange, 52
  prefix transposition, 37
  reversal, 40
    perfect, 69
    poset, 79
    prefix, 70
  RT, 172
  string edit, 123
  syntenic, 182
    linear, 186
  TDRL, 59
  translocation, 165, 166
  transposition, 26
  Ulam's, 31
DNA, 1
  complementarity, 1
  segment, 1
  strand, 1
Double cut-and-join, 73, 170
Dual genome representation, 183

Duplicate, 92
Duplication, 3, 91
 block, 94
 tandem, 3

Edge, 235
 desire, 65
 extremities, 235
 interleaving, 83
 reality, 65
Edit
 block edit, 93
 character edit, 123
Element of a chromosome, 169
Elimination graph, 85
Eulerian
 cycle, 237
 path, 237
Evolution, 2
Extremity
 edge, 235
 gene, 161
 interval, 163, 164
 path, 236

Fission, 4, 159, 165, 182
Forest, 238
Function
 goal, 246
 objective, 246
Fusion, 4, 159, 165, 182

Gene, 1, 13, 92
 duplicated, 200
 extremity, 161
 family, 91
 head, 65, 161
 signed, 161
 tail, 161
 unsigned, 161
Generator, 16
Genome, 1
 compact representation, 182
 doubled, 200, 203
 duplicated, 200, 203
 halving, 201
 linear, 162
 median, 193
 set system, 203
 unsigned, 161
Genome rearrangement problem, 6
Graph
 Γ-, 30
 k-regular, 236
 acyclic, 237
 bipartite, 238
  balanced, 238
  complete, 238
 breakpoint, 65
 Cayley, 17
 complement of a, 235
 complete, 235
 connected, 237
 cycle, 28
 diameter, 237
 digraph, 240
 directed, 240
 disconnected, 237
 interleaving, 83
 loop, 235
 order of a, 235
 overlap, 84
 path, 236
 of a permutation, 14
 simple, 235
 size of a, 235
 synteny, 184, 203
 undirected, 235
Group
 alternating, 15
 hyperoctahedral, 15
 symmetric, 14
Grouping, 138
Guided halving, 202

Hamiltonian
 cycle, 237
 path, 237
Hasse diagram, 77
Homology, 4, 91

Identity genome, 166
In-degree, 240
 maximum, 241
 minimum, 240
Incidence, 235
Independent
 edges, 235
 vertices, 235
In-neighborhood, 240
Insertion
 block, 94
  exclusive, 102
 of a nucleotide, 3
Interleaving
 cycles, 83
 edges, 83
 graph, 83
Interval
 2-interval, 110
 active, 44
 break, 69
 common, 21, 164
  of two strings, 113
 conserved, 64, 113, 164
  of a set of permutations, 115
 extremities of an, 21
 genome, 163

Interval (cont.)
linear, 23
(maximal) location of, 112
nested, 69
passive, 44
of a permutation, 21
prime, 23
separable, 70
of a string, 113
Inversion, 3

Kendall's tau, 51

Leaf, 238
Left code
of an element, 31
of a permutation, 31
Left-invariance, 18
Legal
chromosome, 169
rearrangement, 169
Length of a cycle in a graph, 236
Linear
rearrangement, 172
set, 23
Linear extension
of a permutation, 20
of a poset, 76
of a string, 136
Linearly ordered set, *see* Totally ordered set
Local complementation, 85
game, 85
Longest common prefix, 93
Longest common subsequence, 93
Longest common substring, 93
up to a reversal, 93
Longest common suffix, 93

Matching, 238
t-dimensional, 115
exemplar, 99
full, 99
intermediate, 99
maximal, 238
maximum, 238
perfect, 238
Matrix adjacency, 239
Measure
exemplar, 99
full, 99
intermediate, 99
Median
genome, 193
problem, 193
score, 193
Metric, 6
space, 6

Model
exemplar, 98
full, 98
intermediate, 98
MPBE, 218
Mutation point, 2

Neighborhood, 235
Node, vertex of a graph, 235
Noncomparability, 76
NP, 243
Nucleotide, 1
Number
MAD, 118
SAD, 118

Oriented
component of a genome, 173
component of an overlap graph, 86
component of a permutation, 66
cycle, 84
edge, 84
element, 85
point, 66
reversal, 68
vertex, 84, 85
Orthology, 91
Out-degree, 240
maximum, 241
minimum, 240
Out-neighborhood, 240
Overlap
graph, 84
intervals, 22

P, 243
Paralogy, 91
Parameterized complexity, 250
Partially ordered set, 76
antichain, 76
chain, 76
down-set, 77
height, 76
linear extension, 76
maximal element, 77
minimal element, 77
up-set, 77
weak order, 76
width, 76
Partial order, 75
Partition
common, 134
common possibly reversed, 137
Path
directed, 241
in a graph, 236
Hamiltonian, 237

Perfect
 matching, 238
 scenario, 69
Permutation, 13
 2-permutation, 32
 3-permutation, 32
 γ-permutation, 34
 k-permutation, 83
 circular, 19
 even, 15
 extended, 19
 genomic circular, 19
 Gollan, 40
 identity, 14
 inverse, 14
 linear, 13
 linearization, 19
 odd, 15
 pattern, 57
 reduced, 27
 reversed, 33
 signed, 15
 simple, 32
 spin of a, 43
 stack-sortable, 56
 subsequence, 31
  increasing, 31
 toric, 32
Phylogeny, 191
 breakpoint, 210
 reversal, 211
Plateau, 31
Point
 belonging to a component, 66
 in a permutation, 20
Polynomial time
 reducibility, 244
 reduction, 244
Poset. *See* Partially ordered set
PQ-tree
 components, 66
 intervals, 23
 P-node, 23, 66
 Q-node, 23, 66
Prefix transposition, 152
Prime interval, 23
Problem
 decision, 243
 optimization, 246
Pruning
 $\mathcal{M}$-pruning of two sets of permutations, 116
 t-dimensional, 116
 exemplar, 99
 full, 99
 intermediate, 99

Random loss, 58
Replacement block, 94
Replication, 2
 origin, 70
 terminus, 70
Retrotransposition, 3
Reversal, 3, 40
 k-reversal, 54, 71
 length, 52
 oriented, 68
 phylogeny, 211
 prefix, 47
 safe, 67
 signed, 64
 symmetric around a point, 70
Revrev, 73
Right code
 of an element, 31
 of a permutation, 31

Segment
 of DNA, 1
 of a permutation, 21
Sequence, 1
 alignment, 2
 sorting, 18
Similarity
 adjacency, 109
 common intervals, 111
 conserved intervals, 114
Singleton, 102
 strip, 43
Sink, 240
Size, genome, 181
Source, 240
Speciation, 91
Spin, canonical, 46
Stack, 56
 full-pop, 56
 pop, 56
 push, 56
 sorting, 56
 top, 56
Steiner
 set, 208
 tree, 208
  full, 208
  terminal, 208
Stirling number of the first kind, 51
String, 92
 h-span, 104
 k-ary, 92
 balanced, 94
 binary, 92
 circular, 93
 pegged, 102
 prefix, 92
 related, 94
 signed, 92
 subsequence, 93

String (cont.)
 substring, 92
 suffix, 93
 ternary, 92
Strip, 27
 move, 56
 strong, 43
  decreasing, 43
  increasing, 43
  long, 43
Strong interval, 22
Subgraph, 236
 induced, 236
 spanning, 236
Substitution of a nucleotide, 2
Synteny, 182
 exact, 188
 graph, 184
 linear, 186
  nested, 187
 uncovering, 188

Telomere, 1, 162
Telomeric
 adjacency, 162
 marker, 161
Tetraploidy, 202
Tight
 directed graph, 87
 graph, 85
 matrix, 85
Topological sorting, 77
Toric
 equivalence, 32
 permutation, 32
Totally ordered set, 76
Transfer
 horizontal, 4
 lateral, 4
Translocation, 164
 reciprocal, 4, 159, 164, 182
Transposition, 25
 k-transposition, 34, 37
 (p, q)-transposition, 55
 prefix, 37
Transreversal, 72
Tree, 238
 ordered, 238
 root of a, 238
 rooted, 238
 Steiner, 208
 strong interval, 22
Tree topology, 212

Vertex, 235
 child, 238
 class, 238
 elimination, 85
 extremity, 236
 isolated, 235
 matched, 238
 parent, 238
 reachable, 241
 sink, 240
 source, 240
 well-separated set, 204